CLINICAL APPLICATIONS OF MONOCLONAL ANTIBODIES

CLINICAL APPLICATIONS OF MONOCLONAL ANTIBODIES

Edited by

Ron Hubbard and **Vincent Marks**

University of Surrey
Guildford, Surrey, United Kingdom

PLENUM PRESS • NEW YORK AND LONDON

Library of Congress Cataloging in Publication Data

Clinical applications of monoclonal antibodies / edited by Ron Hubbard and Vincent Marks.

p. cm.

Proceedings of an international symposium held Sept. 15–17, 1987, at the University of Surrey, Guildford, U.K.

Includes bibliographies and index.

ISBN 0-306-43142-4

1. Antibodies, Monoclonal—Diagnostic use—Congresses. 2. Antibodies, Monoclonal—Therapeutic use—Congresses. I. Hubbard, Ronald. II. Marks, Vincent.

[DNLM: 1. Antibodies, Monoclonal—congresses. QW 575 C6407 1987]

RB46.5.C53 1988

616′.07′56—dc19

DNLM/DLC 89-3451

for Library of Congress CIP

Proceedings of an International Symposium on the Clinical Applications of Monoclonal Antibodies, held September 15–17, 1987, at the University of Surrey, Guildford, United Kingdom

A Division of Plenum Publishing Corporation
233 Spring Street, New York, N.Y. 10013

Printed in the United States of America

PREFACE

Immunology has come a long way in the hundred or so years since the general concepts were first enuciated by Metchnikoff, Ehrlich, Von Behring and others. One of the landmarks in this progress was the invention and development of monoclonal antibody secreting hybridomas by Milstein and his co-workers in Cambridge. Unlike most modern inventions of this importance that of monoclonal antibody production was made available to the scientific community throughout the world unimpeded by patent protection. This may explain the unusual rapidity with which it has been applied to the benefit of mankind in general. This book, representing as it does the proceedings of the first International Symposium to be held on the clinical applications of monoclonal antibodies, shows just how much has been achieved within the space of little more than a decade. The enormous promise of monoclonal antibody technology, which became apparent soon after its discovery, has already progressed a long way towards fulfillment. The contributors to this volume, all of whom are actively engaged in monoclonal antibody development and application, represent the state of the art.

Professor Vincent Marks

INTRODUCTION

It has been some twelve years since the pioneering experiments of Köhler and Milstein led to the discovery of monoclonal antibodies. Single molecular species antibodies with desired specificities could be produced by the fusion of antibody - producing cells with neoplastic cells. The original aims of Köhler and Milstein were to study the size of the antibody repertoire and to assess the contribution of somatic mutation to antibody diversity. The applications of these single monoclonal antibodies were not completely obvious then, but it was clear they would be of significant value.

In fact we now look back on the production of monoclonal antibodies as a most exciting and tremendous discovery which has led to a vast range of applications including analytical, diagnostic and therapeutic spanning all the chemical, biological and medical sciences.

Applications of monoclonals in research are worldwide and they have clearly been extensively studied for use in immunoassay and immunopurification of a wide range of industrially and clinically important chemicals, biochemicals and microbiological materials. Clinical applications also include immunocytochemistry, imaging, sensors, diagnosis and therapy in man and in the veterinary fields.

A number of advances in the technology of the production of monoclonal antibodies have occurred and these are highlighted in this book. The production of human monoclonal antibodies in very large quantities, has become increasingly important since they are required for human therapeutic application. Mouse or rat monoclonals remain perfectly suitable for analytical, diagnostic, and research applications.

We can now also look forward to a new, second generation of monoclonal antibodies which involve new technologies including molecular biology. It is now possible to modify the antibody molecule extensively, for example (1) hybrid antibodies have arms of different specificity thus enabling them to be used as a bridge between two different chemical structures; (2) chimeric antibodies which are produced by genetic engineering allowing the V-region mouse genes to be linked to the C-region human genes; these antibodies are less immunogenic in man and are much less likely to produce side-effects in human therapy; (3) CDR (complementary determining region) graft antibodies have only the amino acid residues from the mouse as the antigen-binding site i.e., the hypervariable amino acids of the V domains, whilst the rest of the molecule can be human. Clearly with these new antibodies come new applications and safer human therapy.

This book represents the state of the art in the research into, and the value of, monoclonal antibodies in clinical and veterinary applications. The book provides background to the development of certain topics and highlights current areas of progress. This book will be of great interest to research workers in the field and clinicians in general, but is accessible to the non-specialist with an interest in these topics and hence it has broad educational value.

We are grateful to the many eminent scientists and clinicians who contributed to the meeting with presentations and discussions, and to the secretariat and graduate students of the University of Surrey Biochemistry Department who helped in a variety of roles to ensure the success of the Symposium. In particular the editors would like to give special thanks to Peter Goldfarb, Jan McCall and Ann Hanson for all their help.

The Symposium organisers are also most grateful to our sponsors including Celltech Ltd., Cetus Corporation and Serono Diagnostics Ltd. Finally the editors wish to record their thanks to the staff of Plenum Press and Medimedia Ltd. who gave us every encouragement to realise this project.

R. Hubbard
Department of Biochemistry
University of Surrey
July 1988

CONTENTS

SECTION C
CLINICAL AND VETERINARY APPLICATIONS OF MONOCLONAL ANTIBODIES

SECTION D
ANTIBODIES IN TUMOR DIAGNOSIS AND THERAPY

SECTION A
PRODUCTION AND PURIFICATION OF MONOCLONAL ANTIBODIES

THE POTENTIAL OF ELECTROFUSION FOR HYBRIDOMA PRODUCTION

U. Zimmermann, J.J. Schmitt, P. Kleinhans

Lehrstuhl für Biotechnologie der Universität Würzburg
Röntgenring 11, 8700 Würzburg
West Germany

INTRODUCTION

The potential of monoclonal antibodies as diagnostic and therapeutic agents has now been widely accepted. Improvements in murine hybridoma formation protocols are being continuously developed by almost every group of investigators actively involved in the generation of monoclonal antibodies. However, the preparation of human hybridomas is still generally assumed to be more difficult than that of murine hybrids. Apart from the chromosomal instability of human-human hybridomas (Foung et al, 1982) the choice of fusion agent may be an important factor (Gravekamp et al, 1985). Polyethyleneglycol (PEG) normally used in cell fusion has a number of disadvantages which might result in drawbacks in the production of a broad spectrum of hybrid cells.

The electrofusion technique pioneered by Zimmermann and co-workers (see review articles of Zimmermann, 1982, 1986, 1987) may therefore be an efficient alternative to PEG-induced cell fusion, in particular in the production of human hybrids. The electrofusion technique is based on the application of physical (vectorial) forces which allow precise control and monitoring of the fusion process. Visualization and identification of the hybrids under the microscope offers the possibility of avoiding the use of selective mutants as fusion partners and, in consequence, employment of selection media. This aspect is of great relevance in human hybrid production because the number of available tumor cell lines that have to be tested for hybrid selection is limited by the requirement of HAT-sensitivity.

Electrofusion can be universally applied to all living cells. The field conditions found optimal for two fusion partners can therefore be applied, with appropriate modifications to the fusion of cells of other species, provided that the biology and the biophysics of these cells are taken into account (Zimmermann and Urnovitz, 1987).

The present article reviews our knowledge on electrofusion and emphasis is given to recent results which may pave the way to the production of human hybrids on a large scale using this new technology.

PRINCIPLES OF ELECTROFUSION

Electrofusion is based on the temporary permeabilization of the cell membrane in response to electrical breakdown. This phenomenon was discovered in 1973 (Zimmermann et al, 1973, 1974) and was first used for the electroinjection of membrane-impermeable substances (such as dyes, drugs, particles, proteins and DNA) into freely suspended cells without deterioration of cellular functions or membrane integrity (see e.g. Zimmermann et al, 1981, Zimmermann, 1986).

Breakdown Conditions

Experimentally, electrical breakdown is observed when cells are exposed to a field pulse of high intensity (some kV/cm) and short duration (in the nano-to microsecond range). Due to charge separation a potential difference is built up across the membrane which is superimposed on the intrinsic membrane potential - or more precisely on the intrinsic electric field within the membrane. Breakdown occurs when the vectorial sum of the intrinsic and the induced potentials exceeds the breakdown voltage of the biological membrane which is of the order of 1 V (at room temperature). The generated potential difference is proportional to the field strength, to the radius of the cell, and to the cosine of the angle between a given membrane site and the field vector (Zimmermann, 1982, 1986) (Fig. 1). The radius-dependence of the generated potential difference has the important consequence that breakdown of differently sized cells requires different critical field strengths a priori. Electrofusion of lymphocytes and myeloma cells suffers from this problem because the average volume of non-activated lymphocytes is about 120 $(\mu m)^3$, whilst that of myeloma cells about 1200 $(\mu m)^3$. Since lymphocytes stimulated by antigens either in vitro of in vivo exhibit larger volumes (up to 600 μm^3) than inactivated ones the breakdown conditions of these lymphocytes will be more comparable to those of myeloma cells with the consequence that fusion and subsequent hybridization is preferentially induced between antibody producing lymphocytes and myeloma cells. The radius-dependence can, therefore, be used for physical selection of stimulated lymphocytes in the presence of a large amount of non-stimulated ones provided that appropriate field and media conditions are used (see below and ref. Jeltsch and Zimmermann, 1979).

The angular dependence of the generated potential difference across the cell membrane determines the size of the permeabilized area of a given cell once the breakdown voltage is exceeded. As the external electric field strength is increased the electrical breakdown potential is first reached at membrane sites oriented in the field direction and then at membrane sites with progressively increasing angle when supracritical field strengths are applied. Membrane breakdown cannot occur perpendicular to the field vector because cosine 90° is zero. The more the critical field strength is exceeded the greater the area of disturbance of the membrane. In electroinjection work larger molecules can be admitted into freely suspended cells under these conditions (e.g. Stopper et al, 1987). On the other hand, in electrofusion large field-induced pertubations in the contact zone of adhered cells facilitate fusion of differently sized cells and accelerate the intermingling process.

The superposition of the generated potential difference on the intrinsic membrane field can lead to an asymmetric breakdown under some circumstances (Zimmermann and Stopper, 1987). As shown schematically in Fig. 2 the vectors of the generated and intrinsic field within the membrane are parallel in one hemisphere and antiparallel in the other one. In this figure it is assumed that the cell interior is negatively charged with respect to the external solution, as is the case for lymphocytes and myeloma cells. If the intrinsic membrane potential or field is

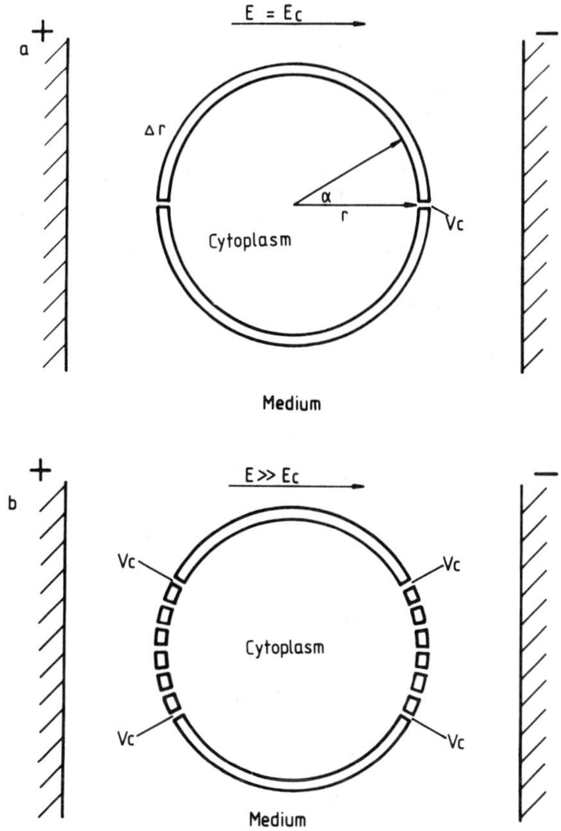

$V_c = 1{,}5 \cdot r \cdot E_c \cdot \cos\alpha$

Fig. 1. Symmetric breakdown
Representation of a spherical cell with radius, r, exposed to an external field, E.
a) The breakdown voltage, Vc, will be attained first at membrane sites oriented in the direction of the field vector because of the angular dependence of the generated membrane potential (see the cosine α term in the equation). The field strength required for breakdown at these sites is donated by Ec. b) When supracritical field strengths E > Ec are applied the breakdown voltage is also exceeded at positions on the membrane oriented at a certain angle, α, to the field vector.
For the purposes of illustration only, the formation of "pores" is shown to indicate where breakdown has occurred even though there is evidence that pores are not generated (Zimmermann, 1986). Note that it is assumed in this figure that the intrinsic membrane potential is zero or negligibly small compared to the potential generated by the external field (taken from Zimmermann, 1986).

sufficiently high, breakdown will occur in the membrane of one hemisphere and only in the other one at much higher field strengths. Asymmetric breakdown related to only one hemisphere minimizes irreversible side effects within the cell and thus increases the viability of the field-treated cells. This is because the current flowing through the cell interior once breakdown in one hemisphere has occurred is restricted by the intact membrane of the other hemisphere. Optimum conditions for asymmetric breakdown can be set up by variation of the ionic strength of the solution and by enzymatic pretreatment of the cells (Zimmermann, 1986). Both procedures change the surface potential of the outer membrane and therefore the intrinsic field within the membrane. Asymmetric breakdown of the cell membrane can therefore be achieved in all living cells. This phenomenon was first seen in plant protoplasts, because of their large size (Mehrle et al, 1985). Fig. 3. represents breakdown-induced asymmetric release of intracellular compounds (not yet identified) from a plant protoplast (Mehrle, Hampp and Zimmermann, unpublished data). The compounds attract chemotactic bacteria which accumulate at that hemisphere which was originally oriented towards the anode. Fig. 4 is an example of double asymmetric uptake of a dye into a plant protoplast. Neutral red was added before the application of the two breakdown pulses to the external solution. In this experiment two breakdown pulses of alternating polarity were applied at an interval of about 1 second. As expected, the uptake of dye now occurs through both hemispheres because each hemisphere is oriented to the anode during this pulse sequence. The angular dependence of the permeabilization is also well demonstrated.

Asymmetric breakdown conditions are useful in electroinjection of foreign membrane-impermeable substances into cells and their application to mammalian cells in the presence of plasmids have led to very high yields of genetically transformed hybrids (Stopper et al, 1987, 1988).

For electrofusion on other hand, asymmetric breakdown conditions have apparently to be avoided. In this process breakdown pulses are applied to the membranes in the contact zone of adhered cells, and asymmetric breakdown may prevent intermingling of the membranes (Zimmermann, 1986). However, pretreatment of one or both fusion partners with digestive enzymes may change the intrinsic field and, in consequence, the external field strengths required for breakdown. Due to changes of the intrinsic field it is, therefore, possible under some circumstances to compensate partly the radius-dependence of the breakdown voltage. Electrofusion of different sized cells is facilitated by pronase pretreatment as was observed for lymphocytes and myeloma cells (Vienken and Zimmermann, 1982). However, it has to be noted that pronase may affect the viability of the hybrids if it is taken up during the fusion process through the permeabilized areas.

Resealing Properties

The electrical breakdown of the cell membrane occurs very rapidly (in less than 50 ns) and is reversible provided that the pulses are not too long and do not exceed a critical intensity which leads to an unfavorable ratio of the permeabilized area to the total surface (Zimmermann, 1986). Under these conditions the pertubations in the membrane introduced by the field pulse reseal, so that the original electrical properties and impermeability of the membrane are restored. Resealing proceeds more rapidly with increasing temperature. In contrast to electroinjection, electrofusion of aligned cells is normally carried out at room temperature, provided that weakly conducting solutions are used during pulse application. At this temperature, the resealing of the field-induced membrane pertubations in the cell chain termini is still fast enough to avoid substantial loss of intracellular substances into the external medium, and slow enough to allow intermingling of the membranes of adhered cells in a chain. Performance of electrofusion at higher temperatures would facilitate the competitive process of resealing of the individual adhered membranes in the contact zone resulting in lower fusion and hybrid yield.

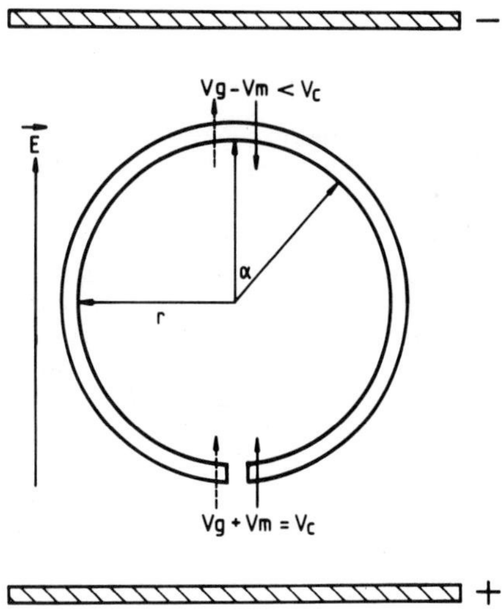

Fig. 2. Asymmetric breakdown
Superimposition of the membrane potential, Vg, generated by an external field, E, on the intrinsic membrane potential, Vm, (which is normally of the order of -100 mV, inside negative). The field vectors are parallel in the hemisphere facing the anode and antiparallel in the other one. As indicated, breakdown will first occur in the membrane of the cell hemisphere oriented to the anode (taken from Zimmermann, 1986).

Membrane Contact

Close membrane contact is a pre-requisite for fusion. Without such contact, only solute exchange between the cell interior and the external medium occurs. This field-induced exchange can be used - as mentioned above - for entrapment of membrane-impermeable substances such as drugs, proteins and DNA (so-called electroinjection, electropermeabilization or electroporation).

Controlled membrane contact can be achieved by various physical forces. In the standard electrofusion technique a non-uniform alternating electric field of low strength (V/cm range) and, therefore, of long duration (seconds to minutes) is applied.

In an inhomogeneous electrical field the forces acting on the induced positive and negative charges within the cells are different due to the field inhomogeneity; thus a net force arises that induces migration of the cells (Pethig, 1979, Pohl, 1979). The movement of a particle in a non-uniform field is independent of the polarity of the field and is therefore also observed in an alternating field (usual frequency range 10 kHz to 5 MHz) provided that poorly conducting media are used in order to avoid turbulence due to local heat development. This phenomenon which has been known since the 1920s (see literature quoted in Zimmermann, 1982, Pethig, 1979) has been termed dielectrophoresis.

Depending on the differences between the dielectric properties of the medium and the cells (which vary with the frequency of the alternating field) movement of the cells or particles into the direction of either highest or lowest field gradients is observed. These effects are therefore termed positive and negative dielectrophoresis, respectively. Particles approaching each other during their movement towards the regions of highest

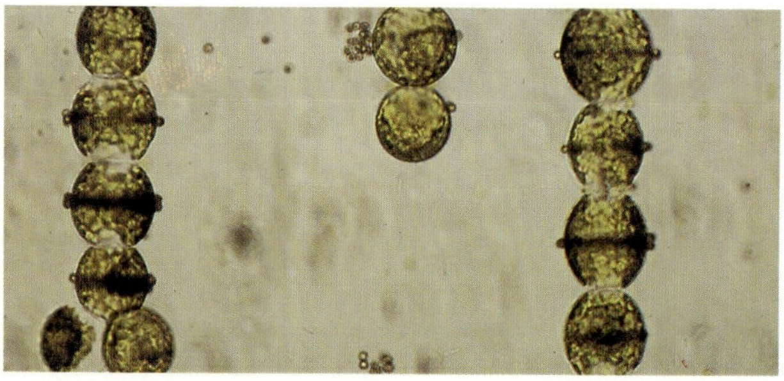

Figure 6

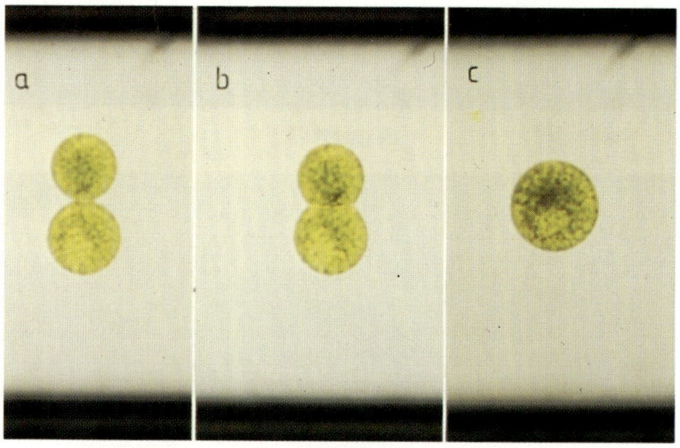

Figure 7

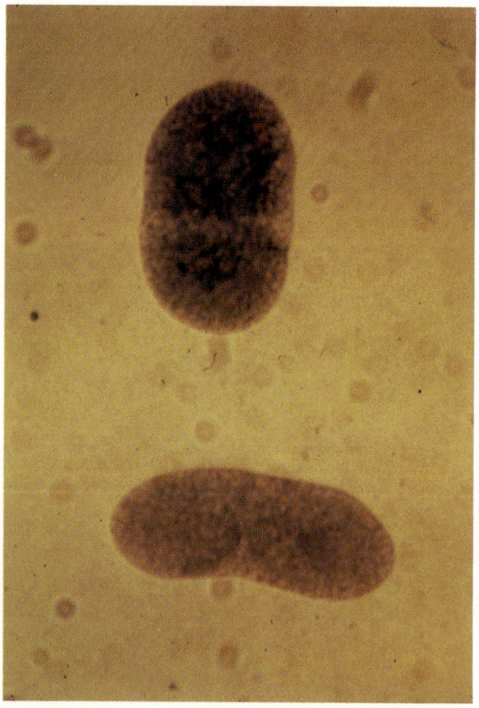

Figure 10

Fig. 6. Formation of concentric rings of polystyrene beads (2.5 μm diameter) around oat protoplasts aligned in the non-uniform alternating field (field strength 240 V/cm, frequency 1 MHz) between two cylindrical electrodes.

In contrast to the bacteria in Fig. 5, the beads show negative dielectrophoresis at this frequency (taken from Mehrle et al, 1988).

Fig. 7. Electrofusion between aligned oat protoplasts.

Field conditions: application of an alternating field of 150 V/cm strength and 1 MHz frequency for 60 s followed by 1 breakdown pulse of 1 kV/cm strength and 20 μs duration. Photographs were taken just before application of the breakdown pulse (a), after 10 s (b) and 10 min after pulse application (c).

Fig. 10. Nuclear membrane fusion in Ehrlich ascites tumor cells after dielectrophoretic alignment and subsequent application of breakdown pulses.

Field conditions: alternating field of 220 V/cm strength and 1.7 MHz frequency followed by 2 pulse trains each of 3 pulses of 5 kV/cm strength and 15 μs duration. Nuclei were Giemsa stained after fixation with Carnoy's fixative. 520 x magnification, oil immersion.

or lowest field gradients attract each other. Provided that the force is sufficiently high to overcome Brownian motion, they move together. The closeness of actual contact between cells will depend on the electrostatic repulsion between the opposed negatively charged cell membranes and on the repulsive hydration forces. The former force will become more effective at a distance determined by the magnitude of the charges and the ionic strength of the medium, whereas the hydration force operates at a separation of about 2 nm. This is the distance required for electrofusion (Stenger and Hui, 1986). Under optimum field conditions particle (cell) chains along the field lines result, in which the cells are in very close contact with each other.

Although a complete theory of dielectrophoresis and chain formation now exists (Sauer 1985), it is in practice almost impossible to predict the detailed response of the cells to a non-uniform alternating field. The reason is - among other things - the pertubation of the field by the introduction of the particles. An idea of the distribution of the field may be obtained by using small particles as field probes (Mehrle et al, 1988). It is assumed that the particles align themselves along the field lines (due to dipole-dipole interaction, in either negative or positive dielectrophoresis (Figs. 5 and 6). Particles more polarizable than the medium such as bacteria will be dielectrophoretically concentrated into volumes of higher field density, whilst particles less polarizable than the medium (such as polystyrene latex particles) will concentrate into areas of lower dielectrophoretic force. As indicated in the two figures, the biological particles are aligned at the "poles" of terminal plant protoplasts (used in these experiments) and as close as possible to the contact zones whereas the artificial beads form concentric rings along the chain of plant protoplasts. The position of the concentric rings around the individual protoplasts in a chain depends on the size of the protoplasts and on their location in the chain (not shown). When differently sized protoplasts are arranged in a chain the smaller protoplasts become enveloped by aligned bacteria. This indicates that the field strength and/or inhomogeneity in such a chain is greater around the smaller than around the larger protoplasts.

These phenomena occur at much lower voltages between the electrodes than those required for alignment of these particles in the absence of the protoplasts, suggesting that the local field strength and the inhomogeneity close to the contact zones and poles of aligned protoplasts may be considerably higher than in the external space. This may explain why symmetric breakdown leading to fusion can occur in both hemispheres of aligned cells despite the fact that isolated cells in the same media show asymmetric breakdown.

The complementary arrangements of positive and negative dielectrophoretic probes clearly indicates that the high density of field lines is restricted to the cell contact and pole areas. Assuming that only moderately supracritical field pulse are used, then breakdown will also be restricted to these areas. This has the important consequence that release of intracellular solutes must be very small under electrofusion conditions. Such a release, which might affect the viability of the fused products, should be restricted to the pole areas of terminal cells in the chain. This selective membrane permeabilization is one reason why electrofusion results in higher yields of viable hybrids than the conventional chemical and viral fusion techniques which modify the whole membrane area.

Application of this technique for field lines illustration, to the lymphocyte/myeloma cell system will undoubtedly permit more accurate prediction of the field parameters to be made, and so achieve optimum membrane contact and subsequent gentle fusion.

General Remarks

The present overview of the experimental conditions for electrofusion is necessarily condensed. For further detailed information with regard to the field and physical system parameters which are involved in electrofusion and have to be considered for successful hybridization, interested readers are referred to other review articles of one of the the authors. However, the above considerations may help to show investigators in this field the intricate, but predictable interactions between the field and system parameters on one hand and the cellular properties on the other hand. Changes in one of the parameters, e.g. of the suspension density of the cells between the two electrodes or of the surface charge of the outer membrane by enzyme pre-treatment, may completely alter the optimum conditions and the kinetics of fusion. Failures in electrical field-mediated hybridization (particularly in hybridoma formation) can arise if the relationships between the various parameters are left out of consideration or if inappropriate equipment is used. This will also become obvious from the following section which deals with the electrofusion of lymphocytes and myeloma cells as well as with mammalian cell model systems. Particular emphasis is given to the influence of media composition and the pre-incubation time of the cells before fusion on hybrid yield.

ELECTROFUSION

Hybridoma Production

Electrofusion can be achieved between cells of the same species (intraspecies) or of different species (interspecies). The fusion of bacteria, invertebrate eggs, yeast and plant protoplasts (Fig. 7) as well as of mammalian cells have demonstrated the general applicability of the electric field pulse technique (Zimmermann, 1986, Zimmermann and Urnovitz, 1987). The use of electrofusion for immortalization of antigen-specific B-cells has been investigated by many researchers because of the apparent limitations of the PEG induced cell fusion, which is commonly used in hybridoma formation. Electrofusion of murine myeloma cells with B-lymphocytes suspended in isotonic mannitol solution was first reported by Vienken and Zimmermann (1982). They used a special electro-hydraulic procedure in order to obtain high yields of pronase-treated myeloma/B-lymphocyte pairs. Production of human hybridoma cells by electrofusion was reported at the same time by Bischoff et al, (1982). Clones were obtained in these experiments; however, the capability of the clones to secrete monoclonal antibodies was not investigated.

Karsten et al, (1985) described the production, by electrofusion, of mouse hybridomas producing monoclonal antibodies to antigens of a human mammary carcinoma cell line. The yield of hybridomas was ten times higher than that in PEG-mediated fusion. Brown et al, (1986) electrically manufactured hybridomas producing antibodies specifically against the glucose transport carrier and against ankyrin from human erythrocytes. In comparative experiments with PEG these authors found that electrofusion only resulted in slightly higher yields. Gravekamp et al, (1985, 1987) electrically fused cytotoxic human T cells from different sources with different types of human T-lymphoma cells and mouse B-myeloma cells. The field conditions were similar to those reported by Vienken and Zimmermann (1982, 1985) with the interesting exception that the frequency of the alignment field was adjusted to 5 MHz (for explanation, see below). Gravekamp et al, (1985, 1987) found that electrofusion resulted in proliferating hybridomas. The yield of growing hybridoma colonies depended on the tumor fusion partner used. For one tumor cell line (HSB-1) considerably more hybridomas were obtained with electrofusion than with the PEG fusion performed in parallel.

In all the above experiments the fusion partners were pretreated with pronase and fused in the absence of ions in the medium. Apart from the specific fusogenic properties of the tumor cell type used for immortalization (Gravekamp et al, 1987), the lack of electrolytes in the fusion medium may also explain the variations in yield reported by these laboratories.

Lo et al, (1984) eliminated the dielectrophoretic collection because it was at that time suspected that the prolonged treatment of the cells by an alternating field may also harm the viability of the fusion partners. These authors were able to establish contact between lymphocytes and myeloma cells by synthesizing a selective cross-linking agent. Immunogen-avidin complexes were used to bridge the specific lymphocytes bearing immunoreactive surface immunoglobulin to biotinylated myeloma cells. Although antibody-secreting hybridoma cells were obtained this method has been criticized for several reasons. A considerable improvement to this avidin-biotin method was recently published by Wojchowski and Sytkowski (1986). In the first step of their approach, biotin was attached to immunogen and to the surface of murine myeloma cells (via N-hydroxy-succinimidobiotin). The cellular molecular complex was fused by electric field pulses resulting in a high frequency of hybridomas producing specific immunoglobulin. This approach may prove to be very powerful although in the meantime it has turned out that the simple standard electrofusion technique (which uses dielectrophoresis for membrane contact) also leads to high frequencies of hybridomas.

Vienken and Zimmermann (1985) showed that, as in the case of yeast cells (Schnettler and Zimmermann, 1985, Broda et al, 1987), divalent cations played an important role in the fusion of myeloma cells and lymphocytes and that the viability of the fusion products is considerably enhanced if the osmoticum used in the fusion medium is inositol or sorbitol. Optimum hybridoma yields which are orders of magnitude larger compared to PEG-mediated fusion, are obtained if 0.1 mM calcium ions and 0.5 mM magnesium ions are present in the fusion medium. The acetates of these cations seem to be superior to the chlorides. The reason may be that acetate uptake into the fusion partners decreases the intracellular pH which, in turn, favors the intermingling of the membranes of the fusion partners by indirect changes in membrane fluidity. Some evidence for a specific influence of acetates on the fusion kinetics has been obtained by experiments in which myeloma cells and activated B-lymphocytes could be fused by addition of high concentrations of potassium or sodium acetate, to the medium (Schmitt and Zimmermann, unpublished data). This "acetate-mediated" fusion resulted in very high yields of antibody-secreting hybridomas.

The presence of the divalent cations in electrofusion media permits the omission of enzyme pre-treatment of the cells, probably because of changes in the surface charge of the fusion partners due to binding of Ca^{2+} and Mg^{2+} and, correspondingly, of electrofusion-promoting changes in the intrinsic field within the membrane (see above). The importance of low concentrations of divalent cations in the fusion medium was also recently confirmed by Ohnishi et al. (1987). These authors showed that the electric-field-mediated generation of antibody-producing hybridomas from either mouse or human fusion partners pre-treated with DNase is about one magnitude larger than in comparable PEG-mediated fusion. Ohnishi et al, (1987) also reported that addition of calmodulin and glycerylmono-oleate (a fusogenic agent) to the fusion medium can slightly enhance the yield of hybridomas. Considerably higher yields of antibody-producing hybridoma cells were obtained very recently by Schmitt et al, (1988). These authors fused mouse myeloma cells with B-lymphocytes taken from mice which were immunized against dinitrophenol. The dinitrophenol was coupled either to bovine serum albumin or to haemocyanine. In these experiments the lymphocytes and myeloma cells (in a ratio 5:1) were dielectrophoretically aligned by an alternating field of 250 V/cm strength and 1.5 MHz frequency for 30 s. Fusion was initiated by injection of three breakdown pulses of 2 kV/cm strength and 15 μs duration. The time interval between the consecutive pulses was adjusted to 1 s. The field strength of 2 kV/cm corresponds to the breakdown voltage at the "pole" of the small lymphocytes. For the larger myeloma cells the "critical" field strength would be 1 kV.cm indicating that a larger area of the membrane of these cells is broken down under these conditions.

For high yields of hybridoma cells it was important that the fusion process was performed at a temperature of 25°C and that the alternating field was applied for further 30 s after application of the pulse train. Furthermore, the fused cells must be kept for about 30 min in the fusion medium before they are transferred to nutrition medium and are subsequently subjected to selection in HAT medium. Most of the hybridoma cells obtained showed antibody production. The yield of 2.68×10^{-4} was higher than that found in comparable PEG-mediated fusion or indeed any other value reported in the literature for hybridoma yields.

Using the improved fusion protocol of Vienken and Zimmermann (1985) Foung and co-workers (personal communication) were able to produce human monoclonal antibodies to cytomegalovirus (CMV) by fusion of a human-mouse cell line with B-lymphocytes isolated from CMV seropositive individuals. The fusion efficiency achieved was eighty fold greater than PEG-induced cell fusion.

Model Systems

Further progress in the prediction (and therefore in the long-term improvement) of hybridoma yield is expected from electrofusion of other cell systems which have appropriate properties to study the influence of certain biological and biophysical parameters on fusion and hybrid yield. Considerable insight into the influence of the size distributions of the fusion partners as well as of growth and pre-incubation conditions on hybrid yield was obtained by electrofusion of yeast cells (Broda et al, 1987) and of hybridoma cell lines with myeloma cells (Urnovitz, Zimmermann, Schmitt, Kleinhans and Shimizu, unpublished data).

In particular, the fusion of hybridoma cell lines with myeloma cells (which themselves are hybridoma cells, see e.g. Stopper et al, 1988) has a decisive advantage in electric field-mediated hybridoma production: i.e., the nearly identical volume of these two fusion partners allows a detailed study of the role of other biochemical and biophysical cellular and media parameters on hybrid yield because of the elimination of the radius-dependence of the breakdown voltage (see above).

Urnovitz, Zimmermann and co-workers electrically fused myeloma (SP2/O, $HGPRT^-$) and G8 (TK^-) cells (a hybridoma produced by chemical fusion of SP2/O and CB6-mouse immune spleen cells) and studied the dependence of the hybrid yield on the pre-incubation time at 25°C. The fusion media contained the appropriate amounts of divalent cations, bovine serum albumin and iso-osmolar concentrations of sorbitol. As indicated by Fig. 8 the hybrid yield decreases nearly exponentially with pre-incubation time suggesting that the entire fusion process has to be performed within about 30 minutes in order to avoid a significant reduction in yield. Preliminary experiments in which the intracellular ATP content of the fusion partners was measured as a function of pre-incubation time showed that the decrease in yield is paralleled by a decrease in intracellular ATP (Fig. 9). Addition of glucose to the external medium resulted in a very rapid reduction of

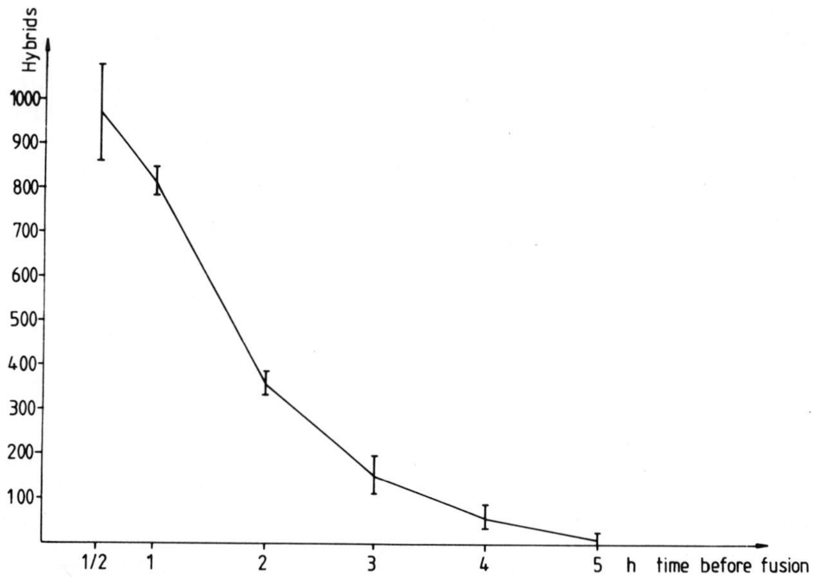

Fig. 8. Hybrid yield from electrofusion between mouse myeloma (SP2/0) and G8 cells (a hybridoma produced by chemical fusion of SP2/0 and CB6 immune spleen cells) as a function of the pre-incubation time. Both cell lines are HAT-sensitive. Incubation was carried out at 25°C in a medium consisting of 0.28 M sorbitol. 0.5 mM magnesium acetate, 0.1 mM calcium acetate and 1 mg/ml bovine serum albumin. Field conditions: alternating field of 250 V/cm strength and 2 MHz frequency for 30 s followed by 3 pulses of 2.3 kV/cm strength and 15 μs duration. After fusion, clones were selected by their HAT-resistance and counted.

intracellular ATP concentration, whereas the presence of additional potassium acetate led to an almost constant ATP level over the time measured.

It is therefore conceivable that the control and maintenance of the ATP level in the fusion partners plays an important role in hybridoma yield. This is also suggested by experiments performed on electrofused plant protoplasts (Verhoek-Köhler et al, 1983). These data also suggest that a second investigation of the effects of acetate and divalent cations on fusion and hybrid yield should be carried out.

These brief considerations clearly indicate that we are still far away from understanding all the numerous steps involved in electrofusion and that studies of model systems may contribute considerably to further improvements in this new fusion technology.

IMPACT OF ELECTROFUSION

Although electrofusion has become an ubiquitous tool in the hands of scientists with a wide array of interests and specialities, the physical background of this technology still raises some major difficulties, particularly to researchers working in the field of hybridoma production. So far results obtained with this technique are very encouraging. However, more combined effort from biophysical and immunological laboratories is required to develop the wide potential of electrofusion. We believe that apart from the ability to predict the fusion conditions of any cell system and to control the fusion process, electrofusion offers some unique features which make this technique superior in comparison to the other conventional fusion methods. The justification for this statement lies partly in recent results from our laboratory.

Zimmermann et al, (1984) and Bertsche et al, (1988) found that in electrofused mammalian cells, a very high frequency of nuclear membrane fusion is observed after fusion of the plasma membranes of dielectrophoretically aligned cells has occurred (Fig. 10). In chemically or virally induced fusion, nuclear membrane fusion has seldom been reported and then only after synchronization. Bertsche et al, (1988) found evidence that high frequencies of nuclear membrane fusion can be achieved by pulling the nuclei of dielectrophoretically aligned cells close to the contact zone between them. It is obvious that this repositioning process facilitates nuclear membrane fusion once the plasma membranes of the adhered cells coalesce. The repositioning of the nuclei is apparently achieved by intracellular dielectrophoresis which can be induced by frequencies of the alternating field in the range of 1 to 5 MHz. At these frequencies the plasma membrane (which shields the cell interior against the field at lower frequencies) becomes conductive so that a potential difference can be built up across the double nuclear membrane.

Intracellular dielectrophoresis can only operate in a very narrow frequency range for several physical reasons (Zimmermann et al, 1988). The absolute value of optimum frequency for this phenomenon in the range given above depends on the specific properties of the fusion partners. High frequencies of nuclear membrane fusion may therefore be one of the main reasons for the high yield of hybridomas in the experiments of Gravekamp et al, (1987) although these workers pretreated the individual cells with pronase (see above). The work of Bertsche et al, (1988) also yielded another result which clearly indicated differences between electrofusion and the conventional fusion methods. High yields of hybrids are obtained only when fusion is predominantly between two cells. Fusion of more than two cells does not lead to proliferating hybrids. However, dielectrophoretic chains almost always consist of more than two cells, except at very low suspension densities. Nevertheless, it is found that fusion products formed from more than two cells are rare events provided that the

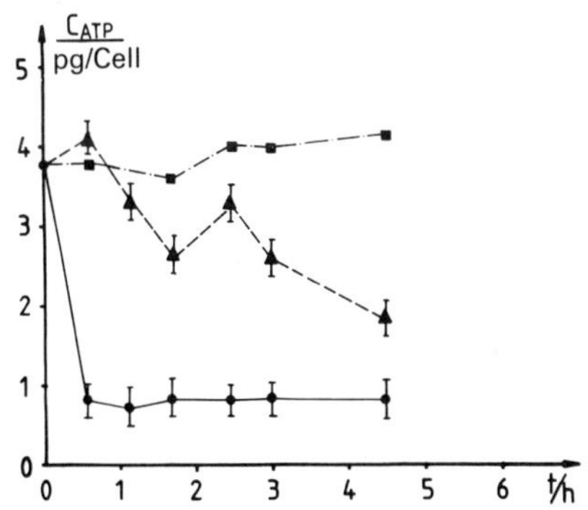

Fig. 9. Intracellular ATP concentration in G8 cells as a function of pre-incubation time in media of different composition. Triangles: medium composition as in Fig. 8; circles: in the presence of additional 10 mM glucose and squares: in the presence of additional 10 mM glucose and 30 mM potassium acetate.

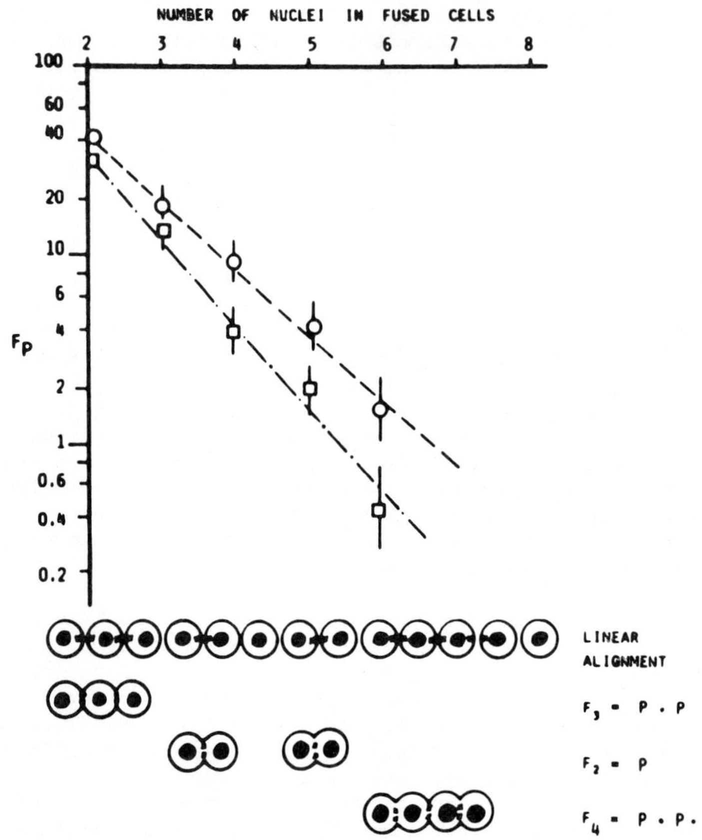

Fig. 11. The partial fusion index Fp (p = 2, 3...) of heterogenous cell systems analyzed by using Giemsa staining of the cell nuclei (see Fig. 10).
Circles represent fusion between Chinese hamster cells (V79-S181) and Ehrlich ascites tumor cells (EAT-F2). Squares represent fusion between EAT-F2 and Muntjac cells. Broken lines through measured points represent a fit to the theoretical relationship between the fusion index and the number, n, of the nuclei. The lower part shows a graphic representation of the probability of fusion between different numbers of cells aligned in a chain according to the theory (see text); the partial fusion index F_3 is the product of p x p with p = probability for fusion of two adjacent aligned cells; correspondingly, F_4 = p x p x p is the probability of fusion of four cells. For further explanation, see Bertsche et al, (1988).

suspension density of the fusion partners in the electrode gap is still so low that the chains are separated from each other. Otherwise, multiple fusion can occur leading to the formation of giant cells (e.g. Zimmermann, 1982, 1986). Bertsche et al, (1988) showed that the relative rates of occurrence of two-cell, three-cell etc. fusion events in a long cell chain is controlled by the probability for fusion occurring between any two neighboring cells and that this probability is constant over the entire chain. These authors defined a partial fusion index, Fp, which was measured on cells having Giemsa stained nuclei. Fp is the ratio of the number of (fusing or fused) cells containing "n" nuclei to the total initial number of cells. According to this definition a semi-logarithmic relationship between this partial fusion index and the number of nuclei is expected provided that a constant fusion probability, P, between two neighboring cells exists. This was indeed found for mammalian cells (Fig. 11) indicating that electrofusion is controlled by a simple law of probability (in contrast to the highly variable chemically or virally induced fusion). Taking these findings together, it is obvious that a high yield of hybrids is an inherent feature of the electrofusion technique.

CONCLUSIONS

The aim of this review was not only to summarize recent results on the production of hybridoma cells by means of electrofusion, but also to point out the numerous aspects of this field-mediated process which have to be investigated in detail before fusion protocols can be developed which allow electrofusion of a small number of lymphocytes with myeloma cells. This is indeed required if production of various human monoclonal antibodies is considered on a large scale. It is open to question whether this result can be achieved by electrofusion. However, a spectacular possibility recently came into the range of reality. This may give further impetus to the application of this technology to human hybridoma production. Performance of electrofusion under microgravity (Zimmermann and Urnovitz, 1987) may pave the way to reach hybridoma yields of the order of 1 to 10%, a figure which is undoubtedly required for the production of human hybridomas. The possibility of simple positioning of the fusion partners as well as the elimination of sedimentation and convection under microgravity conditions allows arrangements of dielectrophoretically aligned cell pairs of the desired combination and a reduction of the field strength and duration of the applied alternating field. The latter possibility could enhance the hybrid yield, because rearrangement of membrane components due to the field is limited (see Zimmermann, 1982). In addition, the altered metabolic conditions of cells under microgravity (Halstead and Dutcher, 1987) may lead to results which are not predictable at the present stage of information. Space biotechnology therefore seems to be a promising line of research that should be performed in parallel to the necessary terrestrial work.

Acknowledgements

This work was supported by grants of the Deutsche Forschungsgemeinschaft (SFB 176) and of the Bundesministerium fuer Forschung und Technologie (DFVLR, no. 01QV3451) to UZ.

REFERENCES

Bertsche, U., Mader, A., and Zimmermann, U., 1988, Nuclear membrane fusion in electrofused mammalian cells, Biochim.Biophys.Acta., 939:509.

Bischoff, R., Eisert, R. M., Schedel, I., Vienken, J., and Zimmermann, U., 1982, Human hybridoma cells produced by electrofusion, FEBS Letters, 147:64.

Broda, H., G., Schnettler, R., and Zimmermann, U., 1987, Parameters controlling yeast hybrid yield in electrofusion: the relevance of pre-incubation and the skewness of the size distributions of both fusion partners Biochem.Biophys.Acta, 899:25.

Brown, S. M., Ahkong, Q. F, Sage, D. A., and Lucy, J. A., 1986, Electrically induced cell fusion in the production of monoclonal antibodies, Biochem.Soc.Trans., 14:289.

Foung, S. H. K., Sasaki, D. T., Grumet, F. C., and Engleman, E. G., 1982, Production of functional human T-T cell hybridomas in selection medium lacking aminopterin and thymidine, Proc.Natl.Acad.Sci.USA., 79:7484.

Gravekamp, C., Bol., S. J. L., Hagemeijer, A., and Bolhuis, R. L. H., 1985, Production of Human T-Cell Hybridomas by Electrofusion in: "Human Hybridomas and Monoclonal Antibodies," E.G. Engelman, S.K. Foung, J.Larrick, eds.

Gravekamp, C., Santoli, D., Vreugdenhil, R., Collard, J. G., and Bolhuis, L. H., 1987, Efforts to produce human cytotoxic T-cell hybridomas by electrofusion and PEG fusion, Hybridoma, 6:121.

Halstead, T. H., and Dutcher, F. R., 1987, Plants in Space, Ann.Rev.Plant. Physiol., 38:317.

Jeltsch, E., and Zimmermann, U., 1979, Particles in a homogeneous electrical field: A model for the electrical breakdown of living cells in a coulter counter, J.Electroanal.Chem., 104:349.

Karsten, U., Papsdorf, G., Roloff, G., Stolley, P., Abel, H., Walther, I., and Weiss, H., 1985, Monoclonal Anti-cytokeratin antibody from a hybridoma clone generated by electrofusion, Cancer Clin.Oncol., 21:733.

Mehrle, W., Hampp, R., Zimmermann, U., and Schwan, H. P., 1988, Mapping of the field distribution around dielectrophoretically aligned cells by means of small particles as field probes, Biochem.Biophys.Acta., 939:561

Mehrle, W., Zimmermann, U., and Hampp, R., 1985, Evidence of asymmetrical uptake of fluorescent dyes through electropermeabilized membranes of Avena mesophyll protoplasts, FEBS LETT., 185:89.

Lo, M. M. S., Tsong, T. Y., Conrad, M. K., Strittmatter, S. M., Hester, L. D., and Snyder, S. H., 1984, Monoclonal antibody production by receptor-mediated electrically induced cell fusion, Nature, 310:792.

Ohnishi, K., Chiba, J., Goto, Y., and Tokunaga, T., 1987, Improvement in the basic technology of electrofusion for generation of antibody-producing hybridomas, J.Immunol.Methods., 100:181.

Pethig, R., 1979, "Dielectric and Electronic Properties of Biological Materials," John Wiley and Sons, Chichester, New York.

Pohl, H. A., 1979, "Dielectrophoresis," Cambridge University Press, Cambridge.

Sauer, A. F., 1985, Interaction forces between microscopic particles in an external electromagnetic field in: "Interaction Between Electromagnetic Fields and Cells," A. Chiabrera, C. Nicolini, and H.P. Schwan, eds., Plenum Press, New York, p.181.

Schmitt, J. J., Zimmermann, U., and Neil, G., 1988, Efficient generation of stable antibody forming hybridoma cells by electrofusion, Hybridoma, in press.

Schnettler, R., and Zimmermann, U., 1985, Influence of the composition of the fusion medium on the yield of electrofused yeast hybrids, FEMS Microbiology Letters, 27:195.

Stenger, D. A., and Hue, S. W., 1986, Kinetics of ultrastructural changes during electrically induced fusion of human erythrocytes, J.Membrane Biol., 93:43.

Stopper, H., Jones, H., and Zimmermann, U., 1987, Large scale transfection of mouse L-cells by electropermeabilization, Biochim.Biophys.Acta., 900:38.

Stopper, H., Zimmermann, U., Neil, G., 1988, Increased efficiency of transfection of murine hybridoma cells with DNA by electro-permeabilization, J.Immun.Meth., 109:145.

Verhoek-Köhler, B., Hamp, R., Ziegler, H., and Zimmermann, U., 1983, Electrofusion of mesophyll protoplasts of Avena sativa: determination of the cellular adenylate-level of hybrids and its influence on the fusion process, Planta, 158:199.

Vienken, J., and Zimmermann, U., 1982, Electric field-induced fusion: electro-hydraulic procedure for production of heterokaryon cells in high yield, FEBS Letters, 137:11.

Vienken, J., and Zimmermann, U., 1985, An improved electrofusion technique for production of mouse hybridoma cells, FEBS Letters, 182:278.

Wojchowski, D. M., and Sytkowski, A. J., 1986, Hybridoma production by simplified avidin-mediated electrofusion, *J.Immun.Meth.*, 90:173.

Zimmermann, U., 1982, Electric field-mediated fusion and related electrical phenomena, *Biochim.Biophys.Acta.*, 694:227.

Zimmermann, U., 1986, Electrical breakdown, Electro-permeabilisation and electrofusion, *Rev.Physiol.Biochem.Pharmacol.*, 105:175.

Zimmermann, U., 1987, Electrofusion of cells *in*: "Methods of Hybridoma Formation," A.H. Bartal and Y. Hirshaut, eds., Humana Press, p.97.

Zimmermann, U., Arnold, W. M., and Mehrle, W., 1988, Biophysics of Electroinjection and Electrofusion, *J.Electrostat.*, in press.

Zimmermann, U., Pilwat, G., and Riemann, F., 1974, Reversible dielectric breakdown of cell membranes by electrostatic fields, *Z.Naturforsch.*, 29:304.

Zimmermann, U., Scheurich, P., Pilwat, G., and Benz, R., 1981, Cells with manipulated functions: new perspectives for cell biology, medicine and technology, *Angewandte Chemie., Int.Ed.Engl.*, 20:325.

Zimmermann, U., Schultz, J., and Pilwat, G., 1973, Transcellular ion flow in E. coli B and electrical sizing of bacteria, *Biophys.J.*, 13:1005.

Zimmermann, U., and Stopper, H., 1987, Electrofusion and electropermeabilization of cells *in*: "Biomembrane and Receptor Mechanisms," E. Bertoli, D. Chapman, A. Cambria, and U. Scapagnini, eds., Liviana Press, Padova and Springer Verlag, Berlin, Heidelberg, New York, p. 371.

Zimmermann, U., Vienken, J., Pilwat, G., and Arnold, W. M. 1984, Electrofusion of cells: principles and potential for the future *in*: "Cell fusion (Ciba symposium 103)," D. Evered and J. Whelan, eds., Pitman, London, p.60.

Zimmermann, U., and Urnovitz, H., 1987, Principles of electrofusion and electropermeabilisation, *Meth.Enzym.*, 151:194.

PRODUCTION OF HUMAN MONOCLONAL ANTIBODIES USING EPSTEIN-BARR VIRUS

Elizabeth V. Walls, Kitty Lam and Dorothy H. Crawford

Department of Virology
Royal Postgraduate Medical School
London

INTRODUCTION

The technique of somatic cell fusion described by Köhler and Milstein (1975) has rapidly become a very successful procedure for the *in vitro* production of rodent monoclonal antibodies (MAbs). In comparison with the rodent system, however, the development of human MAbs has been slow due to difficulties in obtaining ideal material from human subjects and to technical problems which have only been partially resolved.

Human MAbs can be generated by various methods from a) human lymphoblastoid B cell lines (BLCL) obtained by direct immortalization of antigen-specific lymphocytes with Epstein-Barr virus (EBV) (Steinitz, 1977); b) human x human hybridomas made by fusing human lymphocytes with immortal human B cells such as myeloma, lymphoma or lymphoblastoid cell lines (Olsson and Kaplan, 1980); c) human x mouse hybridomas made by fusing human lymphocytes with a mouse myeloma cell line (Nowinski, 1980; Schlom, 1980). These methods have been used to generate human MAbs specific for a wide range of antigens with similar success rates (Larrick and Bourla, 1986).

In certain situations, such as *in vivo* diagnostic or therapeutic procedures in man, the benefits gained from the use of human MAbs make the time and effort required to generate them worthwhile.

PRODUCTION OF HUMAN MONOCLONAL ANTIBODIES BY EBV-IMMORTALIZATION OF B LYMPHOCYTES

Biological Properties of EBV

EBV is a human B lymphotropic herpes virus which is the causative agent of infectious mononucleosis (Henle, 1968) and is also aetiologically associated with two human tumors: African Burkitt's Lymphoma (Epstein and Achong, 1979) and anaplastic nasopharyngeal carcinoma (Epstein, 1978). Primary infection of adolescents and adults with EBV can cause acute infectious mononucleosis whereas childhood infection is generally asymptomatic. The majority of the adult population worldwide (up to 90%) is seropositive for antibodies (Ab) to EBV (Henle, 1968). The virus, however, is not eradicated from the body after primary infection and biologically active EBV can be detected in saliva and throat washings taken from seropositive individuals. Replication of EBV occurs in epithelial cells lining the pharynx and a few virus-infected B lymphocytes are present in the circulation. In this carrier state a balance is maintained between the level of virus infection and the cellular and humoral immune responses which keep the infection controlled.

In vitro, EBV can immortalise B lymphocytes (Pattengale, 1973) giving rise to long-term cell lines which carry the viral genome and express restricted viral genes. B lymphocytes infected with EBV may also be polyclonally activated to proliferate and produce immunoglobulins (Ig) (Rosen, 1977). Polyclonal B lymphocyte activation and B lymphocyte immortalization by EBV are both T lymphocyte independent processes.

Lymphoblastoid B cell lines

Lymphoblastoid B cell lines established from normal blood or tissues, whether infected in vitro with EBV or spontaneously generated from lymphocytes infected in vivo, are usually of polyclonal derivation and associated with EBV. Characteristically, the cells have a "hand-mirror" shape and often grow in clumps (Fig. 1). The cells are usually diploid and the cytoplasm contains many free ribosomes with poorly developed rough endoplasmic reticulum and Golgi apparatus. Phenotypically BLCL express some "normal" B cell markers including HLA class I and class II antigens: CD19: CD20 and surface Ig, with or without cytoplasmic Ig and Ig secretion. Antigens characteristic of activated lymphocytes (not necessarily B cell restricted) are present on BLCL such as CD23; PCA-1; an interleukin-2 receptor similar to that of T cells and receptors for transferrin (Crawford and Ando, 1986). BLCL express EBV encoded antigens including the latent membrane protein and the nuclear antigen complex (EBNA) (Reedman and Klein, 1973), other viral proteins namely the early antigen and viral capsid antigen complexes are only expressed during lytic infections. Some of the phenotypic changes which occur when B lymphocytes are activated and immortalized by EBV are represented diagrammatically in Fig. 2. Under standard culture conditions not more than 5% of EBV-immortalized cells are undergoing lytic infection, thus virus production by such cultures is at a low level.

B Lymphocyte Targets for Immortalization

The receptor for EBV on B lymphocytes is a 140kDa glycoprotein (CD21) which also binds the complement component C3d, albeit at a different site from the virus (Nemerow, 1986). Recently, using limiting dilution experiments, it has been shown that although most peripheral blood B lymphocytes may be infected by EBV, only a subset (from 1-20%) are activated or immortalized in vitro (Tosato and Blaese, 1985; Katsuki, 1977). This may be a reflection of the inherent resistance of certain subpopulations of B lymphocytes to immortalization by EBV or of differences in growth requirements of these subsets after EBV-infection (Aman, 1984; Chan, 1986). The frequency of B lymphocyte immortalization by EBV has been shown to decrease with progress through the cell cycle (Roome and Reading, 1987) and this must be taken into account when using in vitro stimulation to enrich for antigen specific cells.

Enrichment of Antibody-Producing Cells

A supply of antigen specific B lymphocytes is a basic requirement for the generation of long-term Ab-secreting cell lines. Splenic or lymph node lymphocytes of repeatedly immunized animals are routinely used for the generation of rodent MAbs whereas peripheral blood is the most accessible source of human lymphocytes and only 10% of these are B cells. Since

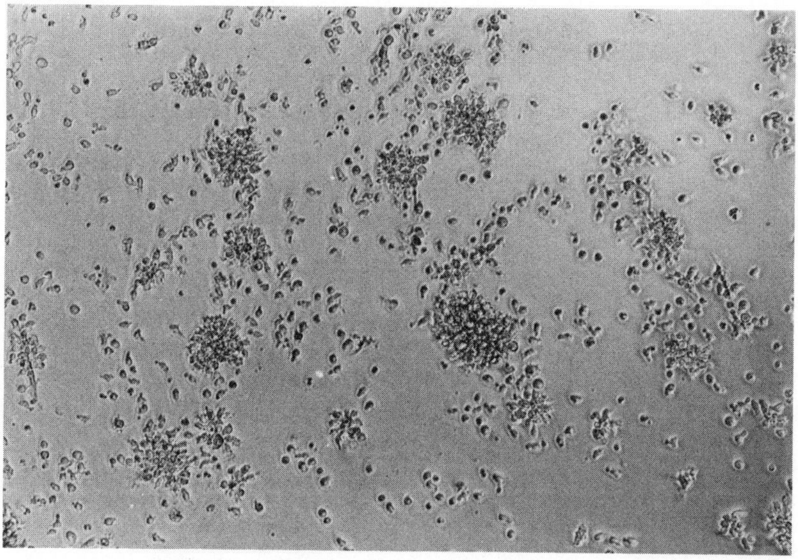

Fig. 1. Photomicrograph of a human lymphoblastoid B cell line generated by infection of cord blood mononuclear cells with EBV.

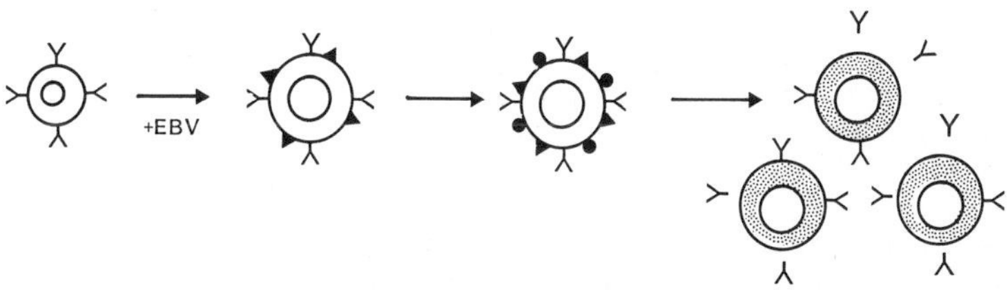

B cell activation:	Small resting B cell	Lymphoblast RNA synthesis 24 hours	Lymphoblast DNA synthesis 72 hours	Cell division
	CD23 - IL-2R - (CD25)	CD23 + ▲ IL-2R - (CD25)	CD23 + IL-2R + ● (CD25)	
B cell maturation:	Small resting B cell surface Ig	Activated B cell surface Ig →	Activated B cell cytoplasmic Ig	Ig secreting lymphoblast cytoplasmic and secreted Ig
EBV antigen expression:	None	EBNA+	EBNA+	EBNA+

(stippled) cytoplasmic Ig

Fig. 2. Activation and immortalisation of B lymphocytes by EBV.

specific Ab producing B lymphocytes are detectable in the blood only for a short period after immunization (Stevens, 1979), donors are restricted to individuals who have recently been immunized (by natural infection or deliberate vaccination) with the antigens of interest. When available, surgical specimens of tonsils, lymph nodes and spleen provide richer sources of B lymphocytes than peripheral blood, however immune donors are still required.

Lymphocyte preparations may be enriched for antigen specific cells either by rosetting with antigen-coupled erythrocytes: "panning" on an antigen coated solid phase or by flow cytometry following binding of antigen labelled with a fluorescent dye (Boyd, 1984). Purification of lymphocytes producing Ab specific for red blood cell surface antigens is most easily achieved by rosetting with the erythrocytes (Crawford, 1983a). Papain or neuraminidase treatment of erythrocytes increases the number of rosettes by removing negatively charged surface sialic acid residues thereby allowing close contact between the red cells. Chromic chloride can be used to couple other antigens to red cells prior to rosetting (Steinitz, 1984). Another method that has been used to enrich for specific Ab-producing cells involves coating a solid phase (for example multiwell plates or petri dishes) with antigen and then incubating the cell preparation in the vessel. Antigen-binding cells adhere to the solid phase and may be harvested after removing the non-adherent cells. The conditions for coating the solid phase vary with the nature of the antigen and the surface properties of the vessel, but generally entail incubation with antigen in an alkaline buffer. This procedure has been used successfully by Winger (1983) to select cells specific for sperm antigens prior to EBV-immortalisation.

Casali (1986) has used flow cytometry to isolate B lymphocytes specific for tetanus toxoid, thyroglobulin, insulin and B-galactosidase, these cells could subsequently be infected with EBV giving rise to BLCL secreting specific Ab. Lymphocyte preparations were first incubated with biotinylated antigens and after a second incubation with an avidin-fluorescein isothiocyanate conjugate, the antigen-binding (fluorescent) cells could be separated using a flow cytometer. Since biotin is a relatively small molecule (approximately 341 daltons) biotinylation of antigens, including short peptides, is unlikely to interfere with their binding to cell surface Ig.

Another, indirect, procedure to purify antigen-specific cells has been developed (Kozbor and Roder, 1983) in which the lymphocytes were first incubated with soluble antigen which bound to, and was subsequently capped from, Ab-bearing B cells, leaving them without surface Ig. Those cells that still expressed surface Ig after this procedure were rosetted with rabbit anti-human Ig coupled to sheep erythrocytes, separated and discarded, leaving an enriched population of antigen specific cells.

An alternative approach to the selection procedure is to stimulate specific Ab-producing cells <u>in vitro</u> by culturing mononuclear cells from a seropositive donor with antigen. A method for stimulation of memory B cells specific for influenza virus was developed (Callard, 1979) and used in the generation of EBV-immortalized BLCL secreting anti-influenza Ab (Crawford, 1983b). Time course studies using this system showed that although maximum Ab responses were obtained 6 to 7 days after stimulation with antigen, by this time the ability of the B lymphocytes to be immortalized by EBV was much reduced. This is presumably due to the previously noted fact that activated B lymphocytes are resistant to EBV immortalisation. Thus, a 3-day incubation period with antigen prior to EBV infection was the time that led to the successful production of MAb. The specific antigen stimulation described by Callard (1979) is a T-lymphocyte-dependent system, and therefore T cells must be removed after the antigen incubation prior to EBV immortalisation in order to prevent activation of cytotoxic T lymphocytes specific for EBV-immortalized cells (Moss, 1978). Protocols have been developed for the <u>in vitro</u> expansion of B lymphocyte populations secreting Ab specific for various antigens. Larrick (1985) and Winger (1983) have reported the successful <u>in vitro</u> stimulation of primary immune responses to a variety of antigens and immortalisation of these cells by EBV to give Ab-producing cell lines.

Production of VIrus and Infection of B Lymphocytes

Infectious, wild-type EBV is present in the saliva or throat washings of infectious mononucleosis patients (Miller, 1973) and can be obtained intermittently from the same sources in normal seropositive individuals (Golden, 1973). In most laboratories, however, the lymphoblastoid cell line B95-8 is used to produce virus (Miller, 1972). The cell line was derived by *in vitro* infection of cotton-topped marmoset peripheral blood mononuclear cells with wild-type virus obtained from an infectious mononucleosis patient. Under standard culture conditions around 1-10% of cells are undergoing a lytic viral infection, but this percentage can be increased up to 80% following treatment with inducing agents such as phorbol esters (zur Hausen, 1978) and sodium-n-butyrate (Luka, 1979). Supernatant medium harvested from cultures at or approaching the stationary phase of their growth curves usually has an immortalizing titre of around 10^{-3} when used to infect human cord blood lymphocytes. It has been shown (Zerbini and Ernberg, 1983) by analysis of virus dose-response curves that increasing the virus dose increases the number of cells infected, thus if a high yield of infected cells is required it is advisable to concentrate the virus by high-speed centrifugation. Stocks of virus preparations can be stored in aliquots at -70°C without loss of viral activity.

Infection of cells with EBV is routinely carried out by suspending pellets of up to 10^7 cells (depending on the virus titre) in 1 ml of undiluted, filtered B95-8 culture supernatant and incubating at 37°C for one hour, agitating the cells occasionally to keep them in suspension. Longer incubation periods (up to 24 hours) can be used, however, there may be a loss of viability of cells incubated at high density in the spent B95-8 culture medium. Following incubation with EBV, the cells are washed and cultured as required.

Cell Culture

When EBV-infected B lymphocytes are cultured the effects of regulatory T lymphocytes in the preparations and the requirements of the B lymphocytes for feeder cells must be taken into consideration. Individuals seropositive for EBV possess circulating memory T lymphocytes which recognize autologous EBV-carrying B lymphocytes. These cells probably control growth of EBV-infected B lymphocytes *in vivo* and give rise to specific cytotoxic cells when stimulated *in vitro* (Rickinson, 1979). Thus, in cultures of EBV-infected blood mononuclear cells from seropositive donors seeded at high density (1-2 x 10^6/ml), proliferating foci of B lymphocytes regress after 3-4 weeks and no cell lines are obtained. The problem of regression may be overcome in one of four ways; cultures may be set up at low cell densities (1-5 x 10^5/ml; mononuclear cells may be depleted of T lymphocytes by rosetting with sheep erythrocytes; the T cells may be inhibited by cyclosporin, (Crawford, 1981) alternatively, phytohaemagglutinin may be added to the medium at the initiation of culture to inhibit specific T lymphocyte activation.

It has been shown that activation of B lymphocytes by EBV to produce Ig is T cell-independent (Rosen, 1977), however, immortalisation of B lymphocytes by the virus does require accessory cells (Pope, 1974). Monocytes present in peripheral blood mononuclear cell preparations are efficient feeders for outgrowth of EBV-infected B lymphocytes. When purified B lymphocyte populations are infected with EBV, feeder cells must be added to the culture for optimal cell growth. The cellular interactions involved in this phenomenon are not fully characterized but probably involve growth factor production. They are not genetically restricted since many allogeneic cells have been used successfully. Sources of feeder cells include foetal fibroblasts; autologous or allogeneic peripheral blood mononuclear cells and cord blood mononuclear cells, all of which are irradiated prior to culture. There is evidence that BLCL, once established, produce their own growth factors (Blazar, 1983; Gordon, 1984).

Cloning of B Lymphoblastoid Cell Lines

Following EBV infection of peripheral blood B lymphocytes from an individual who has serum Abs to a specific antigen, it is usually possible to detect specific Ab in the medium after 2-3 weeks of culture. On continuous culture, however, specific Ab levels decline and become undetectabe by 12-20 weeks after infection. There is some evidence that BLCL are heterogeneous with respect to proliferation, expression of EBV-encoded antigens, B cell differentiation and Ab secretion (Crawford and Ando, 1986). Thus the cells with the slowest growth rates produce the most Ig (Wendel-Hansen, 1987) and may be overgrown by non-producing cells. The factors regulating growth and Ab production by EBV-immortalized B lymphocytes are not fully characterized, although, as mentioned earlier, endogenous growth factors are made (Blazar, 1983) which are essential for cell growth. Repeated cloning and screening for specific Ab secretion should therefore be carried out early in the culture period in order to establish stable cell lines. BLCL have been cloned by limiting dilution in microtitre plates (Crawford, 1983a) and in soft agar (Steinitz, 1977; Kozbor, 1979), both methods require feeder cells to support the growth of cells plated at low concentrations. The same sources of feeder cells described earlier for establishing BLCL may be used for cloning and must be irradiated before use. EBV-immortalized BLCL have a cloning efficiency of around 1-5% which is much lower than that of rodent hybridoma lines (over 10%). After cloning two or three times, stable cell lines may be obtained which continue to grow and produce Ab in culture for years. Occasionally, however, cloned cell lines have grown for 3-6 months in culture and then died (Melamed, 1985). The reason for this cell death is not clear but it may result from senescence of a particular subpopulation of B cells and emphasises the heterogeneity among B lymphocytes with respect to their immortalisation by EBV.

Antibody Production by Cloned Cell Lines

The majority of EBV-immortalized BLCL produce Ab of the IgM class although lines secreting Ab of the IgG and IgA classes have been generated. It was suggested that the predominance of IgM producing cells in immortalized B lymphocyte populations indicated a preference of the virus for infecting IgM-bearing B lymphocytes (Katsuki, 1977). There is, however, little difference in EBV receptor density among different B lymphocyte subpopulations (Aman, 1984) thus the preponderance of lines secreting IgM may just be a reflection of their high frequency in the starting population.

Cloned BLCL have been reported to produce levels of Ab ranging from 10ng/ml up to 10-20μg/ml, the latter being equivalent to the amounts secreted by rodent hybridomas. Comparison of values for Ab production by various cell lines is hindered since cell numbers and culture conditions are not standardized among different laboratories. Attempts have been made to increase the rates of Ig secretion by EBV-immortalized cell lines including the addition of Abs to HLA-class II determinants (Palacios, 1983) or by adding culture supernatants from T lymphocytes stimulated either by mitogens or specific antigen. As yet, however, no reliable strategy has been developed.

SOMATIC CELL HYBRIDISATION OF EBV-IMMORTALIZED CELL LINES

EBV-immortalized BLCL secreting specific Ab have been fused with plasmacytoma cell lines of human or murine origin in order to stabilize Ab production (Kozbor, 1982; Kozbor and Roder, 1983). This approach has, however, been hindered by the lack of suitable cell lines which will fuse efficiently and, after fusion, will support growth and Ab production. Initially it was thought that the immortal fusion partner should have some phenotypic characteristics of plasma cells, in order to support Ab secretion by the hybrids. In practice, however, the fusion frequency, cloning efficiency and growth rate of the fusion partner are more important for the generation of stable hybrids since the fusion efficiency of even the best human cell lines is around 10-fold lower than that of rodent myeloma cell lines (Edwards and O'Hare, 1986; Kozbor and Roder, 1983). Human plasmacytomas, the most differentiated B lymphoid neoplasia, are very difficult to maintain in culture and, as alternatives, EBV-carrying BLCL and mouse myeloma cell lines have been used.

It must be possible to select against the parental cell lines such that only hybrids will grow after fusion. Several procedures are available for this (reviewed by Edwards and O'Hare, 1986) of which hypoxanthine-aminopterin-thymidine (HAT) selection, developed by Littlefield (1964) is generally used. Cell lines rendered deficient in the purine salvage enzyme, hypoxanthine-guanine phosphoribosyl transferase are sensitive to HAT selection medium. When such lines are fused with EBV-immortalized, antigen-specific BLCL counter-selection against the latter cells must be carried out. This may be achieved by making the HAT-sensitive partner resistant to ouabain and since HAT sensitivity is a recessive trait while ouabain resistance is dominant, only the hybrids will grow in medium containing HAT plus ouabain.

One drawback to the somatic cell hybridization procedure is the production of Ig heavy and light chains by the drug-marked fusion partners resulting in dilution of the specific Ab produced by the hybrids. Thus, further selection of the fusion partners for Ig non-producer variants must be undertaken, mouse myeloma cell lines which do not produce Ig heavy or light chains are already available. The main disadvantage of mouse-human fusions is the preferential loss of human chromosomes by the hybrids. This loss of chromosomes is not random, it has been shown that although human chromosome 2 carrying the Kappa light chain genes is lost, chromosomes 14 and 22, carrying the heavy and Lambda light chain genes, respectively, are retained (Croce, 1979; Erikson, 1981). However, retention of the chromosomes with the relevant Ig genes by the hybrids, whether human-human or mouse-human, still may not result in Ab secretion unless other, as yet undetermined, criteria are met (Raison, 1982).

An alternative approach to finding cell lines suitable for fusion with human B lymphocytes has been the development of mouse-human heteromyelomas (Teng, 1983). These cells which are the products of fusions between mouse and human myeloma cells, do not produce mouse or human Ig and are reported to fuse with human B lymphocytes with the efficiency of the mouse cell parent whilst retaining the ability of the human parent to form stable, Ab secreting hybrids.

USES OF HUMAN MONOCLONAL ANTIBODIES

There are several purposes for which human MAbs are potentially useful and would be better suited than rodent MAbs. Analyses of immune responses to infectious agents using human MAbs may reveal which antigenic determinants elicit protective responses and thus aid the development of subunit vaccines. This has been pioneered using human MAbs to tetanus toxoid and Streptococcus pneumoniae in experimental animal models (Tiebout, 1984; Steinitz, 1986). Human immune responses to polymorphic determinants of (human) cell surface molecules often differ from those obtained by xenogeneic immunization. Human MAbs to erythrocyte and histocompatibility antigens would therefore be useful for blood group and tissue typing.

A major application for human MAbs is in the diagnosis, prevention and treatment of a variety of diseases (Engleman, 1985; Larrick and Bourla, 1986) where human antisera and mouse MAbs are already of value. The main disadvantages of using immune sera or pooled serum Ig *in vivo* are the increasing demand which has led to a shortage of suitable donors, the lack of reproducibility among different batches and the possibility of contamination by infectious agents. Mouse MAb preparations on the other hand, are often of limited use in humans due to the production of Ab to mouse Ig (Shawler, 1985). Also, species-specific carbohydrates on Ab molecules may be important for their effector functions such as complement activation, Ab dependent cellular cytotoxicity and phagocytosis. Examples of the therapeutic uses of human Abs include the prevention of Rhesus haemolytic disease of the newborn; neutralisation of bacterial toxins and snake venom; elimination of circulating drugs (overdoses); passive immunisation of "high-risk" individuals against infections such as hepatitis, herpes simplex, varicella zoster, rubella and measles and the prevention and treatment of cytomegalovirus infections in immunocompromised patients. Ab specific for tumor cell surface antigens could be radiolabelled or conjugated with toxins and used for location or eradication of tumors, respectively, although in these situations it is essential that there is no cross-reaction with normal cell surface components.

The therapeutic use of human MAbs is not, however, without problems arising from the procedures by which they are generated. Apart from the practical difficulties associated with bulk cultures of lymphoid cell lines, the MAb preparations must be free from infectious, transforming and pyrogenic agents. Concern has been expressed that MAbs prepared from EBV-immortalized cell lines will be contaminated with the virus or viral DNA. In this context it must be remembered that around 90% of the adult population worldwide has been infected with the virus (Either subclinically or manifested as infectious mononucleosis) and thereafter carry the virus in B lymphocytes as well as secreting it intermittently into the saliva. MAbs produced by mouse-human hybrids may be contaminated with murine retroviruses which are secreted by mouse myeloma cells and can infect human cells in culture (Weiss, 1982). To overcome these problems, rigorous procedures for purification of Ig from immortalized B lymphocyte cultures have been developed and the products are tested for the presence of infectious viruses and viral DNA (Crawford, 1983c; Crawford, 1985). A purified human MAb to the Rhesus D antigen produced by an EBV-immortalized B cell line has been tested *in vivo* and found to be safe (Crawford, 1986).

FUTURE PROSPECTS

There is much potential for improving the methods by which human MAbs of varied antigen specificities and effector functions may be generated.

One area of practical difficulty still to overcome is associated with the production of large quantities of human MAbs that may be required especially for therapeutic use. Rodent hybridomas can be grown as ascites tumors *in vivo* yielding Ab levels one thousand fold greater than can be achieved in culture. At present human cell lines cannot be grown this way, although it may become feasible by growing them in immuno-deficient mice. However, this is expensive and may lead to contamination of the product with mouse Ig or mouse viruses. Most *in vitro* culture techniques use foetal calf serum which is becoming increasingly expensive and scarce and also adds to the time and cost of purification of the Ab. This has led to the development of serum free media (Iscove and Melchers, 1978) and the possible use of surplus fractions of human plasma as another source of protein. In recent years, there has been a major commitment to developing new mammalian cell culture methodologies that increase both cell density and product concentration. Several techniques including rotating perfusion filter (Feder and Tolbert, 1983) and hollow fibers developed by Bio response (Lyderson, 1985) have proven useful in supporting the growth of cells to higher densities. Another technique, originally described by Lim and Sun (1980) has been patented by Damon Biotech where the cells are encased in a semi-permeable polymer membrane. Of major significance is their claim that human-human hybridomas can produce MAbs in the mg/ml range within the capsules, clearly a significant increase over conventional systems.

Apart from increasing the efficiency of the current procedures, the techniques of molecular biology are being utilized to produce "designer Abs" with specified properties of antigen binding, effector function, distribution and susceptibility to degradation. Chimeric Abs, in which mouse antigen binding regions are coupled with human constant regions have been produced using recombinant DNA technology (Morrison, 1985) and hybrid molecules of Ab and enzymes or toxins could also be constructed. Investigations to find oncogenes or viruses other than EBV which could directly transform human B lymphocytes are underway and there have been some encouraging results using oncogenes (Kennett, 1984). As the availability and variety of human MAbs increases, so the range of applications for these reagents will broaden. This is dependent not only on technological advances, but also on improved understanding of the processes of B lymphocyte activation and differentiation as well as the mechanisms of cellular transformation.

REFERENCES

Aman, P., Ehlin-Henriksson, B., and Klein, G., 1984, Epstein-Barr virus susceptibility of normal human B lymphocyte populations, *J.Exp.Med.*, 159:208.

Blazar, B. A., Sutton, L. M., and Strome, M., 1983, Self-stimulating growth factor production by B-cell lines derived from Burkitt's lymphomas and other lines transformed *in vitro* by Epstein-Barr virus, *Cancer Res.*, 43:4562.

Boyd, J. E., James, K., and McClelland, D. B. L., 1984, Human monoclonal antibodies - production and potential, *Trends in Biotechnology*, 2:70.

Callard, R. E., 1979, Specific *in vitro* antibody response to influenza virus by human blood lymphocytes, *Nature*, 282:734.

Casali, P., Inghirami, G., Nakamura, M., Davies, T. F., and Notkins, A. L., 1986, Human monoclonals from antigen-specific selection of B lymphocytes and transformation by EBV, *Science*, 234:476.

Chan, M .A., Stein, L. D., Dosch, H-M., and Sigal, N. H., 1986, Heterogeneity of EBV-transformable human B lymphocyte populations, *J.Immunol.*, 136:106.

Crawford, D. H., 1981, Lymphomas after cyclosporin A treatment *in*: "Transplantation and Clinical Immunology XIII, 2, J.T. Touraine, J. Traeger, H. Betuel, J. Brochier, J.M. Dubernard, J.P. Revillard, and R. Triau, eds., Excerpta Medica, Amsterdam.

Crawford, D. H., Barlow, M. J., Harrison, J. F., Winger, L., and Huehns, E. R., 1983a, Production of human monoclonal antibodies to rhesus D antigen, *Lancet*, i:386.

Crawford, D. H., Callard, R. E. Muggeridge, M. I., Mitchell, D. M., Zanders, E. D., and Beverley, P. C. L., 1983b, Production of human monoclonal antibody to X31 influenza virus nucleoprotein, J.Gen. Virol., 64:697.

Crawford, D. H., Huehns, E. R., and Epstein, M. A., 1983c, Therapeutic use of human monoclonal antibodies, Lancet, i:1040.

Crawford, D H., 1985, Production of human monoclonal antibodies using Epstein-Barr virus in: "Human Hybridomas and Monoclonal Antibodies," E.G. Engleman, S.K.H. Foung, J. Larrick, and A. Raubitschek, eds., Plenum, New York.

Crawford, D. H., and Ando, I., 1986, EB virus induction is associated with B cell maturation, Immunology, 59:405.

Crawford, D. H., Azim, T., Daniels, G. L., and Huehns, E. R., 1986, Monoclonal antibodies to the Rh.D. antigen in: "Progress in Transfusion Medicine," J. Cash, ed., Churchill Livingstone, Edinburgh.

Croce, C. M., Shander, M., Martinis, J., Cicurel, L., D'Ancona, G. G., Dolby, T. W., and Koprowski, H., 1979, Chromosomal locations of the genes for human immunoglobulin heavy chains, Proc.Natl.Acad.Sci., USA, 76:3416.

Edwards, P. A. W., and O'Hare, M. J. 1986, The production of human monoclonal antibodies in: "Monoclonal antibodies," P.C.L. Beverley, ed., Churchill Livingstone, Edinburgh.

Engleman, E. G., Foung, S.K.H., Larrick, J., and Raubitschek, A. 1985, "Human hybridomas and monoclonal antibodies," Plenum, New York.

Epstein, M. A., 1978, Epstein-Barr virus - discovery, properties and relationship to nasopharyngeal carcinoma in: "Nasopharyngeal Carcinoma, Aetiology and Control," G. de Thé, Y. Ho, and W. Davis, eds., IARC, Lyon.

Epstein, M. A, and Achong, B. G., 1979, The relationship of the virus to Burkitt's lymphoma in: "The Epstein-Barr Virus," M.A. Epstein and B.G. Achong, eds., Springer, Berlin.

Erikson, J., Martinis, J., and Croce, C. M., 1981, Assignment of the genes for human λ immunoglobulin chains to chromosome 22, Nature, 294:173.

Feder, J., and Tolbert, W. R., 1983, Large scale cultivation of mammalian cells, Sci.Am., 248:36.

Golden, H. D., Chang, R. S., Prescott, W., Simpson, E., and Cooper, T.Y., 1973, Leukocyte-transforming agent: prolonged excretion by patients with mononucleosis and excretion by normal individuals, J.Infect Dis., 127:471.

Gordon, J., Ley, S. C., Melamed, M. D., English, L. S. and Hughes-Jones, N. C., 1984, Immortalized B lymphocytes produce B-cell growth factor, Nature, 310:145.

Henle, G., Henle, W., and Diehl, V., 1968, Relation of Burkitt's tumor-associated herpes-type virus to infectious mononucleosis, Proc.Natl. Acad.Sci.USA., 59:94.

Iscove, N. N., and Melchers, F., 1978, Complete replacement of serum by albumin, transferrin, and soyabean lipid in cultures of lipopolysaccharide-reactive B lymphocytes, J.Exp.Med., 147:923.

Katsuki, T., Hiruma, Y., Yamamoto, N., Abo, T., and Kumagi, K., 1977, Identification of the target cells in human B lymphocytes for transformation by Epstein-Barr virus, Virology, 83:287.

Kennett, R., Jonak, Z. and Ikegaki, N., 1984, Monoclonal antibodies and molecular genetics: oncogenes and oncogene products in: "Monoclonal antibodies and functional cell lines progress and application," R. Kennett, K. Bechtol and T. McKearn, eds., Plenum, New York.

Köhler, G., and Milstein, C., 1975, Continuous cultures of fused cells secreting antibody of predefined specificity, Nature, 256:495.

Kozbor, D., Steinitz, M., Klein, G., Koskimies, S., and Makela, O., 1979, Establishment of anti-TNP antibody producing human lymphoid lines by preselection for hapten-binding followed by EBV-transformation, Scand.J.Immunol, 10:187.

Kozbor, D., Lagarde, A. E., and Roder, J. C., 1982, Human hybridomas constructed with antigen-specific Epstein-Barr virus-transformed cell lines, Proc.Natl.Acad.Sci, USA., 79:6651.
Kozbor, D., and Roder, J. C., 1983, The production of monoclonal antibodies from human lymphocytes, Immunol. Today, 4:72.
Larrick, J. W., Dyer, B. J., Senyk, G., Hart, S. M., Moss, R., Lippman, D., Jahnsen, M. C., Wang, J., Weintraub, H., and Raubitschek, A. A., 1985, In vitro expansion of human B cells for the production of human monoclonal antibodies, in: "Human hybridomas and Monoclonal Antibodies," E.G. Engleman, S.K.H. Foung, J. Larrick, and A. Raubitschek, eds., Plenum, New York.
Larrick, J. W., and Bourla, J. M., 1986, Prospects for the therapeutic use of human monoclonal antibodies, J.Biol.Resp.Modifiers, 5:379.
Lim, F., and Sun, A. M., 1980, Microencapsulated islets as bioartificial endocrine pancreas, Science, 210:908.
Littlefield, J. W., 1964, Selection of hybrids from matings of fibroblasts in vitro and their presumed recombinants, Science, 145:709.
Luka, J., Kallin, B., and Klein, G., 1979, Induction of the Epstein-Barr virus (EBV) cycle in latently infected cells by n-butyrate, Virology, 94:228.
Lynderson, B. K., Pugh, G. C., Paris, M. S., Sharma, B. P., and Noll, L. A., 1985, Ceramic matrix for large scale animal cell culture, Bio Technol, 3:63.
Melamed, M. D., Gordon, J., Ley, S. J., Edgar, D., and Hughes-Jones, N. C., 1985, Senescence of a human lymphoblastoid clone producing anti-Rhesus (D), Eur.J.Immunol., 15:742.
Miller, G., Shope, T., Lisco, H., Stitt, D., and Lipman, M., 1972, Epstein-Barr virus: transformation, cytopathic changes and viral antigens in squirrel monkey and marmoset leucocytes, Proc.Natl.Acad.Sci,USA., 69:383.
Miller, G., Niederman, J. C., and Andrews, L. L., 1973, Prolonged oropharyngeal excretion of Epstein-Barr virus after infectious mononucleosis, New Engl.J.Med., 288:229.
Morrison, S., 1985, Transfectomas provide novel chimaeric antibodies, Science, 229:1202.
Moss, D. J., Rickinson, A. B., and Pope, J. H., 1978, Long-term T-cell-mediated immunity to Epstein-Barr virus in man I. Complete regression of virus-induced transformation in cultures of seropositive donor leucocytes, Int.J.Cancer, 22:662.
Nemerow, G. R., Siaw, M. F. G., and Cooper, N. R., 1986, Purification of the Epstein-Barr virus/C3d complement receptor of human B lymphocytes: antigenic and functional properties of the purified protein, J.Virol, 58:709.
Nowinski, R., Berglund, C., Lane, J., Lorton, M., Bernstein, I., Young, W., Hakomori, S., Hill, L., and Corney, M., 1980, Human monoclonal antibody against Forssman antigen, Science, 210:537.
Olsson, L., and Kaplan, H., 1980, Human/human hybridomas producing monoclonal antibodies of predefined antigenic specificity, Proc.Natl. Acad.Sci, USA., 77:5429.
Palacios, R. Martinez-Maza, O., and Guy, K., 1983, Monoclonal antibodies against HLA DR antigens replace T helper cells in activation of B lymphocytes, Proc.Natl.Acad.Sci, USA., 80:3456.
Pattengale, P. K., Smith, R. W., and Gerber, P., 1973, Selective transformation of B lymphocytes by EB virus, Lancet, ii:93.
Pope, J. H., Scott, W. and Moss, D. J., 1974, Cell relationships in transformation of human leucocytes by Epstein-Barr virus, Int.J.Cancer, 14:122.
Raison, R. L., Walker, K. Z., Halnan, C. R. E., Briscoe, D., and Basten, A., 1982, Loss of Secretion in mouse-human hybrids need not be due to the loss of a structural gene, J.Exp.Med., 156:1380.
Reedman, B. M., and Klein, G., 1973, Cellular localization of an Epstein-

Barr virus (EBV)-associated complement-fixing antigen in producer and non-producer lymphoblastoid cell lines, Int.J.Cancer, 11:499.

Rickinson, A. B., Moss, D. J., and Pope, J. H., 1979, Long-term T-cell mediated immunity to Epstein-Barr virus in man. II Components necessary for regression in virus infected leucocyte culture, Int.J.Cancer, 23:610.

Roome, A. J., and Reading, C. L., 1987, Frequency of B-lymphocyte transformation by Epstein-Barr virus decreases with entry into the cell cycle, Immunology, 60:195.

Rosen, A., Gergely, P., Jondal, M., Klein, G., and Britton, S., 1977, Polyclonal Ig production after Epstein-Barr virus infection of human lymphocytes in vitro, Nature, 267:52.

Schlom, J., Wunderlich, D., and Teramoto, Y. A., 1980, Generation of human monoclonal antibodies reactive with human mammary carcinoma cells, Proc.Natl.Acad.Sci, USA 77:6841.

Shawler, D., Bartholomew, R., Smith, L., and Dillman, R., 1985, Human immune response to multiple injections of murine monoclonal IgG1, J.Immunol., 135:1530.

Steinitz, M., Klein, G., Koskimies, S., and Makela, O., 1977, E. B. virus-induced B lymphocyte cell lines producing specific antibody, Nature, 209:420.

Steinitz, M., Tamir, S., and Goldfarb, A., 1984, Human anti-pneumococci antibody produced by an Epstein-Barr virus (EBV)-immortalized cell line, J.Immunol, 132:877.

Steinitz, M., Tamir, S., Ferne, M., and Goldfarb, A., 1986, A protective human monoclonal IgA antibody produced in vitro: anti-pneumococcal antibody engendered by Epstein-Barr virus immortalized cell line, Eur.J.Immunol., 16:187.

Stevens, R. H., Macy, E., Morrow, C., and Saxon, A., 1979, Characterization of a circulating subpopulation of spontaneous anti-tetanus toxoid producing B cells following in vivo booster immunization, J.Immunol., 122:2498.

Teng, N. N. H., Lam, K. S., Riera, F. C., and Kaplan, H. S., 1983, Construction and testing of mouse-human heteromyelomas for human monoclonal antibody production Proc.Natl.Acad.Sci, USA., 80:7308.

Teibout, R. F., Stricker, E. A. M., Hagenaars, R., and Ziejlemaker, W. P., 1984, Human lympoblastoid cell line producing protective monoclonal Ig GK anti-tetanus toxin, Eur.J.Immunol., 14:399.

Tosato, G., and Blaese, R. M., 1985, Epstein-Barr virus infection and immunoregulation in man, Adv.in Immunol., 37:99.

Weiss, R. A., 1982, Hybridomas produce viruses as well as antibodies, Immunol Today, 3:292.

Wendel-Hansen, V., Rosen, A., and Klein, G., 1987, EBV-transformed lymphoblastoid cell lines down-regulate EBNA in parallel with secretory differentiation, Int.J.Cancer, 39:404.

Winger, L., Winger, C., Shastry, P., Russell, A., and Longenecker, M., 1983, Efficient generation in vitro of monoclonal Epstein-Barr virus-transformants producing specific antibody to a variety of antigens, without prior deliberate immunization, Proc.Natl.Acad.Sci, USA, 80:4484.

Zerbini, M., and Ernberg, I., 1983, Can Epstein-Barr virus infect and transform all the B lymphocytes of human cord blood? J.Gen. Virol., 64:539.

Zur Hausen, H., O'Neill, F. J., Freese, U. K., and Hecker, E., 1978, Persisting oncogenic herpes virus induced by the tumor promoter TPA, Nature, 272:373.

Abbreviations: Ab: Antibody, BLCL: B-lymphoblastoid cell line,
EBNA: Epstein-Barr virus nuclear antigen, EBV: Epstein-Barr virus
HAT: Hypoxanthine-Aminopterin-Thymidine, Ig: Immunoglobulin
MAb: Monoclonal antibody

HUMANISING OF RECOMBINANT ANTIBODIES

Lutz Riechmann

MRC Laboratory of Molecular Biology
Hills Road
Cambridge CB2 2QH

INTRODUCTION

Immune therapy has been used to treat toxin pathogens and to provide passive immunity. However, there can be problems, as in antisera the "neutralising" antibody is only a small fraction of the total immunoglobulin, human antisera may be infected with viruses such as AIDS and hepatitis, and antibodies from animal sources are recognized as foreign. These problems should not occur if human monoclonal antibodies are used. Unfortunately the production of human monoclonal antibodies has proved to be difficult. Therefore we have attempted, via recombinant DNA technology, to turn mouse or rat monoclonal antibodies into "Human" antibodies.

At its simplest, variable regions from mouse or rat antibodies conferring the antigen specificity can be combined with human constant domains, resulting in a chimeric antibody having predetermined specificity and effector function (Neuberger et al, 1985). Although such antibodies may yet prove to be therapeutically useful, an immune response of the recipient to mouse or rat variable regions seems probable. Therefore, the parts of the animal variable domain conferring the antigen specificity - the complementarity determining regions (CDRs) - have been transplanted into the framework regions of a human variable domain. We have applied this technique to construct "humanised" antibodies binding a small hapten (Jones et al, 1986), to lysozyme (Verhoeyen et al, in preparation) and to a cell surface antigen, CAMPATH-1 (Riechmann et al, in preparation). Such "humanised" antibodies expressed in eukaryotic cells show the same specificity as the original mouse or rat antibody or a conventional chimeric antibody, but can be expected to behave like genuine human monoclonal antibodies. Furthermore they offer a choice of the constant domain and therefore the biologic effect when used therapeutically.

REFERENCES

Jones, P. J., Dear, P. H., Foote, J., Neuberger, M. S., & Winter, G., 1986, Replacing the complementarity-determining regions in a human antibody with those from a mouse, Nature, 321:522.

Neuberger, M. S., Williams, G. T., Mitchell, E. B., Jomhal, S. S., Flanagan, J. G., & Rabbitts, T. H., 1985, A hapten-specific chimaeric

IgE antibody with human physiological effector function, Nature 314:268.
Riechmann, L., Clark, M., Waldmann, H., & Winter, G., (in preparation).
Verhoeyen, M., Milstein, C., and Winter, G., (in preparation).

SECTION B
MONOCLONAL ANTIBODIES IN THE CLINICAL LABORATORY

IMMUNOASSAYS INTO THE 1990s

J. Landon and S. Akman

Department of Chemical Pathology
St. Bartholomew's Hospital
London EC1A 7BE

INTRODUCTION

All immunoassays depend upon the reversible, non-covalent binding of the analyte (the antigen) by the binding sites of specific antibodies in a reaction which obeys the Law of Mass Action. The ideal immunoassay would involve only the addition of sample (preferably saliva or urine) to a single reagent with a visual end-point and would not require the routine use of standards. This is seldom achieved and the criteria which should be applied to immunoassays are that they should be sufficiently sensitive, specific and precise for the purposes to which they will be applied and that they should be simple, robust, rapid and inexpensive and require only widely available instrumentation.

Immunoassays are used for such a wide diversity of purposes that it seems unlikely that there will be a single preferred analytical approach. For example, the most suitable type of immunoassay to quantitate circulating levels of a specific protein in the μmol/L range (such as albumin) may be inappropriate for small peptides in the fmol/L range (such as a hypothalamic releasing hormone). Likewise, the best immunoassay technique to accurately quantitate serum levels of a drug for therapeutic monitoring may be inappropriate to detect low levels of urinary human chorionic gonadotrophin for pregnancy diagnosis.

TYPES OF IMMUNOASSAY

Immunoassays can be categorized into those requiring no labelled reactant, those employing labelled antigen and those based on the use of labelled specific antibodies.

Non-Labelled Immunoassays

Many immunoassays have been developed based on the detection of the formation of large antigen-antibody complexes either as a precipitin line or by turbidometric or nephelometric means. Such assays lack sensitivity and are applicable only to large molecules with multiple epitopes. Nonetheless, their simplicity and suitability for many centrifugal and other automated systems available in most clinical chemistry departments will ensure their widespread and increasing use throughout the 1990s, for example, for the measurement of several specific plasma proteins.

Immunoassays Employing Labelled Antigens

The use of assays employing isotopically labelled antigens by Ekins and by Yalow and Bersen in 1960 greatly enhanced sensitivity and enabled the immunoassay of haptens as well as macromolecules. This approach, albeit with a non-isotopic label, should continue to dominate for the quantitation or detection of small molecules.

Immunoassays Employing Labelled Antibodies

Immunometric assays, especially of the two-site type, seem certain to dominate for the assay of molecules with more than one epitope when sensitivity, specificity and precision are requisites. The development of such assays by Miles and Hales in 1968 ranks in importance with radioimmunoassay, over which they offer considerable advantages in many circumstances.

TECHNICAL ADVANTAGES

Monoclonal Versus Polyclonal Antibodies

The antibodies are the most important reactants in all immunoassays since it is they which in large part determine both the specificity and ultimate sensitivity which can be achieved. The introduction of hybridomas to produce monoclonal antibodies (McAB) has had (and will continue to have) a major impact on immunoassay development and design. At present, such antibodies are unsuitable for non-labelled assays, offer significant advantages (with regard to ease of purification, etc) for at least one of the pair of antibodies required for two-site immunometric assays and provide a marketing advantage for labelled-antigen immunoassays. The 1990s will witness a more rational choice between McAb and their polyclonal counterparts. Thus it remains to be answered why McAb often have relatively low K values, adsorb poorly and inconsistently to assay tubes and lack all the carbohydrate moities of normal immunoglobulin G molecules.

Separation Techniques

Many techniques have been developed to simplify the separation of the antibody bound and free fractions in an immunoassay. The use of coated tubes and magnetisable particles both help to avoid the need for centrifugation, which is a time-consuming step in most assays. However, the continuing aim of most groups will be to develop immunoassays which do not require separation, either because antibody binding alters the nature of the signal or, more commonly, enhances or suppresses the signal.

Non-Isotopic Labels

Both beta- and gamma-emitting isotopes still have much to offer as the label in some immunoassays. However, their eventual replacement by fluorophores, enzymes or some other non-isotope is inevitable in the long term and, again, it is unlikely that one single type of label will dominate in all areas. For example, enzymes are ideal for non-quantitative assays since they enable a visual end-point; fluorophores can be used to attain great precision and a wide range of sensitivities; and chemiluminescent labels enable maximum sensitivity to be achieved. This general topic will probably prove to belong to the 1980s rather than the 1990s. It remains to be seen if the 1990s see the development of simple, sensitive immunosensor techniques.

THERAPEUTIC ADVANCES

One experiences a feeling of deja vu when attending meetings concerned with immunodiagnostics and it is difficult to visualize many future advances of real significance, other than with regard to new fields of application and the eventual introduction of sensitive immunologically based biosensors. Hopefully, the many advantages of specific antibodies, such as their specificity and the avidity with which they bind appropriate antigens, can now be turned to therapeutic, as opposed to diagnostic, use.

Endogenous Antibody Production

During this century the active immunization of the young has virtually eradicated many diseases, such as poliomyelitis and diptheria, in the developed world and succeeded in eliminating smallpox worldwide. The protection conferred by immunization is long-term.

Exogenous Antibodies

Short-term protection against bacteria or their toxins can be conferred by the systemic administration of exogenous antibodies. In 1891 Behring was the first to employ such antibodies, raised in the horse and directed against the diptheria toxin, and subsequently anti-toxins were introduced for the treatment of tetanus and gas gangrene. Injections of foreign proteins often induced an allergic response and, although their severity was reduced by the use of more purified immunoglobulin preparations, human antibodies are preferred.

Recently, this approach has been successfully extended to the treatment of digoxin overdose. The intravenous injection of the Fab of ovine anti-digoxin IgG results in the rapid amelioration of signs and symptoms and, in the majority of subjects, cure. Removal of the Fc fragment, which is not involved in antibody binding, appears to reduce the risk of allergic side-effects, increases the volume of distribution of the antidote and accelerates elimination of the drug, partially via the kidneys. It seems likely that this approach will be extended to other areas, although it must be noted that digoxin is particularly suitable because of its relatively large molecular weight and the small amounts necessary to cause deleterious effects.

Abzymes

Abzymes are antibodies with enzymatic (catalytic) activity. It was stated earlier that immunoassays depend upon the reversible and non-covalent binding of the antigen by specific antibodies. Enzymes also specifically bind their substrate but differ from antibodies in that they then catalyse substrate cleavage. Accepted teaching is that such cleavage occurs because the enzyme encourages the substrate to exist in its transition state. Abzymes are antibodies raised against haptens whose structure is similar to the transition state of another molecule and, during the past eighteen months, two American groups have succeeded in raising monoclonal antibodies with catalytic activity. Recently we have been the first to raise polyclonal antibodies with this ability.

Many groups will enter this exciting and potentially valuable area of research, which may rival in the 1990s the contribution of the monoclonal antibody in the 1980s. Thus, it is possible that the intravenous injection of an appropriate abzyme may become the treatment of choice for some drug overdoses and such exogenous antibodies could even threaten HIV.

METHODS AND LABELS IN IMMUNOASSAY

Roger Ekins

Department of Molecular Endocrinology
University College and Middlesex School of Medicine
University of London, London UK

INTRODUCTION

Immunoassay methods relying on radioisotopic labels have played a major role in medicine and other biologically-related fields (agriculture, veterinary science, the food industry etc) during the past two decades. Their importance has derived from their exploitation both of the "structural specificity" characterizing antibody-antigen reactions and the "detectability" of isotopically-labeled reagents, the latter permitting observation of the binding reactions between exceedingly small concentrations of the key reactants involved. The combination of these features has endowed these methods with unique specificity and sensitivity characteristics, and accounts for their ubiquitous use throughout modern medicine and biology. However, in the past few years, interest has increasingly focussed on so-called "alternative", non-radioisotopic, methods - techniques which are based on essentially the same analytical principles but which differ in the markers used to label the particular reactant (antibody or analyte) whose distribution following the basic analytical reaction constitutes the assay response. The reasons for this interest may be grouped under four headings:

1. Environmental; logistic; economic; practicality and convenience etc. (i.e. "non-scientific");
2. The attainment of higher sensitivity;
3. The development of "immunosensors" and "immunoprobes";
4. The development of "multi-analyte" assay systems.

Though intended as an indication of the use of monoclonal antibodies in an immunoassay context with special reference to non-isotopic immunoassay techniques, this presentation will primarily center on the newer and generally less familiar concepts underlying headings 3 and 4.

"NON-SCIENTIFIC" REASONS FOR THE DEVELOPMENT OF NON-ISOTOPIC IMMUNOASSAYS

Radioisotopes constitute almost ideal markers for labeling either one of the principal reactants (analyte (i.e. antigen) or antibody) involved in the immunological reactions on which immunoassays depend, and which enable the course of these reactions to be observed. Their high "specific activity" (i.e. the number of observable events per unit mass per unit time) implies that reactions between exceedingly small numbers of labeled molecules may be monitored, leading in turn to high assay sensitivity; moreover the invulnerability of radioactive decay to environmental influence, coupled with the low background levels of radioactivity normally present in biological materials, implies that the "signals" emitted by the assay system can be measured with great accuracy albeit using relatively simple equipment requiring minimal technical expertise in its use. The latter confers considerable "ruggedness" - if not to the assay procedure taken as a whole - to the signal measurements on which the assay system relies.

Conversely, the practical disadvantages of the use of radioisotopes in this context are obvious, and have been frequently discussed. They include

a. the limited shelf life of labeled reagents;
b. the necessity for radioisotopic counting equipment (precluding their "home" use by the public, or their "field" use (e.g) in third world contexts, or in agricultural settings, etc);
c. their perceived "hazardous" nature (of real significance only in major institutions faces with a large immunoassay workload and consequent problems of waste disposal, but resulting in legal constraints in many countries);
d. the difficulties of transport of chemically unstable radioactive substances between countries, and
e. the inconveniences deriving from the necessity to physically separate immunological reaction products when these are isotopically labeled. Financial penalties and/or inducements attaching to the diagnostic use of radioisotopically-based methods also influence current pressures to replace them by non-isotopic methodologies.

Such considerations have led to a proliferation of immunoassay techniques - some of great ingenuity - which, though generally lacking the sensitivity of the isotopically-based methods, may obviate particular problems associated with the latter which are of importance to the individual user. Many such methods rely on enzyme labels which, when combined with an appropriate substrate, yield colored reaction products which are observable without the use of expensive instruments, and which therefore readily lend themselves to simple qualitative measurements. Enzyme methods (and certain other techniques, such as those relying on polarized fluorescence) can also be exploited, in the case of substances of small molecular size, in the development of "non-separation" methods (i.e. methods in which the immunological reaction products need not be isolated prior to their individual measurement), and which can therefore, for example, be readily accommodated on conventional autoanalyzer equipment. Many of these methods have been discussed extensively in other publications, and need no further examination here.

THE ATTAINMENT OF HIGHER ASSAY SENSITIVITY

Aside from the practical and logistic problems associated with radioisotopic methods indicated above, a more fundamental and equally compelling reason for their abandonment derives from the current need to develop microanalytical techniques of even greater sensitivity. Radioisotopic methods are, in practice, limited to the measurement of analyte concentrations above about 10^8- 10^9 molecules/ml (i.e. approx 0.15 - 1.5 pmol/L) (Dakubu et al, 1984). However, in certain fields (e.g. virology, tumor detection, etc) there is an increasing requirement to detect and measure molecular concentrations below this level.

The factors which determine immunoassay sensitivity have also been extensively discussed (Dakubu et al, 1984; Ekins, 1985), and need not be reviewed in detail here. Nevertheless, the principal concepts involved may be summarized as follows:

a. Immunoassays relying on reagent "labels" may be subdivided into two classes; i. those relying on labeled "analyte", and observation of the labeled analyte distribution between antibody bound and unbound moieties (e.g. RIA), and ii. those relying on labeled antibody, and analogous observation of the labeled antibody distribution between analyte bound and unbound moieties (e.g. IRMA).

b. Though it has frequently been claimed (or assumed) that labeled antibody methods are intrinsically more sensitive, this proposition is, in a general sense, untrue and the reasons frequently advanced in its support are invalid.

c. Immunoassays may also be conveniently subdivided into "limited reagent" and "excess reagent" methods, depending on whether the optimal concentration of antibody required to maximize assay sensitivity tends towards zero or infinity respectively. (It should be noted that the optimal antibody concentration is essentially determined by the "error structure" of the assay system - i.e. the relationship between the precision of the measurement and the various parameters governing the design of the system. Were an assay to be totally error-free, no antibody concentration would be optimal, and the distinction between "limited" and "excess" reagent systems would be neither relevant nor valid). Limited reagent systems are frequently described as "competitive"; conversely excess reagent systems are often said to be "non-competitive". The reasons for this choice of nomenclature are diverse and poorly understood, primarily deriving from the notion that, in a typical RIA, labeled and unlabeled analyte molecules "compete" with each other for a limited number of antibody binding sites. However, the term is also applied to certain labeled antibody methods (where such a concept is clearly inappropriate), but which are distinguished by the optimal use of very small amounts of antibody, and which are correctly characterized as of "limited reagent" design. Moreover, as discussed below, immunoassays in which no label of any kind is involved can likewise be subdivided into those of "limited reagent" (or "competitive") and "excess reagent" (or "non-competitive") design. In short, the classification of immunoassays into these two categories is unrelated to the nature of the particular reactant labeled (or, indeed, whether or not a label is used); it rests entirely on the concentration of antibody shown, on the basis of statistical considerations, as necessary to maximize the precision of measurement of the target analyte concentration.

d. Nevertheless, in practice, all "labeled analyte" methods are of "limited reagent" (or "competitive") design. Labeled antibody methods are "competitive" when the signal emitted by the antibody unbound to analyte (alone) is detected and measured. Conversely when the signal emitted by the antibody bound to analyte is measured, the assay is generally of "excess reagent" (or "non-competitive") design.

e. Theoretical analysis shows that, making reasonable assumptions relating to the errors arising in immunoassay procedures, and assuming the use of high affinity antibodies (i.e. $\sim 10^{11} - 10^{12}$ L/M) sensitivities yielded by radioisotopically-based techniques (whether relying on labeled antibody (IRMA) or labeled analyte (RIA), or whether of competitive or non-competitive design) are all of closely comparable sensitivity, i.e. in the order of 10^{7} molecules/ml (see Fig. 1).

However, it may also be shown that, in the case of the non-competitive assays, a critical constraint on assay sensitivity is the specific activity of the label used. This constraint does not apply, in practice, to competitive assays relying on radioisotopic labels, whose sensitivity is limited by other factors. The significance of this conclusion is that, only by the use of labels possessing specific activities higher than those of the commonly-used radioisotopes in assays of non-competitive design, can current sensitivity limits be breached. Conversely, the use of a higher specific activity label in a competitive assay would have no significant effect on its sensitivity. Table 1 compares the specific activities of potentially-useful labels with that of ^{125}I.

f. Non-competitive designs also display a number of other advantages deriving from the high reagent (antibody) concentrations on which they generally rely. These include increased reaction speeds (and hence shorter incubation times), decreased vulnerability to environmental effects (causing variations in binding affinity between antibody and analyte), lesser dependence on high antibody binding affinity, etc. Conversely, non-competitive assays are intrinsically less specific, a disadvantage which generally necessitates the use, whenever possible, of an "immunoextraction" procedure such as is incorporated in "two-site" or "sandwich" assay. (Note: the loss of specificity inherent in non-competitive assay designs implies that they are less readily applicable to the measurement of analytes of small molecular size, which cannot be simultaneously bound by two different antibodies directed against different antigenic sites on the molecule. Such

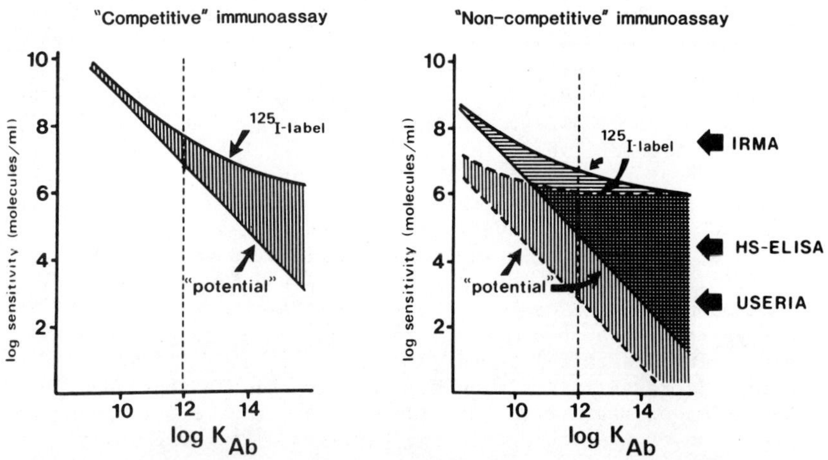

Fig. 1. Curves showing the theoretically-predicted relationship between antibody affinity and the sensitivities achievable using "competitive" and "non-competitive" immunoassay strategies. The "potential" sensitivity curves assume the use of labels of infinite specific activity; the sensitivities achievable using ^{125}I-labeled antigen or antibody are also shown. Shaded areas indicate the loss in sensitivity due to errors in measurement of the label. Curves relating to "competitive" assays assume a 1% error in measurement of the response variable arising from "experimental" errors (i.e. errors other than those arising from measurement of the label per se). Non-competitive curves assume "non-specific binding" of labeled antibody of 0.01% (lower curves) and 1% (upper curves) respectively. Arrows indicate sensitivities claimed for typical non-competitive immunoassay methodologies.

analytes are generally more appropriately measured using competitive assay methodologies.)

These basic concepts have been set out in some detail because they lead to important new insights into immunoassay design which are relevant to assays in which (as indicated above) no label of any kind is used. Fig. 2 summarizes some of these insights and illustrates the following principles:

a. All immunoassays fundamentally rely on observation on the "occupancy" of "specific" antibody binding sites by analyte following reaction between analyte and antibody.

b. "Non-competitive" assays may, in practice, be defined as those in which the assay relies on observation of an observable "signal" emitted by occupied antibody binding sites.

c. "Competitive" assays are those which rely on observation of the signal emitted by unoccupied antibody binding sites.

These concepts, which further emphasize the proposition that the distinction between competitive and non-competitive assays is unrelated to which component of the analytical reaction (analyte or antibody) is labeled, can be extrapolated to immunoassays in which neither antibody nor analyte is labeled. This notion is particularly pertinent to the development of "immunosensors," discussed in the next section.

The perception that the development of "ultra-sensitive" immunoassay systems (i.e. systems surpassing conventional RIA methods in sensitivity) depends on

a. reliance on "excess reagent" or "non-competitive" assay designs;
b. the use of non-isotopic labels displaying higher specific activities than commonly-used radioisotopes;
c. the development of highly efficient separation systems (to ensure minimization of "misclassification" of reaction products, and hence of signal "backgrounds"), and
d. dual or multi-antibody analyte-recognition systems (exemplified by "sandwich" or "two-site" assays) to maintain or increase assay specificity, has formed the basis of my own laboratory's immunoassay development program since the early to mid-1970s. Of the candidate labels for use in this context, the use of fluorescent labels combined with sophisticated time-resolution techniques for their detection (a concept arising from discussions with J. F. Tait in 1970) appeared

Table 1

<u>Specific activity of I-125</u>
1 detectable event/sec/7.5 x 10^6 labelled molecules

<u>Specific activity of enzyme label</u>
determined by enzyme "amplification factor" and detectability of reaction product.

<u>Specific activity of chemiluminescent label</u>
1 detectable event/labelled molecule

<u>Specific activity of fluorescent label</u>
many detectable events/labelled molecule

specially attractive for a number of reasons, including the possibility they offered of providing the enhanced specific activities and high signal to noise ratios required for ultra-sensitive immunoassay as indicated above. They also provide the means of development of "multi-analyte" assay systems of the kind described in a later section.

In pursuance of these ideas, my laboratory began collaboration with LKB/Wallac circa 1976-7 in the development of the instrumentation and measurement technology required to develop such methods. Fortunately a group of fluorescent substances generally known as the lanthanide chelates (and including, in particular, the chelates of europium and tellurium) facilitate such development, possessing prolonged fluorescence decay times (∿10-1000 μs), large Stokes shift (∿290 nm) and other desirable physical characteristics which permit the construction of relatively cheap instrumentation for their measurement (Hemmila et al, 1983). The fluorescent properties of the lanthanide chelates may be compared with those of a conventional fluorophor such as fluorescein which is characterized by a much smaller Stokes shift (∿28 nm), and a fluorescent decay time and emission spectrum which do not permit it to be readily distinguished form fluorescent substances present in blood (such as

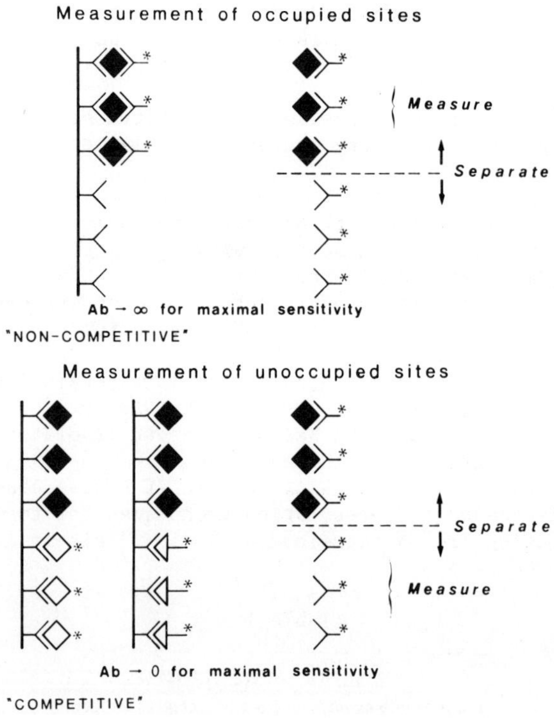

Fig. 2. The distinction between "non-competitive" (above) and "competitive" immunoassays (below) reflects the means adopted to measure occupancy of antibody binding sites. Labeled antibody methods are "non-competitive" if occupied sites of the (labeled) antibody are measured, but are "competitive" (below right) when <u>unoccupied</u> sites are measured.
Methods using labeled antigen (below left) or labeled anti-idiotypic antibody (below center) rely on measurement of sites <u>unoccupied</u> by analyte, and are therefore invariably of "competitive" design.

bilirubin) or in plastic sample holders. The unique fluorescence characteristics of the lanthanide chelates thus permit them to be measured in the presence of a fluorescence background (deriving from extraneous sources) which, in practice, approaches zero. Fig. 3 illustrates the basic concepts involved in pulsed-light, time-resolved, fluorescence measurement, which form the basis of the DELFIA immunoassay system currently marketed by LKB/Wallac.

Though it is inappropriate to pursue this subject in greater detail here, attention should also be drawn to the possibilities offered by phase-resolved fluorometry. This technique permits the separate identification of fluorophors differing in their fluorescence lifetimes by their exposure to a sinusoidally-modulated exciting light source, and observation of their demodulated, phase-shifted, light emission (McGown and Bright, 1984). This technique thus offers the possibility, not only of the development of homogenous assays (relying on a difference in fluorescence decay time of bound and free forms of the fluorescent-labeled molecule), but of distinguishing between two labeled antibody molecules in the context of multi-analyte immunoassay as discussed below.

The fundamental immunoassay design concepts discussed in this section have also subsequently been exploited by a number of other manufacturers using different "enhanced specific activity" labels in "non-competitive", "two-site", assay designs. Such labels include enzymes, which can be used as the basis of enzyme amplification systems of the kind developed by IQ Bio (Stanley et al, 1985), or as catalysts of chemiluminescent reactions (as developed by Whitehead et al, (1983) and marketed by Amersham International) or reactions yielding fluorescent (Shalev et al, 1980) or radioactive (Harris et al, 1979) products; they also include chemiluminescent markers such as the acridinium esters developed by McCapra et al, (1977) and utilized by Woodhead and his colleagues (Weeks et al, 1984), now marketed by Ciba-Corning. Common to all "ultra-sensitive" immunoassay methodologies relying on such alternative labels is their dependence on a non-competitive, labeled antibody, assay strategy when ever this is appropriate; however - for the reasons indicated above - competitive methods are generally employed for the measurement of analytes of small molecular size (such as therapeutic drugs, steroid and thyroid hormones, etc). Nevertheless the convenience (from a manufacturing viewpoint, and for other technical reasons) of generally relying on labeled antibodies has meant that, even in these circumstances, labeled antibody techniques are increasingly preferred.

The advent of the in vitro hybridoma techniques of monoclonal antibody production pioneered by Kohler and Milstein (1975) has clearly been of great importance in the present context. The ability to produce a selected antibody directed against a single antigenic site (or "epitope") in unlimited amounts and in relatively pure form obviously facilitated the exploitation of the concepts discussed above, which, up to this time, had been impeded by the technical difficulties associated with the isolation and purification of the labeled antibodies on which they depend. Though, as indicated above, methodologies relying on labeled antibodies can be of either competitive (limited reagent) or non-competitive (excess reagent) design, the particular importance of the hybridoma techniques has always primarily rested, in my own opinion, on the means they offer for the easier development of non-competitive "sandwich" assay methodologies. Amongst the principal reasons for this view are

a. the requirement inherent in such techniques for relatively large amounts of antibody in comparison with "limited reagent" methods, and
b. the lesser dependence of non-competitive methods on high antibody affinity constants.

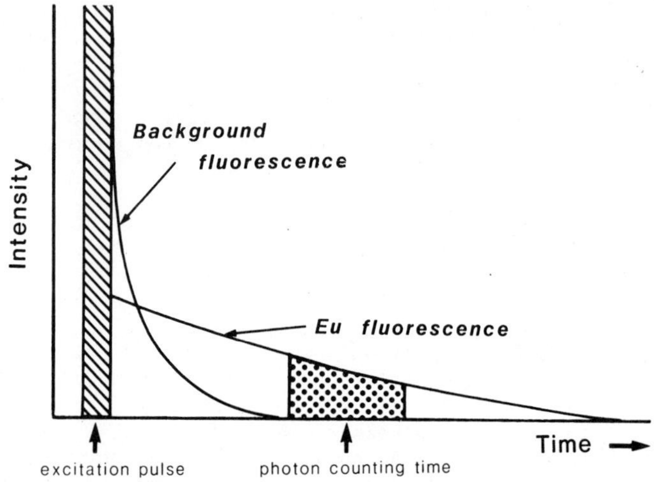

Fig. 3. Basic principles of pulsed-light, time-resolved fluorescence. Fluorescence emitted by the fluorophor (typically a europium chelate) is distinguished from background fluorescence, which decays more rapidly.

Thus, as illustrated in Fig. 4, an immunoassay of equal sensitivity to a competitive method can be constructed using a non-competitive design and an antibody of considerably lower affinity, the permitted affinity difference being dependent, inter alia, on the efficiency of the separation system employed. This conclusion is of considerable practical importance since - given the laborious nature of the techniques involved - the probability of identifying stable antibody-secreting clones yielding antibodies possessing the high affinities demanded by competitive methods (and with appropriate specificity characteristics) is relatively low. For this reason monoclonal antibodies are, in practice, generally of lower affinity than those produced by conventional in vivo techniques, although in time this difference is likely to disappear. Thus hybridoma-based monoclonal antibody production methods have never appeared to me to offer particular advantages in the development of competitive, labeled-antigen, methods such as RIA, although clearly there are, in principle, no reasons precluding their use in this context, nor is it impossible that certain selected monoclonal antibodies might prove to yield improved specificity characteristics (as compared with a polyclonal mixture) when used in assays of conventional competitive design. These reservations are not intended to deny the existence of certain logistic advantages in the use of monoclonal antibodies in competitive labeled antibody designs, such as are now increasingly relied on by manufacturers for the assay of analytes of small molecular size.

It should perhaps be noted that these concepts were specifically presented and openly discussed in all international British Council courses on immunoassay based on my own Department following the first announcement of the development of the hybridoma technique by Kohler and Milstein late in 1975 (the first such occasion being in Feb. 1976); furthermore, the first attempts (to my knowledge) to develop monoclonal antibodies for in vitro immunoassay purposes were made in 1977 at the Middlesex Hospital Medical School (as a collaborative project between Professor Ivan Roitt and myself, with financial help from the WHO, and the generous assistance of Dr. Milstein), whilst the specific reasons underlying the adoption of such techniques were discussed at length in presentations by myself in Berlin in 1977 (Ekins, 1977) and Seattle, March 1979 (Ekins, 1979). It is therefore

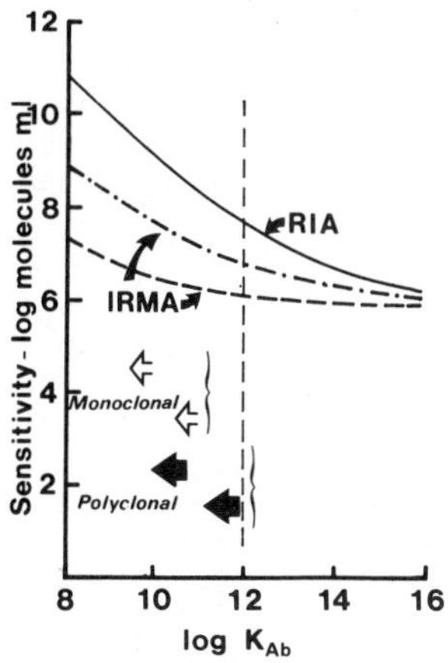

Fig. 4. Use of a labeled monoclonal antibody of low affinity in a non-competitive assay design may yield greater assay sensitivity than obtained with a competitive methodology using a relatively high affinity (polyclonal) antibody.

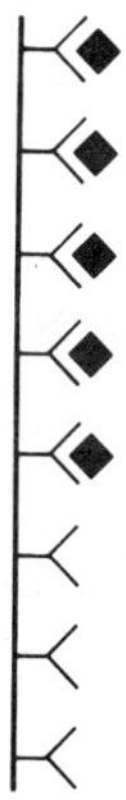

Fractional occupancy: $F = 1 - \frac{1}{1 + K[An]}$

Fig. 5. Basic principle of "ambient analyte" immunoassay. The fractional occupancy (F) of a vanishingly small amount of antibody (of affinity K) is determined by the analyte concentration in the surrounding medium ([An]).

both ironic and disturbing that patent applications relating to the use of monoclonal antibodies in sandwich assays submitted in 1980 by manufacturers in the U.S. and Europe (see Ezzell, 1987) - some of which have been provisionally granted - should now threaten the widespread use of monoclonal antibodies in an immunoassay context.

IMMUNOSENSORS AND IMMUNOPROBES

Speculation has centered for some years on the possibility of constructing viable "immunosensors", i.e. devices which, when immersed in an analyte-containing medium, emit a signal indicative of the amount of analyte present. Although certain practical and logistic advantages might be claimed for an immunosensor which, once used in this manner, must subsequently be either discarded or "reactivated", it is evident that the principal potential benefit offered by such an instrument is the ability continuously to monitor an analyte level in a body fluid such as blood, or in other situations (for example, in a continuous-flow manufacturing process) in which the analyte level may be continuously changing.

By definition, such a device must rely on detection of antibody occupancy by analyte in the surrounding medium without use of a label, i.e. by the recognition of a change in the shape, charge or other "intrinsic" physico-chemical property of the antibody molecule when it is thus occupied. Indeed, it is useful to distinguish between "intrinsic" immunoassay methods which function in this way, and "extrinsic" methods, which depend on exposure of antibody (either concurrently or sequentially) to a chemical "developer", whose specific function is to reveal the extent of antibody binding-site occupancy. Conventional immunoassay methods currently used in the clinical chemistry laboratory are, of course, almost all of the latter type.

It is probably also useful to distinguish between the terms "immunosensor" (which, in subsequent discussion, I will employ when referring to devices relying on the change in an intrinsic property of the occupied antibody molecule using some form of transducer), and "immunoprobe". I propose to use the latter term to describe devices which, following exposure to the analyte-containing medium, are withdrawn and subsequently "developed" - much as a photographic film is developed - in order to reveal binding site occupancy, and which therefore rely on "extrinsic" methods of antibody-occupancy recognition.

Another important concept in the present context is that embodied in the analytical approach I have termed "ambient analyte immunoassay" (Ekins, 1983). This term is intended to describe a type of immunoassay system which, unlike conventional methods, measures the analyte concentration in the medium to which the antibody is exposed, being essentially independent of sample volume. This concept is illustrated in Fig. 5 and relies on the physico-chemical proposition that, when a "vanishingly small" amount of antibody (preferably, but not essentially, coupled to a solid support) is exposed to an analyte-containing medium the resulting (fractional) occupancy of antibody binding-sites reflects the ambient analyte concentration. Clearly the binding by antibody of analyte results in a depletion of the amount of analyte in the surrounding medium, but provided the proportion so bound is small (i.e. less than, for example, 1% of the total), such disturbance can be ignored. (This concept is analogous to that underlying the insertion of a thermometer into a medium possessing a much larger thermal capacity; the temperature disturbance caused by the thermometer itself can, in these circumstances, be disregarded).

Ambient analyte immunoassay methodology exhibits some important and distinctive features. The first is that the system "response" essentially constitutes the ratio of occupied (or, alternatively, unoccupied) to total antibody binding sites. Secondly, within the limits implied above, the amount of antibody used in the system is irrelevant. For example, the introduction of 10, 100, or 1000 antibody molecules into a medium containing millions or billions of analyte molecules will result, in each case, in virtually identical fractional antibody binding site occupancy, this being determined solely by the ambient analyte concentration in the medium and the affinity constant of the antibody used.

These concepts were originally exploited by myself and my colleagues in our original development of what has come to be known as "two-step" free hormone immunoassay (Ekins et al, 1980), but it is clear that they are of far wider application, and can, in particular be utilized in the construction of immunosensors and immunoprobes. One such example is a probe for the measurement of salivary steroids, currently being developed and tested in my Department. Comprising a small antibody-coated plastic "dipstick" comparable in size and shape to a clinical thermometer, this device is intended to permit the measurement of salivary steroid levels without necessitating the collection of saliva.

Following exposure of such a "probe" to a test medium, it is removed, and antibody occupancy subsequently determined. This can be effected in two ways:

a. by exposure of the probe to, for example, a labeled antibody capable of reacting with occupied binding sites, or
b. by its exposure to a labeled substance (such as labeled analyte or labeled anti-idiotypic antibody) capable of reacting with unoccupied sites.

In conformity with the concepts and terminological definitions discussed earlier in this presentation, a probe relying on measurement strategy

a. may be described as "non-competitive"; likewise a probe relying on measurement strategy
b. constitutes a "competitive" probe. Such differentiation between these two forms of probe is important because, inter alia, the particular approach adopted dictates such design features as the amount and affinity constant of the antibody coated on the probe surface.

I have discussed these considerations at some length because they are equally relevant to the design of "immunosensors" (although the concepts involved appear to have been largely disregarded by those engaged in this field). Immunosensors can, as in the case of immunoprobes, be subdivided into two types: "competitive" immunosensors, and "non-competitive" immunosensors. Analogously, the first rely on the measurement of the signal yielded by unoccupied antibody binding sites, ideally yielding zero signal in the presence of an analyte excess in the medium (causing all sites to be occupied (see Fig. 6a)). This proposition may be compared with the concepts governing conventional RIA, which likewise require that the radiolabel bound to antibody falls to zero in the presence of excess analyte. Conversely, a "non-competitive" immunosensor relies on the signal emitted by filled sites (see Fig. 6b), implying that it should yield zero signal in the presence of zero analyte in the medium (in a manner analogous to the operation of a two site IRMA). It is clearly evident that, in principle, such considerations should govern the optimal design of immunosensors just as they do that of conventional immunoassays, determining the optimal amount of antibody coated on the probe surface and other such parameters. In practice, neglect of these fundamental aspects of immunoassay design, coupled with the fact that no transduction system so far revealed possesses characteristics ideally suited for use in either competitive or non-competitive modes of immunosensor operation, has meant that such devices as have been developed possess relatively low sensitivities, and are appropriate only for the measurement of high concentration analytes.

A second important impediment to the construction of useful immunosensors derives from the slow kinetics of antibody/antigen reactions. In general, the dissociation half-lives of the high affinity antibodies conventionally used for immunoassay purposes are in the order of hours or days, and are rarely less than 15-30 minutes. Clearly such characteristics severely limit the ability of an immunosensor to respond rapidly to changes

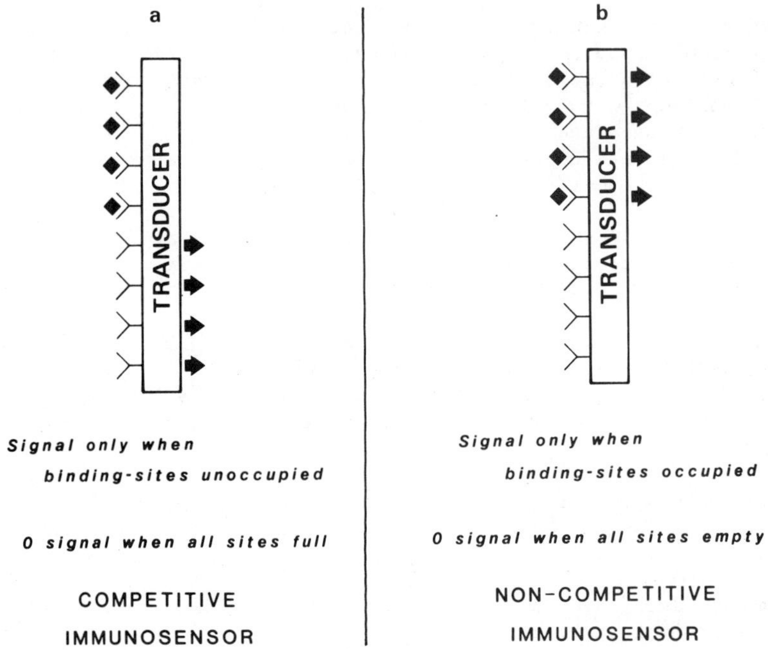

Fig. 6. a. "Competitive" immunosensor, ideally yielding zero signal when all antibody binding sites are filled.
b. "non-competitive" immunosensor, ideally yielding zero signal when all antibody binding sites are empty.

in analyte (antigen) concentration in their immediate environment, which constitutes one of the principal reasons for their development. Whether antibodies can be identified combining the high antigen binding affinity necessary for high assay sensitivity and the high dissociation rate constants required to yield response times of the order of seconds is unknown but appears intrinsically unlikely. (Certain hormone binding proteins such as thyroxine binding globulin (dissociation half time: 39 secs; affinity constant: 2.2 x 10^{10}L/M) might prove to be suitable in this context, but even this protein would yield a transduced signal requiring several minutes to readjust fully to an abrupt change in ambient thyroxine level.) Alternatively, methods might be devised for dissociating antigen at intervals from antibody on the sensor surface (by, for example, electrical means), and deducing, by observation of the antigen rebinding rate, the current ambient antigen concentration. However, such ideas are entirely speculative, and no technique has, as yet, been developed which resolves this fundamental problem.

Finally, in this overview of basic difficulties standing in the way of practical immunosensor development, there exists the problem of analyte specificity. By their nature, immunosensors necessarily rely on single epitope interactions, and are clearly subject to influence by any substance in the environment which reacts with the binding sites of antibody coated on its surface. In addition, other substances are liable to exert "non-specific" effects, either on the antibody itself, or on the transduction system used to signal changes in antibody binding site occupancy.

These fundamental physicochemical problems together constitute a formidable challenge. They are further compounded by the difficulties involved in the development of transducers of sufficient sensitivity to permit observation of changes in occupancy of small numbers of antibody molecules.

Proposed devices of this kind can be broadly divided into two groups; those which depend upon a "non-specific" modification in the surface properties of the immunosensor, such as a change in its electrical or optical properties, resulting from the adsorption of antigen molecules on its surface, and those which endeavour to exploit in a more specific manner any alterations in the conformation of antibody molecules caused by their binding of antigen.

Amongst the former, those which have been relatively extensively studied include field effect transistor (FET), potentiometric, piezo-electric and optically-based systems. An "immunoFET" essentially comprises a transistor, the current through which is modulated by changes in an applied electric field deriving from the interactions of an antigen or antibody on the transistor surface. Devices of this kind have been claimed to respond to antibodies to albumin and syphilis (Janata and Huber, 1980); Collins and Janata, 1982); however, a variety of non-specific effects (caused by ions etc) are also liable to cause alterations in the electric field controlling the current flow. For these and other technical reasons, early hopes centered on immunoFETs appear to have evaporated, and there seems to be little prospect of rapid development in this field until major advances in FET design have occurred (Janata and Blackburn, 1984).

Similar problems of non-specificity arise with "potentiometric" methods of monitoring antigen/antibody reactions. These likewise depend on detection of the charge or potential differences caused when antigen attaches to antibody-coated electrodes, the latter being constructed of glass or amino-silanized titanium (Aizawa et al., 1977). Though claimed to be capable of detecting hCG in pregnancy urine (Yamamoto et al, 1980), the susceptibility of these devices to non-specific effects have caused a diminution of interest in them in recent years.

Piezo-electric systems depend on the ability of certain crystals and plastics to respond to mechanical stress by the generation of electrical signals. When molecules of antigen bind to antibodies covalently linked to the surface of an immunosensor of this type, the resulting change in mass can be detected electrically, either by observations of the change in the resonant frequency of the crystal in an appropriate oscillator circuit, or by observation of the changes in the propogation velocity of acoustic waves across the crystal surface. Surface acoustic wave devices are widely used in conventional electronic applications, and immunosensors exploiting this technology have been constructed capable of measuring human IgG and influenza virus (Roederer and Bastiaans, 1983). Claimed as able to detect amounts of antigen in the nanogram range, piezo-electric sensors are nevertheless still far removed from achieving the sensitivities yielded by conventional immunoassay methods.

Meanwhile interest in optical methods has centered primarily on "evanescent wave" and plasmon resonance systems. The principle of the evanescent wave approach is illustrated in Fig. 7. In simple terms, it relies on alteration of the internal light-reflecting properties of an optical wave guide - for example, an optical fibre - resulting from changes in the composition and reflective properties of the fibre surface (Sutherland et al, 1964). The binding of analyte molecules to antibody coated on such a surface is thus detected as a change in light transmission along the guide. These changes are, in practice, small in relation to the total light transmitted, implying low measurement sensitivity (currently claimed to be in the nanomolar range). Meanwhile surface plasmon resonance effects arise in consequence of light falling on a metallic surface at a

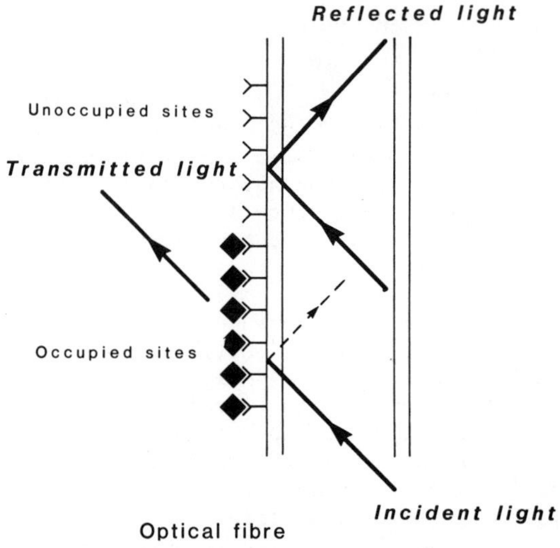

Fig. 7. Basic principle of proposed "evanescent wave" immunosensor. Light reflected along the optical fibre is reduced when antibody binding sites are occupied. Assuming measurement of the reflected light, this (in principle) constitutes a "competitive" immunosensor, the reflected light ideally falling to zero in the presence of very high concentrations of analyte in the medium. In practice this does not occur; the device thus functions in a manner equivalent to a conventional RIA performed using a radioisotope counter with a very high background.

particular wave length and at a particular angle. These effects are manifested as a sharp fall in the reflected light intensity at a critical angle of light incidence, accompanied by the occurrence of electrical phenomena in the metal surface. Exploitation of these phenomena has involved the use of metallized diffraction gratings, antibody being coated in the areas adjacent to the metal deposits. The critical angle of light incidence at which plasmon resonance effects occur is very sensitive to changes in the dielectric constant in the antibody-coated regions, and hence to the binding of antigen by antibody. Model devices of this kind have been constructed, likewise with claimed sensitivities in the nanogram range.

The responsiveness of all these devices to non-specific effects - e.g. alterations in the ionic composition of the medium, non-specific adsorption of proteins to the probe surface, etc - makes their application to the measurement of analytes in the sub-nanogram and picogram range formidably difficult. An alternative approach is to attempt to utilize the change in the structure of the antibody molecule when antigen is bound to generate an observable signal. Unfortunately such changes appear to be restricted to the Fab region of the molecule, and are not transmitted to the Fc region. Speculation has therefore centered on the possibility of inserting "reporter" groups close to the antibody binding site, and whose activity would be affected by binding of antigen (North, 1985). Such groups might, for example, comprise ionophores, fluorophores or luminophores, sensitive to the electronic environment in which they are situated. However, practical implementation of such ideas obviously requires a far more detailed knowledge of the intramolecular events involved in antigen/antibody reactions than is currently available, and the development of sophisticated techniques for the attachment of reporter groups to specific points on the antibody molecule, and measurement of the signals they emit.

To summarize, the development of practical, rapidly-responding, immunosensors possessing sensitivities comparable to those of conventional immunoassay techniques represents an analytical challenge of formidable magnitude. At the present time, a number of model systems have been described, and others are the subject of intensive investigation by large commercial organizations. Though the construction of systems capable of measuring relatively high concentration analytes appears to be feasible, the extension of known methodologies to analytes in the concentration ranges in which immunoassay techniques are generally applied constitutes a problem for which no solution appears to be presently in sight.

MULTI-ANALYTE IMMUNOASSAY SYSTEMS

Though much current research is centered on the development of immunosensors of the kind indicated above, my own Department is directing its efforts towards a different objective: the development of a random access, multi-analyte immunoprobe system capable of measuring, in the same small sample, any number of individual analytes from selected analyte 'menus' (e.g. hormone menu, a viral antigen menu, an allergen menu etc). The possibility of measuring tens or even hundreds of analytes in the same sample which is implicit in some of the concepts discussed earlier in this presentation is, in my own view, both more feasible - using present technology- than the development of useful immunosensors, and potentially of equal or greater importance. Though still at an early stage, the broad concepts underlying our present approach may be briefly indicated.

As discussed above, the notion of ambient analyte assay simultaneously introduces two extremely important and novel concepts:

a. that an estimate of analyte concentration can be based upon the use of an infinitesimal amount of "sampling " antibody, and
b. that such an estimate derives from a direct measurement of the degree of antibody occupancy by analyte, irrespective of the exact amount of antibody used.

It should be emphasized that the latter proposition is valid only in the context of ambient analyte assay, and is not true in current conventional immunoassay systems (in which fractional antibody occupancy depends both upon the amount of antibody in the system, and sample volume). In short, exposure of a small number of antibody molecules (in the form, for example, of a "microspot" located on a solid support) to an analyte-containing fluid will result in an antibody binding-site occupancy which reflects the analyte concentration in the medium. Following such exposure, the antibody-bearing probe may be removed and exposed to a "developing" solution containing a high concentration of an appropriate second antibody directed against either a second epitope on the analyte molecule if this is large (i.e. the occupied site), or against unfilled antibody binding sites in the case of small analyte molecules. (An antibody reacting with unoccupied binding sites is described as a "mirror-image anti-idiotypic antibody"; the use of such an antibody instead of labeled antigen in the present context is convenient but not essential, and is suggested here merely to simplify discussion of the basic concepts involved).

Subsequently, an estimate of binding site occupancy of the "sampling" (solid-phase) antibody may be derived by measurement of the relative numbers of the two antibodies forming the dual-antibody "couplets". This can be conveniently achieved by labeling each of the antibodies used with different markers; for example, a pair of radioactive, enzyme or chemiluminescent markers. Fluorescent labels are particularly useful in this context because, by the use of optical scanning techniques, they permit arrays of different antibody "microspots" distributed over a surface, each directed to a different analyte, to be individually examined, thus enabling multiple assays to be simultaneously carried out on the same small sample. Fig. 8 illustrates these basic ideas, and Fig. 9 such an array.

It is inappropriate to discuss here the technical details of such a system, and sufficient to draw attention to the possibilities it offers. Laser beams can be focussed onto very small areas, so that the number of individual assays potentially containable within a multianalyte array is restricted primarily by the density with which the different sampling antibodies can be packed on a suitable surface. Meanwhile it should be noted that, as the surface area exposed to the laser beam is reduced, the background signal deriving from the solid support, from "non-specifically-bound" antibody and from other similar extraneous sources correspondingly falls. This implies that the ratio of the two fluorescence signals can be successfully measured from very small areas (as, indeed, is often done in other biological applications).

Differentiation of the fluorescent signals yielded by the two fluorophors can, of course, be readily achieved using a combination of physical techniques exploiting the differences in wave length and/or decay time of the two emissions (for example, by the use of a pulsed or sinusoidally-modulated laser source, and time-or phase-resolved detectors combined with appropriate wave-length filters). This aside, the technology involved closely resembles that employed in compact disk recorders and other similar data-storage devices, the obvious difference being that light emitted from each of the discrete zones forming the antibody-array is

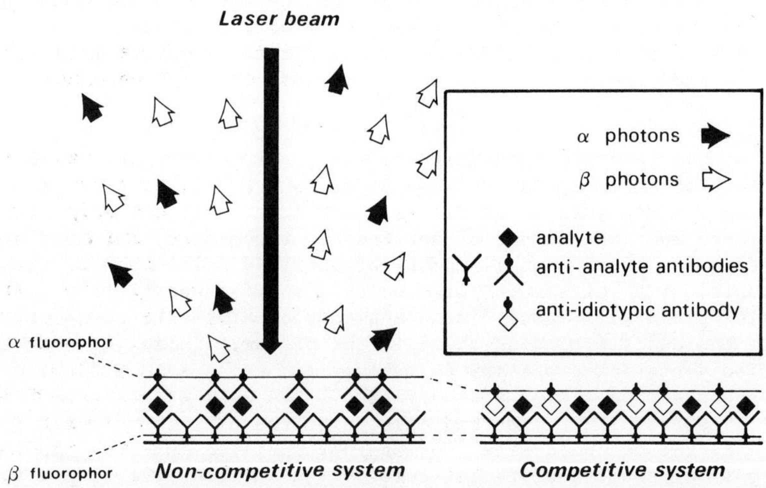

Fig. 8. Basic principle of dual-label, ambient-analyte, immunoassay relying on fluorescent labeled antibodies. The ratio of α and β fluorescent photons emitted by the two labeled antibody species reflects the value of F (see Fig. 5) and is thus solely a function of the analyte concentration in the medium to which the probe has been exposed. The ratio is unaffected by the amount or distribution of antibody coated (as a monomolecular layer) on the surface of the probe.

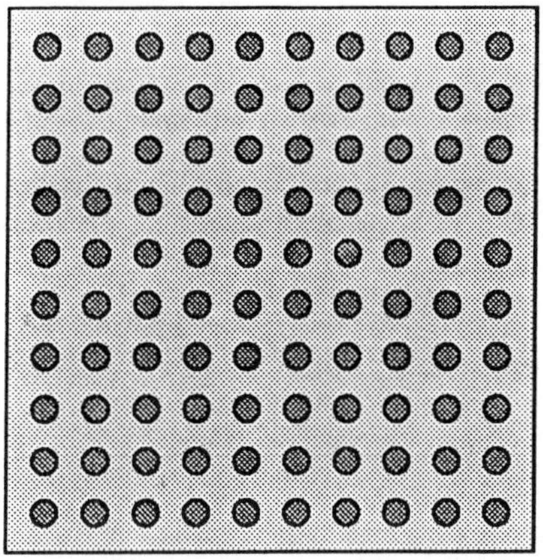

Fig. 9. Multi-analyte immunoarray. Each "microspot" represents a "vanishingly" small amount of (labeled) antibody directed against a specific analyte (antigen). The fractional occupancy of each antibody microspot (following exposure to sample and 'developing' antibody solution) may be determined by measurement of the ratio of the two signals emitted by each member of the resulting (labeled) antibody "couplet". This can be readily effected with fluorescent labels (for example), by "interrogation" of the surface with a scanning laser beam. Dual enzyme label systems based on the same principle may also be readily visualized.

fluorescent rather than reflected, and yields chemical rather than physical information.

The development of multi-analyte assay systems of this kind can be anticipated to bring about fundamental changes in medical diagnosis and many other biologically related areas. For example, it is both conceivable and within present technological reach that immunoprobes will be developed capable of measuring every hormone and other endocrinologically-related substance within a single small sample of blood, providing data which, when analyzed with the aid of computer-based "expert" pattern-recognition systems, will reveal endocrine deficiences only dimly perceived using current "single-analyte" diagnostic procedures. Such systems also provide a possible solution to a need frequently voiced by manufacturers, i.e. the development of "random access" immunoassay methodology, permitting the selection of any desired test or combination of tests from an extensive menu. Clearly the accommodation of a large range of individual immunoassays on a small immunoprobe (comparable in its overall physical dimensions with a few drops of blood) is likely to totally transform the logistics of immunodiagnostic testing. Yet a further likely application of potentially great importance is in situations (in medicine, in the food industry, agriculture, environmental research etc) in which samples must be screened for the presence of one or more substances drawn from a wide range of possibilities.

CONCLUSION

In this brief overview, I have attempted to identify the main trends in current immunoassay development, emphasizing the needs and objectives that underlie such development rather than simply listing the variety of techniques, labels and novel immunodiagnostic devices of varying ingenuity that manufacturers have evolved in the past few years.

The advent of monoclonal antibody production techniques has not, in any fundamental sense, revolutionized the immunoassay field since many of the methods to which they have been applied were previously well established. Nevertheless, as predicted, they have provided the basis for the more widespread use of "non-competitive" immunoassay methodology, and thus to the development of techniques of greater sensitivity, speed and specificity as compared with conventional RIA. Such development represents the route that many manufacturers are currently following, and it is in this area of application that monoclonal antibodies are likely to make their greatest impact. It is likewise within the context of non-competitive immunoassay methodology that the development of new labels and improved separation techniques (whose efficiency is crucial in this context) is likely to prove of key importance in overcoming current assay sensitivity limits.

I have briefly discussed present efforts to develop "immunosensors", not because I entertain any hopes of early success in this field, but because it is one which has clearly attracted the interest of major manufacturers, and in which large financial resources are now being invested. Clearly for any significant progress to be made in this area, fundamental breakthroughs must be achieved in the construction of transducers that are capable of recognizing conformational changes in small numbers of antibody molecules; means must also be found for circumventing the physicochemical laws governing antibody/antigen reactions. Furthermore such devices, if ever developed, appear to be doomed to be of limited specificity.

Finally I have briefly described methodological developments in my own laboratory which aim at objectives I believe to be comparable in importance with those underlying immunosensor development, and which appear to offer a better chance of attainment in a Department (such as my own) with limited financial, technical and intellectual resources. Indeed, the development of multianalyte systems is likely, in my view, to totally transform the immunodiagnostic field. Clearly monoclonal antibodies, though not crucial in this context, greatly assist the development of the technology, permitting, for example, the dissection and measurement of heterogeneous mixtures of molecules now classified as single substances and known by a single name. As is often the case, only when such a methodology is fully developed and widely applied will its impact and usefulness, if any, be fully assessable.

REFERENCES

Aizawa, M., Kato, S., Suzuki, S., 1977, Immunoresponsive membrane I. Membrane potential change associated with an immunochemical reaction between membrane-bound antigen and free antibody, J.Membr.Sci., 2:125.

Collins, S., Janata, J., 1982, A critical evaluation of the mechanism of potential response of antigen polymer membranes to the corresponding antiserum, Anal.Chim.Acta., 136:93.

Dakubu, S., Ekins, R., Jackson, T., Marshall, N. J., 1984, High sensitivity, pulsed light time-resolved fluoroimmunoassay, in: "Practical Immunoassay. The State of the Art," W.R. Butt, ed., Marcel Dekker, Inc., p. 71. New York.

Ekins, R. P., 1983, Measurement of analyte concentration, British Patent no. 8224600.

Ekins, R., 1985, Current concepts and future developments, in: "Alternative Immunoassays, " W.P. Collins, ed., John Wiley & Sons Ltd., p.219. Chichester.

Ekins, R. P., Filetti, S., Kurtz, A. B., Dwyer, K., 1980, A simple general method for the assay of free hormones (and drugs); its application to the measurement of serum free thyroxine levels and the bearing of assay results on the 'free thyroxine' concept, J.Endocrinol., 85:29.

Ezzell, C., 1987, Hybritech wins court injunction over sandwich assays, ed., Nature, 327:5.

Harris, C. C., Yolken, R. H., Krokan, H., Hsu, I. C., 1979, Ultrasensitive enzymatic radioimmunoassay: Application to detection of cholera toxin and rotavirus, Proc.Natl.Acad.Sci.USA 76:5336.

Hemmila, I., Dakubu, S., Mukkala, V.-M., Siitari, H., Lovgren, T., 1983, Europium as a label in time-resolved immunofluorometric assays, Anal.Biochem., 137:335.

Janata, J., Blackburn, G. F., 1984, Immunochemical potentiometric sensors, Ann.NY Acad.Sci., 286.

Janata, J., Huber, R. J., 1980, Chemically sensitive field effect transisters, in: "lon-sensitive Electrodes in Analytical Chemistry," H. Freiser ed, Plenum Press, Vol 2:107.

Kohler, G., Milstein, C., 1975, Continuous culture of fused cells secreting specific antibody, Nature, 256:495.

McCapra, F., Tutt, D. E., Topping, R. M., 1977, Assay method utilizing chemiluminescence. British Patent no. 1, 461,877.

McGown L. B., Bright, F. V., 1984, Phase-resolved fluorescence spectroscopy, Anal.Chem., 56:1400.

North, J. R., 1985, Immunosensors: antibody-based biosensors, Trends in Biotechnology, 3/7:180.

Roederer, J. E., Bastiaans, G. J., 1983, Microgravimetric immunoasay with piezo-electric crystals, Anal.Chem., 55:2333.

Shalev, A., Greenberg, G. H., McAlpine, P. J., 1980, Detection of attograms of antigen by a high sensitivity enzyme-linked immunosorbent assay (HS-ELISA) using a fluorogenic substrate, J.Immunol.Methods., 38:125.

Stanley, C. J., Paris, F., Plumb, A., Webb, A., Johannsson, A., 1985, Enzyme amplification: A new technique for enhancing the speed and sensitivity of enzyme immunoassays, Int.Clin.Products Review July/ August 1985, 44.

Sutherland, R. M., Dahne, C., Place, J. F., Ringrose, A. R., 1984, Immunoassays at a quartz-liquid interface: theory, instrumentation and preliminary application to the fluorescent immunoassay of human immunoglobulin G., J.Immunol.Methods., 74:253.

Weeks, I., Campbell, A. K., Woodhead, S., McCapra, F., 1984, Immunoassays using chemiluminescent labels in: "Practical Immunoassay. The State of the Art," W.R. Butt, ed., Marcel Dekker, Inc. p. 103, New York.

Whitehead, T. P., Thorpe, G. H., Carter, T. J., Groucutt, C., Kricka, L. J., 1983, Enhanced luminescence procedure for sensitive determination of peroxidase-labelled conjugates in immunoassay, Nature, 305:158.

Yamamoto, N., Yoshikatsu, N., Sadanobu, S., Tsubomura, H., Sawai, M., Okumura, H., 1980, Antigen-antibody reaction investigated with use of a chemically modified electrode, Clin.Chem., 26:1539.

MONOCLONAL AND POLYCLONAL ANTIBODIES FOR IMMUNOASSAY

J. Stuart Woodhead, J. Paul Aston,
Richard C. Brown and Ian Weeks

Department of Medical Biochemistry
University of Wales College of Medicine
Cardiff CF4 4XN

INTRODUCTION

It is axiomatic that the key component of an immunoassay system is the antibody. Though other factors such as choice of label or method of separation may be important, it is the antibody which will ultimately determine the performance characteristics of the system. It is also generally accepted that if a monoclonal antibody can be readily produced to a required specification then it would invariably be used in preference to a polyclonal reagent. In reality the introduction of monoclonal antibodies into diagnostic immunoassay systems has been relatively slow, mainly for practical reasons. In this paper we shall consider the relative merits of monoclonal and polyclonal reagents within the context of the fundamental requirements for immunoassay design and optimization.

IMMUNOASSAY FORMAT

High sensitivity immunoassays require labels of high specific activity to enable the binding reaction to be monitored at low analyte concentrations. Two basic assay formats exist. The first is where the analyte is labelled and the labelled and unlabelled forms compete for a limited number of antibody binding sites (competitive immunoassay). The second format is where the binding reagent is labelled and is then used in excess to convert analyte present to a labelled derivative (non-competitive or immunometric assay). These two formats are not simple variants of each other, but represent two fundamentally different approaches to the quantitation of low levels of material in biological samples (Ekins, 1981). Just as the principles on which these two systems are based differ so do their properties and their requirements (Hales and Woodhead, 1980).

In a competitive system, the specificity of the binding reaction being monitored is conferred by the use of pure analyte for labelling. In a non-competitive assay it is the purity of the labelled antibody which gives the system its specificity. It is this requirement for pure antibody (not merely pure immunoglobulin) which, more than any other single factor, has restricted the applications of immunometric methodology in the past. Before monoclonal antibodies were readily available, the only means of achieving the required purity was by affinity chromatography using immobilized antigen, a time consuming, sometimes expensive and frequently wasteful procedure. It is primarily their ease of use which has made monoclonal antibodies such important components of the many immunometric assays in use today. It has long been recognized that such methods can provide improvements in analytical specificity, sensitivity and speed when compared with conventional competitive systems. As we describe later, the realization of such advantages depends on careful optimization which in turn requires an understanding of certain fundamental principles of assay design.

LABELLED ANTIBODY IMMUNOASSAYS OF HAPTENS

The concept of using labelled antibody rather than labelled antigen methodology for the measurement of low molecular weight compounds is in theory attractive. The preparation of labelled antigen is relatively simple only as long as it involves the incorporation of ^{3}H or ^{14}C as an integral part of the molecule. The chemistry of external labelling with ^{125}I may be much more complex, as it is when non-isotopic derivatives are used. One attraction of labelled antibodies is the availability of uniform methods of labelling, irrespective of the chemical nature of the analyte.

In practice, however, the complexity of hapten immunochemistry raises a number of problems regardless of the analtyical procedure. In the first place, there is the need for the analyte to be coupled to protein to render it immunogenic. As a result of 'bridge' recognition, the antibodies produced generally have a higher affinity for the antigen than for the hapten itself. Design of a label for a conventional competitive assay may thus become a complex affair.

The analysis itself is complicated by the fact that the analyte is present in blood primarily bound to carrier proteins with only a small fraction being in free circulation. Thus direct assays of such molecules require a means of displacing the protein bound analyte if the total concentration is to be measured or a procedure which will allow quantitation of the free analyte concentration. This latter is preferable in the case of many hormones since the free concentration is recognized as determining the level of biological activity of such molecules.

Finally, the choice of methodology may be influenced by the reaction kinetics of the hapten-antibody reaction. In many cases, dissociation of the complex is rapid so that the use of reagent excess procedures is precluded. Some years ago, we attempted to develop an immunometric assay for progesterone using affinity purified rabbit antibodies (Woodhead et al, 1975). Apart from the difficulty in obtaining good recoveries of antibody by this procedure, sensitivity was extremely poor. This we attributed to the use of excess solid-phase progesterone to bind unreacted labelled antibody. Presumably the solid phase was also binding antibody which was dissociating rapidly from its complex with free hapten.

More recently, we have developed methods for the measurement of total (Sturgess et al, 1986) and free (Sturgess et al, 1987) thyroxine using acridinium ester labelled antibodies, but where the reagent is limiting rather than in excess. In this system hapten and a hapten-protein conjugate compete for a limited amount of labelled antibody, the conjugate bound fraction being subsequently removed following reaction with a solid phase antibody to the appropriate protein (in this case rabbit IgG). In addition to providing adequate sensitivity for the routine measurement of free T4, this method has been found to yield reliable clinical information where methods based on the use of radioactive hormone analogues are misleading (Sturgess et al, 1987). It is significant that the success of this method derived from the use of a monoclonal antibody raised to thyroxine (Mpoko et al, 1985), our attempts to use affinity purified sheep antibodies being thwarted by inadequate analytical sensitivity.

This procedure thus appears to offer a general approach to the direct measurement of haptens in blood. Since the non-isotopic labelling of antibody is a straight forward and reproducible operation, it is logical to suggest that it is worthwhile devoting more effort to the production of monoclonal reagents for hapten immunoassay.

LABELLED ANTIBODY-IMMUNOASSAYS FOR POLYPEPTIDES

In contrast to the limited developments in labelled antibody methodology for hapten measurement, immunometric assays for peptides have appeared in profusion during recent years. From a theoretical standpoint it could be argued that this is only to be expected in view of the immense potential advantages of such methods (Ekins, 1981; Hales and Woodhead, 1980). However, it is hybridoma technology which has provided the purified reagents needed to convert the potential advantages into practice.

We were among the first to demonstrate the potential advantages of a labelled monoclonal antibody assay in practice (Weeks et al, 1981). We were able to show that a monoclonal antibody to alphafetoprotein (AFP) with an affinity constant of 3.8×10^8 l/mol yielded insufficient sensitivity to be useful in a conventional competitive assay format. In contrast, when used as the labelled component of a two-site immunoradiometric assay, a greatly improved performance could be achieved in what proved to be a simpler and more rapid analytical procedure. However, antibody affinity remains a crucial issue if assays of extremely high sensitivity are to be developed.

A careful theoretical analysis by Ekins and Dakubu (1985) has demonstrated the relationship between antibody affinity and analytical sensitivity for both competitive and non-competitive assay systems, with particular regard to the specific activity of the label used. It is clear that for all practical purposes, increasing the label specific activity to a level greater than 1 atom/molecule of ^{125}I will have no significant effect on the sensitivity of a competitive radioimmunoassay. In contrast, in a non-competitive system it is theroetically possible to achieve improved sensitivity when higher specific activity labels are used and that such improvements can be dramatic with antibodies of high affinity.

The implications of these simple conclusions are profound, particularly since recent developments in non-isotopic labelling have led to the production of reagents with vastly improved detection limits. While it has been recognized for many years that increased incorporation of ^{125}I into antibody molecules is impracticable due to progressive impairment of immunoreactivity (Woodhead et al, 1974) the ability to produce higher specific activities with enzyme, chemiluminescent or fluorescent labels has been widely discussed (Ekins and Dakubu, 1985). In practice we were able to demonstrate a fivefold improvement in sensitivity in the measurement of AFP when the incorporation of a chemiluminescent acridinium ester into a monoclonal antibody was increased from 0.3 mol/mol to 3.0 mol/mol (Weeks et al, 1983).

Many immunodiagnostic procedures stand to benefit from this increased awareness of the principles of immunometric assay design. We have recently described an assay for parathyroid hormone (PTH) which has finally overcome the problems of inadequate specificity and sensitivity which have restricted the usefulness of previous assays for the hormone (Brown et al,

1987). While some of these problems have been technical in nature, others derive from the physiology of PTH secretion itself.

The molecule which is secreted is an 84 amino acid peptide which undergoes rapid metabolism in peripheral blood, with the removal of a biologically active N-terminal sequence and considerably slower breakdown and clearance of biologically inactive mid-region and C-terminal fragments. These latter processes occur in the kidney with the result that in patients with renal impairment there is an accumulation in blood of biologically inert fragments of PTH, which do not reflect parathyroid secretory activity. Though it has been possible to raise high affinity polyclonal antisera to PTH the presence of multiple antigenic determinants in the hormone molecule has led to the production of antisera of mixed specificity (O'Riordan et al, 1972). Fig. 1 shows cross-reactivities of various hormone fragments in a competitive assay using a sheep antiserum raised against purified bovine PTH. The lack of specificity for the biologically important region of the molecule ultimately leads to difficulties of interpretation particularly in patients with renal impairment.

There have been many attempts to improve assay specificity, the most popular being the use of synthetic fragments of human PTH to produce antibodies. The improvement in specificity which can be achieved by this approach is illustrated in Fig. 2 which shows cross-reactivity data obtained with a sheep antibody raised against human (1-34)PTH coupled to keyhole limpet haemocyanin. In this case the antibodies recognize only N-terminal fragments and the intact hormone molecule, fragments such as 28-48 have no cross-reactivity. The affinity constant of these antibodies for human PTH (1-34) is just short of 10^{11} l/mol and while there is almost identical reactivity with human (1-84) PTH, it should be stressed that this is a relatively unusual situation, since many antibodies raised to fragments cross-react poorly with the native peptide. Though the specificity of assays based on this reagent is satisfactory, the sensitivity has proved inadequate to measure basal PTH in normal subjects with the result that it is difficult to resolve the patient with mild hyperparathyroidism using conventional immunoassay procedures. The two-site assay we devised was based on the concept of using antibodies one of which is specific for the N-terminal sequence and the other recognizing a C-terminal sequence. For this to be practicable the latter antibody would have to be present at a sufficiently high concentration to bind both fragments and intact hormone. The scheme by which this was achieved is shown in Fig. 3, where intact hormone is depicted in the presence of C-terminal fragments. The objective was to incubate the sample with high affinity, labelled N-terminal antibody and then react with sufficient solid-phase C-terminal antibody to immobilize all the immune complex so formed together with the fragments, with the result that the quantity of label associated with the solid phase would reflect the concentration of intact PTH in the sample.

After considerable efforts to produce monoclonal antibodies to both N-terminal and mid-region fragments, we identified one reagent directed against (44-68) PTH which appeared potentially useful. Its affinity for intact (1-84) PTH was 5 x 10^8 l/mol and it could be purified in high yield from ascitic fluid by high performance liquid chromatography on a hydroxylapatite column (Fig. 4). Initially, this reagent was coupled to a diazonium salt of reprecipitated aminoaryl cellulose for use as a solid phase reagent. More recently, we have used glutaraldehyde activated paramagnetic particles (Advanced Magnetics, Cambridge, Massachusetts, USA) as the solid support. The rationale was that this reagent could be used in large excess and so overcome the drawback of rather poor affinity.

For the labelled reagent, it was clear that a high affinity was needed. None of the monoclonal antibodies tested had affinities

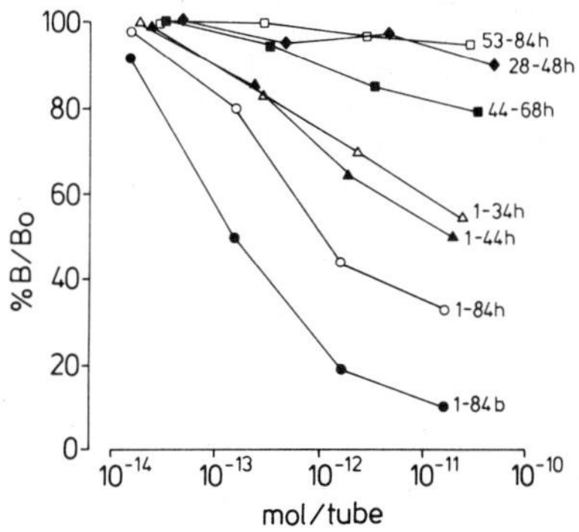

Fig. 1. Cross reactivity of fragments of the human PTH molecule with an antiserum raised in a sheep against bovine (1-84) PTH, seen in a competitive radioimmunoassay. h = human, b = bovine
W71 1:10,000, ^{125}IbPTH (1-84).

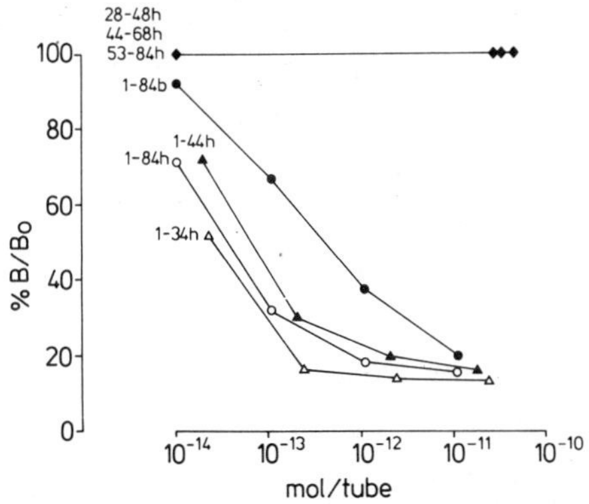

Fig. 2. Cross reactivity of PTH fragments with an antiserum raised in a sheep against human (1-34) PTH, seen in a competitive radioimmunoassay. h = human, b = bovine
LB34 1:20,000, ^{125}IhPTH (1-34).

approaching that of the polyclonal reagent (LB 34). We therefore carried out affinity chromatography on this polyclonal material using a 3 ml column of Sepharose 4B to which had been coupled 4 mg of highly purified bovine PTH. The lower affinity of the antibody for bovine PTH encouraged us to think that recoveries of specific antibody would be high. In practice, using dilute HCl as the eluting agent, it was possible to isolate approximately 3 mg purified antibody from 75 ml of antiserum. The purity of this antibody was confirmed by its elution as a single peak on DEAE high performance liquid chromatography. This material was labelled with an acridinium ester to yield a specific activity of 1 mol ester/mol immunoglobulin.

The assay procedure currently used involves incubation of 100 µl patient serum with approximately 3 ng of this labelled antibody for 18h. Solid phase antibody (≡10µg IgG) is then added and after a further 1 h reaction, the immune complex is separated magnetically and quantified in a luminometer (MAGIC Lite Analyzer: Ciba-Corning Diagnostics, Halstead, Essex, UK) which also carries out the data reduction. The method is sensitive enough to measure PTH in all normal subjects and to discriminate

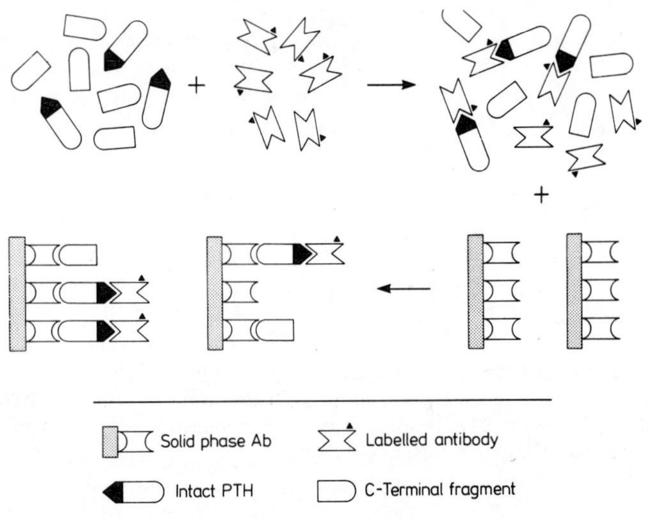

Fig. 3. Scheme depicting the protocol adopted for the specific measurement of intact PTH in the presence of circulating C-terminal fragments.

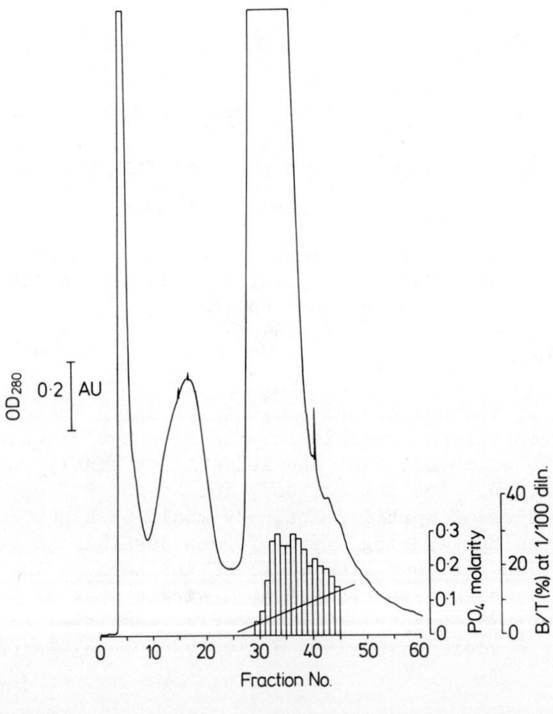

Fig. 4. High performance liquid chromatography on hydroxylapatite of ascitic fluid produced by a hybridoma secreting antibody to human (44-68)PTH. The histogram shows the distribution of immunological activity. The concentration of IgG in this ascitic fluid was approximately 10 mg/ml.

patients with hyperparathyroidism. Moreover it is not influenced by extremely high concentrations of C-terminal fragments as may be found in patients with renal impairment (Brown et al, 1987).

Clearly, it would be convenient to use a monoclonal antibody rather than the affinity purified polycolonal antibody as the labelled agent. However, the obligatory requirement for high affinity has so far precluded this possibility. This should not be taken to imply that because it may be difficult to raise high affinity monoclonal antibodies, it is not worth the effort. The production of high affinity monoclonal antibodies to thyrotrophin has led to the development of assays of a clinical performance which more than justifies the efforts involved in their production (Soos et al, 1984; Weeks et al, 1984).

CONCLUSIONS

Polyclonal antibodies for immunoassay purposes are relatively easy to produce even when high affinity is a necessary requirement. Their main disadvantages are that there is a theoretically limited supply even from a large animal and there are uncertainties over production as evidenced by changes in specificity which may occur with successive bleeds. In addition there is the problem of multiple epitope recognition and possible requirement for expensive ligands to carry out affinity chromatography This step is necessary if the reagent is to be labelled with a view to taking advantages of immunometric assay methodology.

Monoclonal antibodies have the advantages of being available in unlimited supply and their screening and production are easy to control. In addition, there is the enormous advantage of homogeneous reagent with recognition for only one epitope. On the disadvantage side there is the occasional problem of labelling damage particularly with regard to the incorporation of ^{125}I. Finally, there is the question of how much effort is needed to generate antibodies of sufficiently high affinity to capitalize on the potentially high sensitivities to be achieved with certain non-isotopic labels when used in immunometric procedures.

It remains to be seen whether improvements in the technology of monoclonal antibody production will overcome these problems and lead to the disappearance of the polyclonal reagent. It seems likely that this will not happen in the immediate future.

Acknowledgments

Studies reported here were supported by the Medical Research Council, the Welsh Office and Bioanalysis Ltd.

REFERENCES

Brown, R. C., Aston, J. P., Weeks, I., and Woodhead, J. S., 1987, Circulating intact parathyroid hormone measured by a two-site immunochemiluminometric assay, J.Clin.Endoc.Metab., 65:407.

Ekins, R. P., 1981, Towards immunoassays of greater sensitivity, specificity and speed: an overview. In: "Monoclonal Antibodies and Developments in Immunoassay," A. Albertini and R.P. Ekins, eds., Elsevier, Amsterdam, pp. 3-21.

Ekins, R. P., and Dakubu, S., 1985, The development of high sensitivity pulsed light, time-resolved fluoroimmunoassays, Pure & Appl. Chem., 57:473.

Hales, C. N., and Woodhead, J. S., 1980, Labelled antibodies and their use in the immunoradiometric assay, in: "Methods in Enzymology," Vol. 70. H. Van Vanukis and J.J. Langone, eds., Academic Press, New York, pp. 334-355.

Mpoko, C. N., Gordon, D. B., Laing, I., Corbitt, G., and Storey C. C., 1985, The production and characterization of a monoclonal antibody to thyroxine, Clin.Chim.Acta., 146:215.

O'Riordan J. L. H., Addison, G. M., Woodhead, J. S., Keutmann, H. T., and Potts, J. T., 1972, Localization of antigenic sites in parathyroid hormone, in: Endocrinology, 1971, Proceedings of the 3rd International Symposium, S. Taylor, ed., Heinemann, London pp. 386-392.

Soos, M., Taylor, S. J., Gard, T., and Siddle, K., 1984, A rapid sensitive two-site immunometric assay for TSH using monoclonal antibodies: investigation of factors affecting optimisation, J.Immunol.Meth, 73:237.

Sturgess, M. L., Weeks, I., Evans, P. J., Mpoko, C. N., Laing, I., and Woodhead, J. S., 1987, An immunochemiluminometric assay for serum free thyroxine, Clin.Endoc. 27:383.

Sturgess, M. L., Weeks, I., Mpoko, C. N., Laing I and Woodhead, J. S., 1986, Chemiluminescent labelled-antibody assay for thyroxine in serum with magnetic separation of the solid phase, Clin.Chem., 32:532.

Weeks, I., Campbell, A. K., and Woodhead, J. S., 1983, Two-site immunochemiluminometric assay for human α_1-fetoprotein, Clin.Chem., 29:1474.

Weeks, I, Kemp, H. A., and Woodhead, J. S., 1981, Two-site assay of human α_1-fetoprotein using ^{125}I-labelled monoclonal antibodies, Bioscience Rep., 1:785.

Weeks, I., Sturgess, M. L., Siddle, K., Jones, M. K., and Woodhead J. S., 1984, A high sensitivity immunochemiluminometric assay for human thyrotrophin, Clin.Endoc., 20:489.

Woodhead, J. S., Addison, G. M., and Hales, C. N., 1974, The immunoradiometric assay and related techniques, Brit.Med.Bull., 30:44.

Woodhead, J. S., Evans, M. J., Scarisbrick, J. J., Read, C., and Cameron E. H. D., 1975, The preparation and properties of ^{125}I-labelled antibodies to progesterone. In Radioimmunoassay of Steroids, E.H.D. Cammeron, S.G. Hillier and K. Griffiths, eds., Alpha Omega, Cardiff, pp. 269-271.

MONOCLONAL ANTIBODIES IN CHEMILUMINESCENT IMMUNOASSAYS

Ian Weeks and J. Stuart Woodhead

Department of Medical Biochemistry
University of Wales College of Medicine
Heath Park, Cardiff, CF4 4XN

INTRODUCTION

The field of immunoassay has experienced several major advances since the first description of radioimmunoassay (RIA) over 25 years ago (Yalow and Berson, 1960). Undoubtedly one of the most important of these is the development of the immunoradiometric assay (IRMA) by Miles and Hales (1968). This methodology involves the use of excess concentrations of binding reagents relative to the analyte concentration in contrast to the limiting antibody concentrations of RIA. Further, the IRMA methods involve the use of labelled antibody rather than labelled antigen reagents. The classical IRMA involved preliminary incubation of the anaytical sample with excess labelled antibodies following which a solid phase antigen derivative was introduced to bind any labelled antibodies not involved in immune complex formation. Quantitation of radioactivity in either the solid-phase or the liquid thus gave a measure of the amount of analyte originally present. For theroetical and practical reasons this type of assay was not widely accepted and a variant of the basic system became more popular. This is the so-called two-site immunoradiometric assay (Woodhead et al, 1974) in which antigen is bound by labelled antibodies and the subsequent immune complexes bound by solid-phase antibody at other antigenic determinants (epitopes). The radioactivity measured in the separated solid-phase immune complexes is thus directly proportional to the concentration of analyte present.

In spite of its theoretical advantages the two-site IRMA was not widely used because of practical difficulties. Firstly, high performance assays required the use of specific labelled antibodies which had to be prepared by immunoaffinity purification. Secondly, the preparation of solid-phase antibody reagents required the use of large quantities of often expensive antisera. Thirdly, whilst it is an obvious requirement that the analyte must be large enough to bind two antibodies simultaneously, assays for peptides could only be configured using carefully selected, site-specific antibodies even then yielding immunoassays of limited use due to poor performance.

At this point the advent of monoclonal antibody technology provided a second major advance in immunoassay methodology particularly in terms of two-site immunoradiometric assays (Weeks et al, 1981). The use of these techniques enables large quantities of pure, specific antibodies to be produced and thus overcomes the major technical difficulties associated with two-site IRMA. Careful optimisation of the immunochemistry has thus succeeded in enabling IRMA to be engineered to the point at which the label itself becomes the limiting factor in terms of assay performance and ease of use.

In parallel with improvements in immunochemistry the field of immunoassay has seen attempts to replace radioisotope tracers with non-radioactive labels. Many methodologies have been demonstrated but few are capable of matching the levels of performance offered by state-of-the-art IRMA. Only chemiluminescence immunoassay stands out in terms of offering improved performance over ^{125}I-based assays without recourse to complex chemistry or instrumentation for end-point quantitation.

CHEMILUMINESCENCE

In simple terms chemiluminescence is the light emitted during certain chemical reactions, in contrast to fluorescence and phosphorescence which are essentially photophysical processes. Though all of these luminescence phenomena involve the emission of light when an electronically excited state relaxes to its ground state, the generation of the excited states occurs by different mechanisms. In fluorescence an electronically excited state is produced when a molecule absorbs light of a suitable wavelength, this energy rich state loses energy by emission of light during relaxation to the ground state. Phosphorescence is analogous except that intersystem crossing e.g. from a singlet to triplet state occurs prior to emission. In chemiluminescence a chemical reaction occurs in which the reaction product is formed as an electronically excited state which subsequently relaxes to its ground state with photonic emission.

The high energy of the emitted radiation (several hundred kilojoules per mole) can only be achieved by 'channelling' the energy of the metathetical reactions involved in a highly coordinated fashion. In spite of this requirement a large number of chemiluminescent reactions exist ranging from simple atom recombination reactions in the gas phase to the complex enzyme-substrate systems of 'natural' chemiluminescence (bioluminescence).* Further discussion of chemiluminescence phenomena will be restricted to those reactions which proceed under aqueous conditions and lend themselves to coupling into immunochemical systems. Fig. 1. shows some examples of chemiluminescent molecules. Reactions of these molecules involve a variety of 'active' oxygen species and may also require catalysts or cofactors. Classical examples of this are seen in the firefly and bacterial bioluminescence systems where the substrate (luciferin) is acted upon by an enzyme (luciferase) in the presence of cofactors. In these cases the cofactors are ATP and NAD(P)H/long chain aldehydes respectively and many examples exist where a bioluminescent reaction has been coupled into another biochemical system via processes which have a dependency on the cofactors.

In systems involving synthetic chemiluminescent molecules there are potentially many ways of coupling the luminescent system into the biochemical reaction. An example of this is the use of enzyme labelled antibodies or antigens which are capable of generating 'active' oxygen species which subsequently react in a chemiluminescent system. A particular example of this is the use of glucose oxidase labelled reagents to produce hydrogen peroxide which is then quantified using a bis-oxalate ester chemiluminescent system (Arakawa et al, 1985).

---

* See Campbell et al, 1986 for a review of this field.

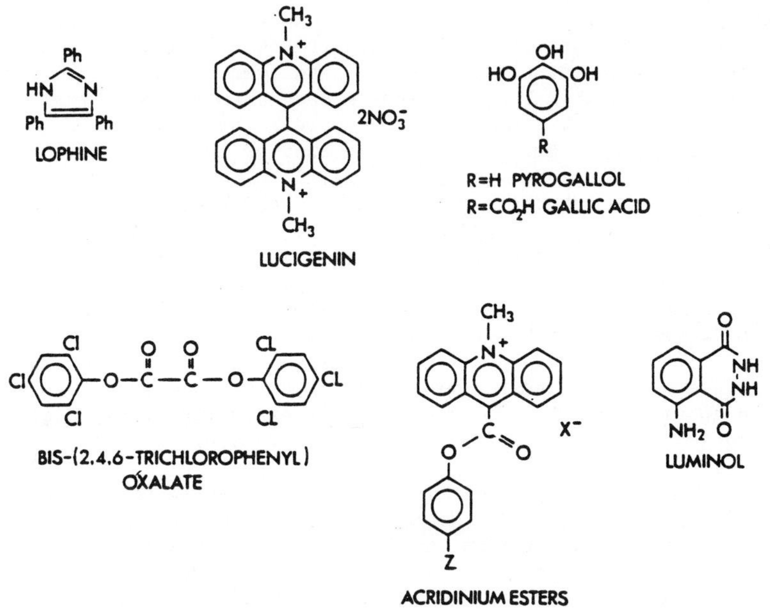

Fig. 1. Examples of chemiluminescent molecules.

A better known system is the one which utilizes the ability of horseradish peroxidase (HRP) to catalyse the luminol reaction (Arakawa et al, 1979). Here an immunoassay is performed using HRP labelled antigen or antibody as the 'tracer' reagent and end-point quantitation is performed by addition of luminol and hydrogen peroxide. More recently this basic system has been improved by the introduction of so-called 'enhancer' molecules such as luciferins or a variety of other small molecules which, by an as yet unknown mechanism, increase the signal/noise ratio of the emission (Whitehead et al, 1983). The most straightforward and logical approach to chemiluminescence immunoassay is to design a simple system where the chemiluminescent molecule is a direct replacement for the radioisotope. Historically, this has proved difficult in many cases for a variety of reasons which are discussed in the next section.

CHEMILUMINESCENCE IMMUNOASSAYS

The earliest work in this field involved the synthesis of chemiluminescent-analyte analogues for use in competitive binding reactions. The most often quoted example is a competitive binding assay for biotin using biotinylisoluminol as the tracer and avidin as the binding reagent (Schroeder et al, 1976). The unusual feature of this system is that it was also a demonstration of a homogeneous system, that is an assay which did not require the separation of bound and free tracer prior to quantitation. This was due to the fact that the intensity of light emission was changed upon binding and was hence a measure of complex formation. Other such examples of homogeneous assays have appeared for steroids but have not been adopted due to the variability of the phenomenon and lack of sensitivity (Kohen et al, 1979). Far more work has been reported on steroid immunoassays based on solid phase, heterogeneous chemiluminescence immunoassays using steroid-isoluminol conjugates as tracers (Collins et al, 1983). These have demonstrated the ability of such systems to replace cumbersome scintillation counting.

In spite of this the use of luminol-based molecules in the development of sensitive immunoassays for proteins has been disappointing. Recently a wide range of such assays has been reported which perform well routinely and offer a genuine alternative to RIA and IRMA (Wood et al, 1984). However the full potential of chemiluminescence has not been realized with these systems in providing assays of improved performance in comparison with ^{125}I based systems. Perhaps some clue as to the reasons for this is to be gained by consideration of earlier work. In the first description of chemiluminescence labelled antibodies, it was shown that coupling luminol to proteins resulted in a massive decrease in light output (quantum yield) from the molecule (Simpson et al, 1979). Even the use of a long chain isoluminol derivative suffered from a decrease in quantum yield of some fourteen-fold, possibly due to microenvironmental quenching, following coupling to protein (Schroeder et al, 1981).

Recent work has shown that certain aryl acridinium esters can be coupled to proteins without impairment of their inherently high quantum efficiencies to yield high specific activity labelled reagents (Weeks et al, 1983). Since we have already seen that the use of monoclonal antibodies permits the development of immunoassays whose performance is limited by the label rather than the immunochemistry, then the importance of this property of acridinium esters can be readily appreciated on theoretical as well as practical grounds. Ekins (1985) has recently shown the theoretical basis for this most strikingly for non-competitive, antibody excess systems where specific activity of labelled antibody and assay non-specific binding are of paramount importance in setting the limit of sensitivity. For an antibody of moderate affinity (10^{10} l/mol) improvements in assay sensitivity of 10 to 100 fold are predicted in going from ^{125}I to a label of considerably higher specific activity provided that assay non-specific binding is low (<0.1% total activity). In going to higher affinities and lower non-specific binding figures a proportional improvement in sensitivity occurs with a high specific activity non-isotopic label in contrast to ^{125}I where the specific activity is the limiting factor. Improvements are also imparted to competitive binding systems but are far less impressive.

Thus on theoretical grounds the use of pure, high affinity monoclonal antibodies of unique specificity which are labelled with a stable acridinium ester, having perhaps a 100 fold greater specific activity than ^{125}I, should yield assays of superior performance to that offered by IRMA. Further the reagents would have all the advantages expected of non-radioactive labels. In practice it is vital that the immunochemistry be fully optimised such that improvements due to the use of a chemiluminescent acridinium ester label can be seen. This is becoming easier as the properties of monoclonal antibodies improve in immunoassay terms with further developments in the methodology. Such improvements in assay performance have recently been demonstrated.

DESIGN OF IMMUNOASSAYS BASED ON CHEMILUMINESCENT ACRIDINIUM ESTER LABELS

In contrast to the luminol reaction, the chemiluminescent reaction of aryl acridinium esters is well understood. The mechanism (Fig. 2) consists of a concerted multiple bond cleavage via a dioxetanone intermediate to yield an electronically excited molecule of N-methylacridone which subsequently relaxes to its ground state with photonic emission at 426 nm. Although a variety of luminometers are commercially available for photon counting few are designed particularly for immunoassay work. The latest machines which are specifically suited for this purpose are the Clini-Lumat (Laboratorium Prof. Dr Berthold, Wildbad, FRG) and the Magic Lite Analyzer

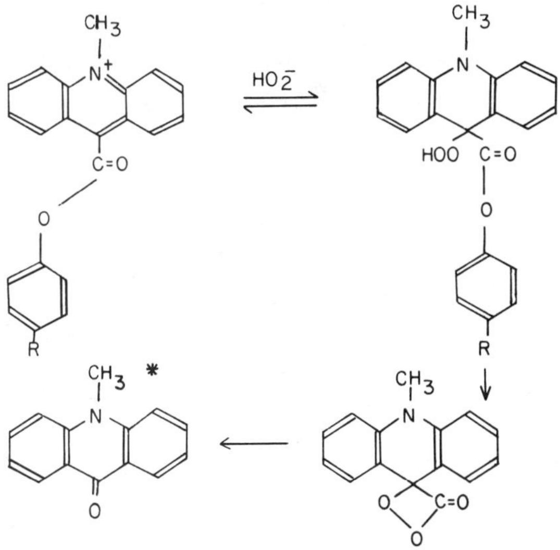

Fig. 2. The mechanism of the chemiluminescent reaction of an acridinium ester.

(Ciba-Corning Diagnostics,Medfield, MA 02052, USA). These are both equipped with sophisticated software for immunoassay data reduction.

The chemistry of the acridinium reactions has important implications in immunoassay. Firstly, it does not require a catalyst and proceeds in the presence of dilute alkaline hydrogen peroxide. This is in contrast to luminol where a vast range of chemical species (many of them present in biological systems) are capable of catalysing the reaction or otherwise affecting the chemiluminescence emission. There exists therefore the problem of high backgrounds, many potential interferences and complex reagent chemistry in the luminol based systems. Secondly, the nature of the acridinium reaction can be exploited in such a way that the emitting species (N-methylacridone) is dissociated from the rest of the original molecule (e.g. that portion coupled to protein) prior to emission of photons such that microenvironmental quenching is minimal in contrast to luminol systems.

Ideally, a labelling system is preferred whereby the chemiluminescent molecule can be reacted with the protein under mild conditions, without the need for other reactants, to yield a stable, easily purified labelled protein. Such a system has recently been reported where an acridinium salt incorporating a N-hydroxysuccinimide ester has been synthesised (Weeks et al, 1983). This compound, 4(2-succinimidyloxycarbonylethyl)phenyl-10-methyl-9-acridiniumcarboxylate fluorosulphate (Fig. 3), reacts with the lysine residues of proteins within 15 minutes at pH 8 at room temperature to yield high specific activity labelled derivatives. In the case of antibodies, specific activities of several million photon counts per nanogram of IgG can be achieved in contrast to ^{125}I-labelled antibodies where the maximum attainable activity is of the order of 30,000 counts per minute per nanogram of IgG.

Since the light emission is generated chemically, it is possible to vary the rate at which photons are emitted (milliseconds-hours). It is therefore desirable to produce a rapid flash of light to maximise throughput at the measurement stage of the assay rather than to produce a long glow. In the former, all the photons are measured within perhaps less than

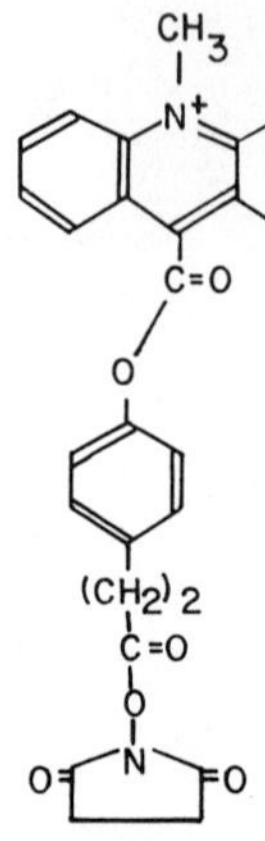

Fig. 3. Acridinium ester for use in protein labelling.

a second without wastage and significant accumulation of background activity. This fact coupled with the high quantum yield of acridinium esters enables high sensitivities of detection to be achieved.

The importance of attaining high specific activities in terms of assay sensitivity has been discussed in theory earlier. In practice this has been demonstrated using a two-site immunochemiluminometric assay for α_1-fetoprotein which utilizes acridinium ester labelled monoclonal antibodies (Weeks et al, 1983). Here the sensitivity of the assay increases when the specific activity of the labelled monoclonal antibody increases provided that the assay non-specific binding is kept at a constant level. Table 1 lists a number of two-site immunochemiluminometric assays based on the use of polyclonal and monoclonal antibodies labelled with chemiluminescent acridinium esters. The solid-phases used in these assays are based on either diazoarylcellulose or paramagnetic particles.

The ICMA for TSH and PTH are worthy of further comment since both analytes require assays of a highly sensitive and specific nature. The TSH-ICMA has a sensitivity of detection of 0.003 mU/l and a working range for inter-assay CV $\leqq$ 10% extending upwards from 0.06 mU/l (Fig. 4). This level of performance permits clear discrimination between thyrotoxic and euthyroid individuals which has previously proved impossible. The PTH-ICMA is specific for intact PTH and therefore suffers no interference from circulating fragments. Since the circulating levels of PTH are very low, particularly in cases of primary hypoparathyroidism, a high sensitivity of detection is necessary to discriminate these from normal concentrations. This assay has a sensitivity of detection of 0.4 pmol/l with an inter-assay CV $\leqq$ 10% at concentrations >2 pmol/l (Fig. 5).

COMPETITIVE BINDING IMMUNOASSAYS BASED ON ACRIDINIUM ESTERS

The constraints of the immunochemistry of competitive assays which use limiting concentrations of antibody mean that such assays exhibit poorer performances than two-site immunometric assays in terms of sensitivity, reaction time and dynamic range. However, the choice of a high specific activity non-isotopic label still offers advantages over radioisotopes even though the full potential of the label is not exploited.

Table 1. Two-Site Immunochemiluminometric Assays Based on Chemiluminescent Acridinium Esters

Analyte	Antibody type * Label	Solid-phase	Reference
α_1-fetoprotein (AFP)	M	P	Weeks et al., 1983
Ferritin	P	P	Weeks & Woodhead, 1984
Complement C9	M	M	Weeks et al., 1985
Tamm-Horsfall glycoprotein	P	P	Horton et al., 1987
Thyrotropin (TSH)	M	P	Weeks et al., 1984
Somatotropin (hGH)	M	P	Weeks & Woodhead, 1986
Parathyrin (PTH)	P	M	Brown et al., 1987

* M = Monoclonal P = Polyclonal

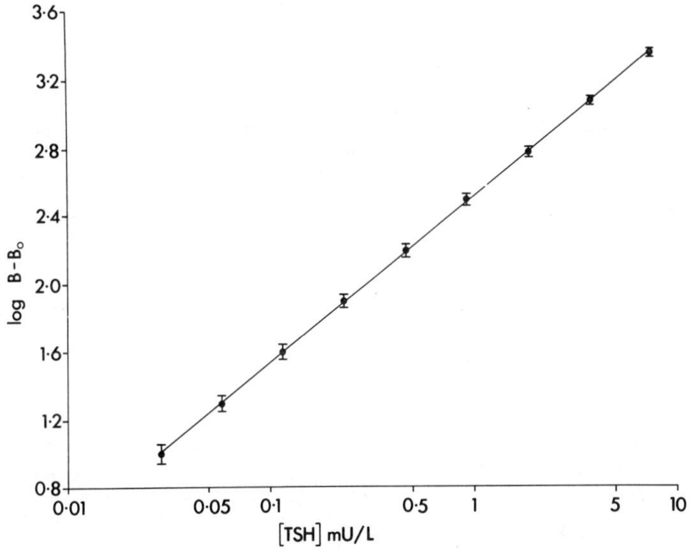

Fig. 4. Dose-response curve of a two-site ICMA for thyrotropin. B = specific bound chemiluminescence, B_o = non-specific binding.

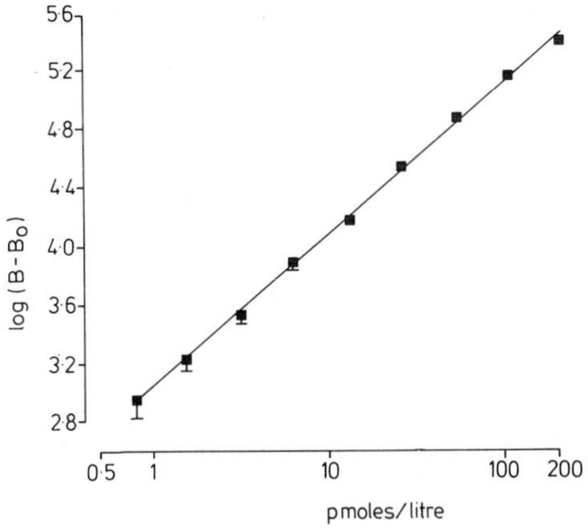

Fig. 5. Dose-response curve of a two-site ICMA for intact parathormone (PTH). B = specific bound chemiluminescence, B_o = non-specific binding.

Two general architectures have been described in the case of chemiluminescent acridinium esters. Firstly, those assays using labelled antigens and secondly, those assays employing labelled antibodies. A list of examples of the former is given in Table 2.

The labelling of proteins and peptides with lysine residues is facilitated by using the acridinium salt/N-hydroxysuccinimide ester as described previously, provided that immunoreactivity is not impaired and suitable purification procedures can be performed (e.g. ion exchange hplc). However for small molecules and peptides lacking suitable amino groups different chemical strategies may be needed for each particular application possibly requiring complex synthetic procedures.

A more universal approach is therefore to use labelled antibodies in the form of competitive binding immunochemiluminometric assays. Here a limiting concentration of labelled antibody is partitioned between the analyte and a solid-phase analyte derivative. Thus the intensity of chemiluminescence associated with the separated solid-phase is inversely proportional to the concentration of analyte present (Fig. 6). Recently a variation of this basic system has been described which has certain practical advantages (Sturgess et al, 1986, Sturgess et al, 1987). Here the analyte derivative is a liquid-phase analyte-protein carrier conjugate which competes with the analyte for binding to the labelled antibody. Immune complexes formed between the conjugate and the labelled antibody are precipitated by the introduction of a solid-phase second antibody directed to the protein portion of the conjugate. Systems described to date have used analyte-rabbit IgG conjugate and a corresponding sheep (anti-rabbit IgG)IgG paramagnetic particle solid-phase. The advantages of this type of system are that the important primary immune-complex formation takes place entirely in solution thus there is no imprecision due to solid-phase heterogeneity. Further, the subsequent reaction with solid-phase anti-protein carrier can be made extremely rapid since a large excess of the antibody can be used to drive the reaction. Table 3 gives examples of this type of assay system using acridinium ester labelled monoclonal antibodies. A review of immunoassays based on acridinium esters has been published recently (Weeks et al, 1987).

CONCLUSION

There is no doubt that further developments in monoclonal antibody technology will result in fully optimized immunochemistry for a wide range of immunoassays. As a consequence the limiting factor in assay performance will be the choice of label. Of the potential alternatives to radioisotopes, chemiluminescent molecules have established themselves as being one of the most useful. More particularly the use of chemiluminescent acridinium esters has provided a basis for the development of immunoassays exhibiting superior performance to conventional assays with the minimum of technical and chemical complexity.

Table 2. Competitive Binding Chemiluminescence Immunoassays Based on Acridinium Ester Labelled Antigens

Analyte	Tracer preparation	Reference
17β-estradiol	Dedicated synthesis	Patel et al., 1982
Progesterone	Dedicated synthesis	Richardson et al., 1986
Calcitonin	NHS-Acridinium *	Hullin et al., 1987
Growth Hormone releasing factor	NHS-Acridinium *	Page et al., 1986

* NHS = N-hydroxysuccinimide ester

LABELLED-ANTIBODY FREE HORMONE ASSAY

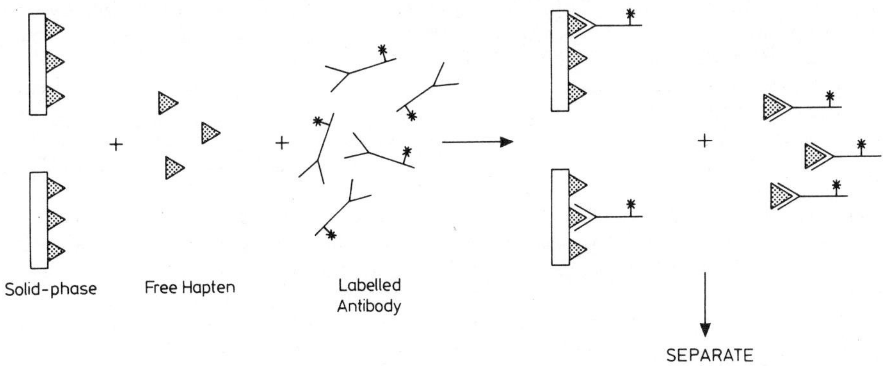

Fig. 6. General architecture for a competitive binding ICMA.

Table 3. Competitive Binding Immunochemiluminometric Assays Based on Chemiluminescent Acridinium Esters

Analyte	Sensitivity of detection	References
Total thyroxine	1.0 nmol/l	Sturgess et al., 1986
Free thyroxine	0.3 pmol/l	Sturgess et al., 1987
Fluphenazine	70.0 pmol/l	Herbert, unpublished work
Prochlorperazine	160.0 pmol/l	Herbert, unpublished work

REFERENCES

Arakawa, H., Maeda, M., and Tsuji, A., 1985, Chemiluminescence enzyme immunoassay for thyroxine with use of glucose oxidase and a bis (2,4,6-trichlorophenyl)oxalate-fluorescent dye, Clin.Chem., 31:430.

Arakawa, M., Maeda, M., and Tsuji, A., 1979, Chemiluminescence enzyme immunoassay of cortisol using peroxidase as label, Anal.Biochem. 97:248.

Brown, R. C., Aston, J. P., Weeks, I., and Woodhead, J. S., 1987, Circulating intact parathyroid hormone measured by a two-site immunochemiluminometric assay, J.Clin.Endocrinol Metab., 65:407.

Campbell, A. K., Hallett, M. B., and Weeks, I., 1985, Chemiluminescence as an analytical tool in cell biology and medicine, in: "Methods of Biochemical Analysis, vol 31', D. Glick, ed., John Wiley and Sons, Chichester, U.K.

Collins, W. P., Barnard, G. J., Kim, J. B., Weerasekera, D. A., Kohen, F., Eshhaz, Z., and Lindner, H. R., 1983, Chemiluminescence immunoassays for plasma steroids and urinary steroid metabolites, in: "Immunoassays for Clinical Chemistry," W.M. Hunter, and J.E.T. Corrie, eds., Churchill Livingstone, Edinburgh, U.K.

Horton, J. K., Davies, M., Woodhead, J. S., and Weeks, I., 1987, The development of an immunochemiluminometric assay for measuring Tamm-Horsfall glycoprotein in serum, in: "Bioluminescence and Chemiluminescence: New Perspectives," J.Scholmerich, R. Anderson, A. Kapp. M. Ernst and W.G. Wood, John Wiley and Sons, Chichester, U.K.

Hullin, D. A., Brown, R. C., and Weeks, I., 1987, Development of a chemiluminescent immunoassay for calcitonin, in: "Bioluminescence and Chemiluminescence: New Perspectives, " J. Scholmerich, R. Anderson, A.

Kapp, M. Ernst and W.G. Wood, eds., John Wiley and Sons, Chichester, U.K.
Kohen, F., Pazzagli, M., Kim, J. B., and Lindner, H. R., 1979, An assay procedure for plasma progesterone based on antibody enhanced chemiluminescence, FEBS Letters 104:201.
Miles, L. E. M., and Hales, C. N., 1968, Labelled antibodies and immunological assay systems, Nature 219:186.
Page, M. D., Dieguez, C., Weeks, I., Woodhead, J. S., and Scanlon M. F., 1987, GRF secretion in man, in: "Growth Hormone and Acromegaly," R. Robbins, ed., Plenum Press, New York, In press.
Patel, A., Woodhead, J. S., Campbell, A. K., Hart, R. C., and McCapra F., 1982, The preparation and properties of a chemiluminescent derivation of 17β - estradiol, in: "Luminescent Assays: Perspectives in Endocrinology and Clinical Chemistry," M. Serio and M. Pazzagli, eds., Raven Press, New York.
Richardson, A. P., Kin, J. B., Barnard, G. J., Collins, W. P., and McCapra, F., 1985, Chemiluminescence immunoassays of plasma progesterone, with progesterone-acridinium ester used as the labelled antigen, Clin.Chem 31:1664.
Schroeder, H. R., Vogelhut, P. O., Carrico, R. J., Boguslaski, R. C., and Buckler, I. R. T., 1976, Competitive protein binding assay for biotin monitored by chemiluminescence, Anal.Chem., 48:1933.
Schroeder, H. R., Hines, C. M., Osborn, D. D., Moore, R. P., Hurtle, R. L., Wogoman, F. F., Rogers, R. W., and Vogelhut, P. O., 1981, Immunochemiluminometric assay for hepatitis B surface antigen, Clin.Chem., 27:1378.
Simpson, J. S. A., Campbell, A. K., Ryall, M. E. T., and Woodhead, J. S., 1979, A stable chemiluminescent labelled antibody for immunological assays, Nature 279:646.
Sturgess, M. L., Weeks, I., Mpoko, C. N., Laing, I., and Woodhead, J. S., 1986, Chemiluminescent labelled antibody assay for thyroxine in serum with magnetic separation of the solid-phase, Clin. Chem., 32:532.
Sturgess, M. L., Weeks, I., Evans, J. P., Mpoko, C. N., Laing I and Woodhead, J. S., 1987, Immunochemiluminometric assay for serum free thyroxine, Clin.Endocrinol, 27:383.
Weeks, I., Kemp, H. A., and Woodhead, J. S., 1981, Two-site assays of human α_1-fetoprotein using ^{125}I-labelled monoclonal antibodies, Biosci Rep., 1:785.
Weeks, I., Beheshti, I., McCapra, F., Campbell, A. K., and Woodhead, J. S., 1983, Acridinium esters as high specific activity labels in immunoassay, Clin.Chem., 29:1474.
Weeks, I., Campbell, A. K., and Woodhead, J. S., 1983, Two-site immunochemiluminometric assays for α_1-fetoprotein, Clin.Chem., 29:1480.
Weeks, I., Sturgess, M., Siddle, K., Jones, M. K., and Woodhead, J. S., 1984, High sensitivity immunochemiluminometric assay for thyrotrophin, Clin.Endocrinol., 20:489.
Weeks, I., Morgan, B. P., Campbell, A. K., and Woodhead, J. S., 1985, Measurement of C9 concentrations using an immunochemiluminometric assay, J.Immunol Meth., 80:33.
Weeks, I., and Woodhead, J. S., 1984, Two-site immunochemiluminometric assay for serum ferritin, Clin.Chim.Acta., 141:275.
Weeks, I., and Woodhead, J. S., 1986, Measurement of human growth hormone (hGH) using a rapid immunochemiluminometric assay, Clin.Chim.Acta., 159:139.
Weeks, I., Sturgess, M., Brown, R. C., and Woodhead, J. S., 1986, Immunoassays using acridinium esters, Methods in Enzymology 133:366.
Whitehead, T. P., Thorpe, G. H. G., Carter, T. J. N., Groucutt, C., and Kricka, L. J., 1983, Enhanced luminescence procedure for sensitive determination of peroxidase labelled conjugates in immunoassay, Nature 305:158.

Wood, W. G., 1984, Routine luminescence immunoassays for haptens and proteins, *in*: "Analytical Applications of Bioluminescence and Chemiluminescence," L.J. Kricka, P.E. Stanley, G.H.G. Thorpe and T.P. Whitehead, eds., Academic Press, London.

Yalow, R. S., and Berson, S. A., 1960, Immunoassay of endogenous plasma insulin in man, *J.Clin.Invest.*, 39:1157.

MONOCLONAL ANTIBODIES TO HUMAN ESTROGEN AND PROGESTERONE RECEPTORS

Geoffrey L. Greene

The Ben May Institute, The University of Chicago
5841 S. Maryland Avenue, Chicago, Illinois 60637

INTRODUCTION

Monoclonal antibodies directed against the hormone-binding forms of human estrogen receptor (ER) and progesterone receptor (PR) have been used to develop quantitative and histochemical immunoassays for these proteins in reproductive tissues and related cancers (Greene and Press, 1987; Greene and Jensen, 1982; King and Greene, 1984; King et al, 1985; Nolan et al, 1984; Press and Greene, 1984). Such assays make use of very specific and sensitive probes that do not depend upon the ability of receptor protein to bind its cognate radiolabeled hormone in tissue extracts. In addition, these antibodies can be used to detect and quantify ER and PR directly in fixed tissue sections and cell dispersions, thereby permitting an evaluation of the type, proportion and distribution of cells that contain immunoreactive ER and/or PR. For metastatic breast cancer, a major potential clinical application of receptor immunoassays is the prediction of response to endocrine therapy and assessment of prognosis (disease-free interval and survival). The clinical utility of ER and PR steroid-binding assays (SBA) for evaluating these parameters in patients with advanced breast cancer has been well documented (DeSombre et al, 1979; DeSombre, 1984; Osborne et al, 1981). However, major limitations of current receptor assays are:

1. their inability to predict response to endocrine therapy in 30-40% of the breast cancers defined as ER- or PR-rich and in 5-10% of those that are receptor-poor, and

2. the difficulty of performing such assays on small tumors, metastasies and needle biopsies.

In addition, assays performed on extracts of breast tumors cannot provide information about the identity, proportion and distribution of receptor-containing cells. Receptor-specific monoclonal antibodies can be used for all of these purposes, as described here.

ESTROGEN RECEPTOR

Monoclonal antibodies to human ER were derived from male Lewis rats immunized with partially purified MCF-7 cytosol [^{3}H] estradiol-receptor complex (E^*R_c) eluted from an estradiol-Sepharose column (5-10% pure). A total of 13 monoclonal antibodies, most or all of which appear to recognize distinct regions of the receptor molecule, have been characterized. These antibodies have high affinity ($K_d=10^{-9}-10^{-10}M$) for both steroid-occupied and unoccupied estrogen receptor and recognize nuclear as well as cytosol forms of the receptor molecule. Although they vary in their cross reactivity with estrogen receptors from various animal species, each antibody appears to be completely specific for the 65-KDa steroid-binding subunit of the estrogen receptor complex, as judged by extensive sucrose gradient and immunoblot analyses of cytosol and nuclear extracts from a variety of tissues and cell lines. Cross reactivity patterns (Greene et al, 1984) indicate both sequence homology and heterogeneity among mammalian and nonmammalian ERs, consistent with the variable homology predicted from cDNA sequences for human, rat, mouse and chicken ERs. Some epitopes (e.g. H222 and H226) are common to all tested estrogen receptors, whereas others (e.g. D547 and D58) are present only in mammalian receptors, and one (D75) appears to be restricted to primate ER. The D75 epitope lies at the extreme carboxy terminus of the receptor, which is a region of low homology among the four characterized ERs. The most well conserved epitopes are associated with the highly homologous DNA and steroid-binding regions of the receptor molecule.

ER monoclonal antibodies have been used to devise a simple enzyme immunoassay (ER-EIA) and an immunocytochemical assay (ER-ICA) for detecting and measuring ER in reproductive tissues and related cancers, especially breast cancer. Two antibodies (D547, H222) were used to develop the current ER-EIA (Nolan et al, 1984). In this assay, polystyrene beads coated with D547 adsorb either occupied or unoccupied receptor present in a tissue or tumor extract. Peroxidase-conjugated H222 is then used to measure the amount of receptor bound to the bead. In recent U.S. and European multicenter studies involving more than 1,000 breast cancers, a comparison of ER-EIA results with steroid-binding results revealed an excellent correlation (>90% concordance) over a broad range of ER concentrations (Cancer Research Supplement, 1986). For seven of the European studies, the slope of the correlation line was very close to 1.0, and the Y intercept was not significantly different from zero in five laboratories. In the U.S. study, the slope of the correlation was generally greater than 1.0 (0.95-2.3), which may reflect an underestimate of ER concentration by the dextran-coated charcoal assays used to measure steroid-binding capacity at the U.S. sites. Inter-and intra-laboratory variation coefficients for ER-EIA were low, probably due to the standardization of this assay. The sensitivity of the D547/H222 combination used in the current ER-EIA is about 2 fmol of ER per milliliter of cytosol (1-2 fmol/mg protein). Although further testing is warranted, the ER-EIA should provide an accurate and sensitive means of measuring ER in tumors from pre-and post-menopausal women. A particular advantage of the EIA over SBA is the ability of this assay to measure ligand-occupied ER, a situation which can occur in tumors from premenopausal women and women who are on estrogen replacement therapy, as well as women who are on tamoxifen therapy.

The immunocytochemical assay for ER (ER-ICA), which affords direct visualization of ER in tumor sections and needle biopsies, by indirect immunoperoxidase staining with the H222 antibody, shows excellent promise as complementary, and possibly alternative, approach to receptor assessment in breast cancers and endometrial cancers. In a study of 117 human breast cancers by the ER-ICA method, the presence or absence of specific nuclear staining for ER was significantly associated with the concentration of cytosolic estrogen receptor determined by steroid-binding assay (King et al, 1985). A more recent analysis of primary breast tumors from post-menopausal patients at high risk for recurrence has shown that several quantifiable features of ER-ICA determinations correlate significantly with

ER concentrations measured in cytosols (DeSombre et al, 1986). The two parameters that contribute the most to this correlation are the staining intensity and the proportion of positively stained cells in the cancer. The overall ER-ICA status for these breast cancers also differentiated two populations of patients with different disease-free experience and survival. The proportion of patients surviving four years after mastectomy is significantly higher (P = 0.001) for ER-ICA positive tumors than for ER-ICA negative tumors. Although some of this difference may be related to the improved prognosis of patients who received adjuvant tamoxifen therapy, patients with ER-ICA positive lesions have a better prognosis whether treated with tamoxifen or not. An additional, and potentially important, clinical application of ER-ICA would be the ability of this assay to identify breast cancer patients with advanced disease who are likely to respond to endocrine therapy. Thus far, four independent studies with clinical response data have been published (Cancer Res. Suppl., 1986). Overall, approximately 74% (56/76) of ER-ICA positive patients responded objectively to endocrine manipulation (complete remission or arrested disease), whereas only 8% (5/66) of ER-ICA negative patients responded. It is of interest that the results among the four unrelated groups are very similar despite the use of different, and subjective, criteria for interpretation of ER-ICA status. Clearly, the number of patients for which response data and ER-ICA evaluations are available is still very low. However, the data suggest that ER-ICA status, and possibly one or more semiquantifiable features of immunocytochemical staining for ER, may be a useful predictor of response to endocrine therapy for patients with advanced breast cancer. The ER-ICA method has also proved useful for the evaluation of breast tumor needle biopsies, ovarian cancers and endometrial cancers.

PROGESTERONE RECEPTOR

The cytosol and nuclear forms of PR from T47D cells have been partially purified by steroid affinity chromatography or by immunoadsorption to a monoclonal PR antibody (JU601) coupled to Sepharose 4B. Purified PR consists of two steroid-binding components which, when photoaffinity-labeled with ^{3}H-R5020, migrate at 109-119 kDa (B) and 88-93 kDa (A) in reducing SDS gels. More than 14 monoclonal antibodies to the A and B steroid-binding components of the PR complex have been prepared and characterized. Of these, 13 recognize epitopes shared by the A and B components. One antibody, KC146, is specific for the B component. Like the ER monoclonal antibodies, the PR antibodies have high affinity for steroid-occupied as well as unoccupied receptor and recognize both nuclear and cytosol forms. The epitope for the KC146 antibody is present on all PRs tested, including the B form of PR from chicken oviduct, whereas nine other antibodies recognize only human PR and the remaining four cross react with rabbit PR.

In regard to the development of quantitative immunoassays for PR, and enzyme immunoassay similar to the one developed for ER (PgR-EIA) has been developed for PR analysis in extracts of hormone-responsive cancers. Several combinations of antibodies have been tested on polystyrene beads and at least two combinations appear to have the necessary sensitivity and specificity for accurate PR measurements (JZB39 and KD68). When 28 human breast tumor cytosols were analyzed by a prototype PR-EIA, the results correlated well (R = 0.96) with values determined by the NEN progestin receptor assay kit (Weigand et al, 1986). The sensitivity of the PR-EIA is <1 fmol of PR per ml (<1 fmol/mg protein) of extract. In multicenter studies carried out during the past two years, the overall concordance of PgR-EIA with SBA is greater than 90%.

For immunocytochemical localization of PR in various tissues and tumors, the JZB39 and KD68 IgGs have proved most useful. As observed for ER-ICA, specific nuclear staining for PR is observed, under the same fixation conditions and with the same reagents used for ER-ICA, in serial sections from PR-rich breast carcinomas. Both ER and PR can be visualized in adjacent sections of the same tumor. Several studies are in progress to evaluate these antibodies as probes for PR in breast tumors, especially in comparison to biochemically determined PR in cytosols and to ER measurement and localization with the ER-ICA method. In a preliminary evaluation of 20 breast tumors by the PR-ICA method, a good correlation was observed between measured PR in tumor cytosols and the presence or absence of specific nuclear staining for PR in frozen, fixed tumor sections.

DISCUSSION

From the above results, it is clear that immunochemical evaluations of ER and PR content and distribution in breast lesions are feasible and informative. The quantitative sandwich enzyme immunoassays are simple, rapid, sensitive and less likely to be affected by chemical and conformational changes in receptor than steroid-binding assays. Monoclonal antibodies prepared against both ER and PR have been shown to be completely specific for their corresponding receptor, with the exception of two PR antibodies that recognize epitopes present on nuclear proteins present in non-target tissues. In addition, all ER and PR antibodies used in these assays recognize occupied as well as unoccupied receptor, permitting measurement of total receptor in cytosol as well as nuclear extracts of tumors. As a complementary, and possibly alternative, approach to receptor assessment in breast cancers, direct localization of ER and PR in tumor cells by immunoperoxidase staining offers several advantages. First, it is possible to distinguish ER- and/or PR-containing tumor cells from negative tumor cells and normal tissue. In addition, cytological preparations from primary tumors or from metastases at different sites can be readily evaluated for ER or PR on relatively few cells. Problems with sample storage and handling are also minimized since no homogenization is required. Finally, as already shown for ER-ICA, several quantifiable features of immunoperoxidase staining, including staining intensity and the proportion of tumor cells stained, correlate with the ER content, patient prognosis for recurrence and length of survival, and response of patients with metastatic breast cancer to hormone therapy. Although clinical data is not yet available for PR-ICA, preliminary data from this and other laboratories indicate that immunocytochemical evaluations of PR in breast cancers correlate well with the PR content of these tumors. It is not clear yet whether the ER-ICA and PR-ICA will provide information that is more useful than current steroid-binding assays for ER and PR. However, the direct assessment of ER and PR in frozen, and possibly paraffin, sections of tumors as well as in needle aspirates of lesions will certainly simplify tumor evaluation and repeat sampling. Finally, if tumor heterogeneity is an indicator of cancers that are less likely to respond to hormone therapy, due to the presence of autonomous, ER- and/or PR-negative neoplastic cells, the ability to identify such cells may be important in determining the appropriate therapy and course of the disease.

REFERENCES

DeSombre, E. R., Carbone, P. P., Jensen, E. V., McGuire, W. L., Wells, S. A., Wittliff, J. L., and Lipsett, M. B., 1979, Special report: steroid receptors in breast cancer, N.Engl.J.Med., 301:1011.

DeSombre, E. R., 1984, Steroid receptors in breast cancer, in: "The Breast," R.W. McDivitt, H.A. Oberman, L. Ozello and N. Kaufman, eds., Williams and Wilkins, Baltimore.

DeSombre, E. R., Thorpe, S. M., Rose, C., Blough, R. R., Anderson, K. W., Rasmussen, B. B., and King, W. J., 1986, Prognostic usefulness of estrogen receptor immunocytochemical assays for human breast cancer, Cancer Res., 46:4241s.

Greene, G. L., and Jensen, E. V., 1982, Monoclonal antibodies as probes for estrogen receptor detection and characterization, J.Steroid Biochem., 16:353.

Greene, G. L., Sobel, N. B., King, W. J., and Jensen, E. V., 1984, Immunochemical studies of estrogen receptors, J.Steroid Biochem., 20:51.

Greene, G. L., and Press, M. F., 1987, Immunochemical evaluation of estrogen receptor and progesterone receptor in breast cancer, in: "Immunological Approaches to the Diagnosis and Therapy of Breast Cancer," R.L. Ceriani, ed., Plenum, New York, p. 119.

King, W. J., and Greene, G. L., 1984, Monoclonal antibodies localize estrogen receptor in the nuclei of target cells, Nature, 307:745.

King, W. J., DeSombre, E. R., Jensen, E. V., and Greene, G. L., 1985, Comparison of cytochemical and steroid-binding assays for estrogen receptor in human breast cancer, Cancer Res., 45:293.

Nolan, C., Przywara, L. W., Miller, L. S., Suduikas, V., and Tomita, J., 1984, A sensitive solid-phase enzyme immunoassay for human estrogen receptor, in: "Current Controversies in Breast Cancer," F.C. Ames, G.R. Blumensheim, and E.D. Montague, eds., Univ. of Texas Press, Austin, p. 433.

Obsorne, C. K., Yochmowitz, M. G., McGuire, W. A., and W. L., 1980, The value of estrogen and progesterone receptors in the treatment of breast cancer, Cancer, 46:2884.

Press, M. F., and Greene, G. L., 1984, An immunocytochemical method for demonstrating estrogen receptor in human uterus using monoclonal antibodies to human estrophilin, Lab.Invest., 50:480.

Weigand, R. A., Noland, C., Kinders, R. J., Przywara, L. W., Cotter, D. L., Dunn, R., Moore, B. S., Suduikas, V. V., and Greene, G. L., 1986, Measurement of progesterone receptor in human breast tumors by monoclonal enzyme immunoassays, 68th Annual meeting of the Endocrine Society, Anaheim, Abstr. 1005.

MONOCLONAL ANTIBODIES TO THE INSULIN RECEPTOR

Kenneth Siddle, Maria A. Soos, Richard O'Brien,
Rosalind H. Ganderton and T. S. Pillay

University of Cambridge, Department of Clinical Biochemistry
Addenbrooke's Hospital, Hills Road, Cambridge CB2 2QR

INTRODUCTION

THE INSULIN RECEPTOR: STRUCTURE AND ACTIVITY

Insulin exerts multiple effects on carbohydrate, lipid and protein metabolism, and on cell growth, in a variety of mammalian tissues. All of these effects are believed to be initiated by the binding of insulin to its receptor in the plasma membrane. Many of the metabolic actions of insulin ultimately involve increases or decreases in the level of phosphorylation of key regulatory enzymes such as glycogen synthase, pyruvate dehydrogenase and triacylglycerol lipase, or the translocation between subcellular pools and the cell surface of membrane components such as the glucose transporter (Kahn 1985; Denton 1986). However, the nature of the signals generated by the receptor which mediate these intracellular events has remained obscure in spite of intense investigation. Other aspects of receptor function which have still to be elucidated include mechanisms regulating cellular sensitivity to insulin, either by acute modulation of receptor activity or by longer term control of receptor levels.

There have been considerable advances in recent years in our understanding of the structure of the insulin receptor (Jacobs and Cuatrecasas 1983; Czech 1985). The receptor is a disulphide-linked heterotetrameric molecule of the structure β-α-α-β, in which the α subunit (Mr 135000 Da) is wholly extracellular, while the β subunit (Mr 95000 Da) has both extracellular and intracellular domains. The subunit contains the insulin binding site, although the binding of insulin to both intact membranes and solubilized receptors displays complex kinetics suggestive of site heterogeneity or negative cooperativity (Gammeltoft, 1984) and it appears that only a single molecule of insulin binds with high affinity to each native $\alpha_2\beta_2$ receptor molecule (Pang and Shafer, 1984). The β subunit possesses intrinsic tyrosine-specific protein kinase activity in its intracellular domain and is itself susceptible to reversible phosphorylation, not only on tyrosine residues as a result of an autophosphorylation reaction, but also on serine and threonine residues following the action of unidentified kinases (Gammeltoft and Van Obberghen, 1986). A further covalent modification of the β subunit is its fatty acylation, probably on cysteine residues (Magee and Siddle, 1986; Hedo et al, 1987) although it is not yet known whether this fatty acyl group turns over independently of the receptor as a whole. The effects of these modifications on receptor function are unclear although there is evidence that tyrosine kinase activity may be regulated by receptor phosphorylation (Bollag et al, 1986; Roth and Beaudoin, 1987; Obermaier et al, 1987).

The two subunits of the insulin receptor are derived from a common polypeptide precursor by proteolytic processing, and the cDNA coding for this precursor has been cloned and sequenced (Ullrich et al, 1985; Ebina et al, 1985). The deduced amino acid sequence suggests there is only a single membrane-spanning hydrophobic helix within each β subunit. The sequence also shows that there is considerable homology between the tyrosine kinase domains of the insulin receptor, the epidermal growth factor receptor and a family of tyrosine kinases encoded by oncogenes such as _src_ and _ros_.

The association of tyrosine kinase activity with multiple growth factor receptors and transforming proteins is itself strong circumstantial evidence for an important role of this activity in the growth-regulating action of all these molecules. Evidence has been obtained that the tyrosine kinase activity of the insulin receptor is also essential for expression of acute metabolic effects of the insulin, particularly by using site-directed mutagenesis to produce receptors lacking kinase activity or autophosphorylation sites (Ellis et al, 1986; Ebina et al, 1987; Chou et al, 1987). Abnormalities of kinase activity have been demonstrated in insulin-resistant metabolic states (Le Marchand Brustel et al, 1985; Caro et al, 1987; Friedenberg et al, 1987), although there is no evidence for a genetic abnormality in the receptor itself in these cases. It is not yet clear what are the physiologically important substrates for the insulin receptor kinase. Several polypeptides have been described which show increased tyrosine phosphorylation after exposure of cells to insulin (Bernier et al, 1987; Haring et al, 1987; Perotti et al, 1987; White et al, 1987). However the identity and function of all of these is unknown, and their role in a putative phosphorylation cascade, perhaps ultimately involving serine kinases or phosphatases, remains to be established. The autophosphorylation of insulin receptors results in activation of receptor kinase which is then independent of insulin binding (Yu and Czech 1984). Whether there are further consequences of autophosphorylation, perhaps affecting interaction of receptors with other proteins, is unclear.

Evidence has also been presented that guanine nucleotide-binding regulatory proteins (Houslay, 1985), soluble, glycolipid-derived compounds (Saltiel, 1987), and protein kinase C (Cooper et al, 1987) have a role in mediating at least some aspects of insulin action. These hypotheses are not necessarily exclusive of a role for tyrosine kinase activity (Houslay et al, 1986) and might reflect sequential steps in the overall mechanism, or parallel pathways acting synergistically or specifically on different targets of insulin action.

The binding of insulin to its receptor is followed by rapid internalization of the complex, after which hormone and receptor appear to follow separate pathways, insulin being delivered predominantly to lysosomes while the receptor is largely recycled to the plasma membrane (Jacobs and Cuatrecasas, 1983; Carpentier et al, 1986). There is no good evidence that receptor internalization is a necessary step in the mechanism of insulin action, although autophosphorylated receptor might well remain active after internalization. Rather, the process seems to provide a mechanism for insulin degradation and/or regulation of cell surface expression of insulin receptor, either by redistribution in the short term, or by down-regulation involving accelerated receptor degradation in the longer term. These mechanisms might play a significant role in controlling cellular sensitivity to insulin, and are thus relevant to understanding the molecular basis of states of insulin resistance. However, the signals which trigger receptor internalization following insulin binding, or which modulate this process in response to other hormones, are unknown.

ANTIBODIES TO THE INSULIN RECEPTOR

The insulin receptor is a minor membrane protein, normally present at only a few thousand or tens of thousands of copies per cell (or approx. 0.001-0.01% of cell protein). Because of this, purification of receptor for structural and functional analysis, and characterization of receptor in situ under different conditions, presents considerable practical difficulties. Anti-receptor antibodies have already proved to be invaluable reagents for such studies, most notably human sera obtained from a few patients with a rare autoimmune syndrome of insulin resistance (Kahn et al, 1981; Van Obberghen and Kahn 1981) and also antisera raised against purified receptor in rabbits (Jacobs and Cuatrecasas 1981). Such polyclonal antibodies have been used for instance in the immunoprecipitation of receptor from cell lysates following biosynthetic or metabolic labelling. The insulin-like and insulin-inhibitory properties of the antibodies on intact cells have also been studied. However, as with all polyclonal antisera, the supply of such reagents is limited, the titres are often low, and the antibodies are heterogeneous within a single serum (De Pirro et al 1984) and variable between sera (Kahn et al, 1981).

Several strategies have been employed in an attempt to produce more specific reagents. Polyclonal antisera have been obtained by immunization with synthetic peptides corresponding to defined regions of receptor sequence (Herrera et al, 1985) and, as anti-idiotypes, following immunization with insulin (Schechter et al, 1984). Monoclonal antibodies to insulin receptors were first obtained by Roth et al, (1982), Kull et al, (1983) and Ganguly et al (1985), and studies employing these antibodies have been reviewed (Roth and Morgan 1985). Subsequently, larger panels of monoclonal antibodies have been described by Soos et al (1986), Morgan and Roth (1986a) and Forsayeth et al (1987a).

Production and Characterization of Monoclonal Antibodies

All the monoclonal antibodies described to date have been raised in mice against human receptor, using as antigen either solubilized, purified placental receptor (Kull et al, 1983; Morgan and Roth 1986a; Forsayeth et al 1987a), or IM-9 lymphocytes, a cell line expressing a good level of insulin receptors (Roth et al, 1982). We employed a combined approach using IM-9 cells to prime a large number of mice and reserving the limited amounts of purified receptor for boosting just those mice which showed a significant primary response (Soos et al, 1986). We were not successful in our initial attempts to induce an immune response in mice to insulin receptor purified from rat liver (M.A. Soos, M. D. Baron and K. Siddle, unpublished).

Screening of antisera and hybridoma culture supernatants has been performed by inhibition of ^{125}I-insulin binding to cells or membranes (Roth et al, 1982; Forsayeth et al 1983a; Soos et al, 1986) or by variants of an immunoprecipitation assay using solubilized receptor with bound ^{125}I-insulin (Kull et al, 1983; Morgan and Roth 1985; Soos et al, 1986). It is unlikely that either assay detects all types of antibody, though some antibodies do react in both assays (Soos et al, 1986). We also used a screening assay with directly iodinated purified receptor in the expectation that all antibodies would react in such a system, but identified one antibody which bound receptor only weakly unless insulin was also present.

We carried our 4 successful fusions, using mice selected as good responders in test bleeds. It was notable that among the 28 antibodies obtained we observed only a limited number of distinctly different specificities within each fusion, and that each fusion produced different types of antibody. This emphasizes the value of using as many animals as is convenient in attempting to produce a large bank of antibodies with distinct properties. Interestingly, all the antibodies we obtained reacted with extracellular domains of the receptor, while Morgan and Roth (1986a) concluded that the main immunogenic regions were localized to the intracellular kinase domain. Our failure to obtain antibodies reacting with intracellular domains may reflect the use of intact cells for primary immunization and/or partial proteolysis of the purified receptor preparations used as booster antigen and for screening assays. On the other hand Morgan and Roth (1986a) obtained 35 of their 37 antibodies from a single fusion, and the corresponding specificities therefore may reflect only the dominant epitopes recognized by one particular mouse.

Initial characterization of antibodies, in terms of identifying the number of distinct epitopes recognized, has been performed by competitive binding studies in which ^{125}I-receptor or ^{125}I-insulin/receptor complex was preincubated with soluble antibody and the extent of inhibition of subsequent binding to an immobilized antibody then determined (Soos et al, 1986; Morgan and Roth 1986a). In principle, pairs of antibodies reacting at the same or closely adjacent sites will show complete mutual inhibition of binding while those reacting at spatially well separated sites will show no interaction. In practice the situation is much more complex, given the relatively large size and bivalency of antibodies themselves (monovalent Fab fragments would be preferable for this type of analysis), the symmetrical, antigenically bivalent structure of the receptor, and the potential for allosteric effects dependent on receptor conformation rather than direct steric interaction. However, inspection of the overall patterns of interaction among large groups of antibodies allows a clear identification of major epitope groups. Moreover, these assignments are supported by studies of other properties of the antibodies, such as their species specificity and functional effects. Thus we identified at least 9 epitopes or antigenic regions on extracellular domains, some of which could be localized to the α or β subunits by immunoblotting (Soos et al, 1986). On the intracellular kinase domain, 4 distinct antigenic regions were proposed (Morgan and Roth 1986a).

The cloning of receptor cDNA and determination of the nucleotide sequence and deduced amino acid sequence opens up the possibility of generating antibodies to predetermined sites on the receptor. We have obtained a monoclonal antibody to the C-terminal penta-decapeptide of the insulin receptor, following immunization with a synthetic peptide conjugated to carrier protein (R.H. Ganderton, M.A. Soos and K. Siddle, unpublished). Alternatively site-specific antibodies might be produced by immunizing with fusion proteins generated in bacteria by the use of expression vectors incorporating fragments of receptor cDNA (Tooze and Stanley 1986). Either approach must take account of the largely empirical rules indicating which short peptide antigens are likely to elicit antibodies capable of recognizing the corresponding sequences in the native protein molecule (Lerner 1984; Palfreyman et al, 1984).

Use of Antibodies for Receptor Isolation and Assay

Methods of insulin receptor purification have generally been based on affinity chromatography employing immobilized lectins, insulin or polyclonal antibodies. By virtue of their availability in quantity and ease of coupling to a variety of support matrices, monoclonal antibodies are an attractive alternative. We developed a method for the rapid, high yield, immunoaffinity purification of active receptor using monoclonal antibody coupled to aminocellulose (O'Brien et al, 1986). Various antibodies were screened in order to select one of relatively low affinity from which receptor could be eluted under mild conditions (1M NaCl, 50 mM acetate, pH 5) which do not adversely affect its kinase or insulin-binding activities.

This is important when the purified receptor is to be used for functional studies, for instance to investigate interactions with other purified components such as guanine nucleotide-binding proteins, or in assessing the susceptibility to, and consequences of, phosphorylation by specific protein kinases. Other procedures have been described for the immunoaffinity purification of insulin receptors (Roth and Morgan 1985; Roth et al, 1986) or IGF-1 receptors (Le Bon et al, 1986) but these generally necessitated the use of potentially denaturing elution conditions.

Immunoaffinity techniques may also be used for subcellular fractionation (Richardson and Luzio, 1986). We hope to develop methods for the isolation of receptor-containing vesicles by using antibodies specifically recognizing the cytosolic domain of the receptor, in order to study the endocytosis and subsequent intracellular processing of insulin/receptor complexes. The antibody recently obtained for the C-terminal peptide of the receptor appears eminently suitable for this purpose, as it reacts well with both phosphorylated and non-phosphorylated forms of the receptor, does not itself inhibit autophosphorylation, and also reacts with both rat and human receptors (R.H. Ganderton, M.A. Soos and K. Siddle, unpublished).

We and others have found monoclonal antibodies to be excellent reagents for the specific immunoprecipitation of receptor from cell lysates, as when studying the biosynthesis or phosphorylation of receptors (Jacobs et al, 1983; Jacobs and Cuatrecasas 1986; Ellis et al, 1986; Whittaker et al, 1987). Although monoclonal antibodies offer the theoretical advantage of high specificity, not all are as efficient as polyclonal sera at precipitating low concentrations of receptor. This may reflect the need for high affinity multivalent binding between antigen and antibody if specific immunoprecipitates are to withstand the relatively stringent washing conditions commonly applied to reduce non-specific adsorption.

Monoclonal antibodies have also been used to establish immunoassays for the insulin receptor, capable of quantifying receptor in tissue extracts independently of insulin binding. Roth and Beaudoin (1986) described a competitive radioimmunoassay, using monoclonal antibody adsorbed to microtitre plates and ^{125}I-receptor. We have developed a two site immunometric assay using two different monoclonal antibodies respectively coupled to aminocellulose solid phase and labelled with ^{125}I (R.O'Brien and K. Siddle, unpublished). Either type of assay might in principle be made specific for different forms of receptor (phosphorylated/non-phosphorylated, intact/proteolyzed) by choice of suitable monoclonal antibodies.

Antibodies as Probes of Receptor Structure

Studies of immunological cross-reactivity are of value, in conjunction with molecular cloning and sequence analysis, in comparing the structures of insulin receptors and related proteins from different sources. It appears that there is only a single insulin receptor gene in man (Ullrich et al, 1985; Ebina et al, 1985) and yet there is structural and functional evidence of receptor heterogeneity within and between tissues (Gammeltoft 1984; Gammeltoft and Van Obberghen, 1986). We could detect no immunological difference between receptors from human placenta, liver and IM-9 cells (Soos et al, 1986), although Roth and Morgan (1985) reported that one of their antibodies displayed differences in relative potency of reaction with lymphocytes, placenta and adipocytes. This antibody reacted only very poorly with receptor from brain, which probably reflects the different glycosylation pattern in this tissue (Roth et al, 1986).

Polyclonal anti-receptor antibodies show extensive cross-reaction between species, and the conformation of the insulin binding site is also believed to be highly conserved. Nevertheless monoclonal antibodies detect substantial structural differences in receptor from different species. None of the antibodies so far described for extracellular domains of the human receptor cross-reacts with rat receptors, although some react with monkey, bovine or rabbit receptors (Roth et al, 1982; Kull et al, 1982; Ganguly et al, 1985; Soos et al, 1986). This may in part reflect a lack of antigenicity in mice of epitopes conserved between human and rodent receptors. However, antibodies to epitopes on intracellular domains show more extensive species cross-reaction, most reacting with rat, and some even with chicken and frog receptors (Morgan et al, 1986). It may be concluded that the structure of the kinase domain has been more highly conserved in evolution than that of extracellular domains.

The receptor for IGF-1 has a high degree of sequence homology with the insulin receptor (Ullrich et al, 1986), and indeed insulin interacts weakly with the receptor for IGF-1 and vice versa (Kahn, 1985). Some monoclonal antibodies have been described which are quite specific for insulin receptors or IGF-1 receptors, relative potencies of cross-reaction being less than 1% (Roth et al, 1983a; Jacobs et al, 1986). Such antibodies have been used to study the importance of IGF-1 receptors in mediating the metabolic and mitogenic effects of insulin (Verspohl et al, 1984; Van Wyk et al, 1985). We found more complex cross reactions with our antibodies. Most of these, representative of several different epitopes on the insulin receptor, react well with a subfraction of IGF-1 binding sites in human cells (M.A. Soos and K. Siddle, unpublished). This is seen whether reaction is tested by inhibition of IGF-1 binding, or by precipitation of solubilized receptors with bound IGF-1, as appropriate for different antibodies. The fraction of total IGF-1 binding sites which reacts in this way varies between different cell types (approx 30% with HEP G2 hepatoma cells, 60% with placenta, 90% with IM-9 lymphocytes). Most surprisingly, the cross-reaction of antibodies with IGF-1 receptors is abolished by a brief pretreatment of receptors with dithiothreitol, conditions which do not affect immunoreactivity of insulin receptors. Similar observations have been reported by other groups, involving a single monoclonal antibody (Morgan and Roth 1986b) and a polyclonal antiserum (Jonas and Harrison 1986). It appears therefore that there exist two subtypes of IGF-1 receptors, one of which is much more closely related to the insulin receptor than the other. The molecular basis and physiological significance of these immunologically defined receptor subtypes remains to be determined.

The mapping of antigenic sites in relation to the known primary structures of the insulin and IGF-1 receptors may shed further light on the structural similarities between these receptors, and between insulin receptors of different species. The identification of regions of primary structure accessible to antibodies in native receptors will also contribute to a picture of receptor conformation, and of functionally important domains. We are attempting such epitope mapping by cloning short random fragments of receptor cDNA in the bacterial expression vector pEX and then screening the resulting library for reaction of the expressed fusion proteins with antibodies, using a colony blot technique (Stanley et al, 1985). This approach will however only detect those epitopes which are predominantly sequential, these being identified by sequencing of cDNA inserts from immunoreactive clones.

ANTIBODIES AS PROBES OF RECEPTOR FUNCTION

One of the most interesting aspects of our studies has been the diversity of biological effects displayed by different antibodies. We initially tested for effects on insulin binding (Soos et al, 1986; Siddle et al, 1987), and identified antibodies for three distinct epitopes which strongly inhibited ^{125}I-insulin binding to intact cells, membranes and solubilized receptor. Scatchard analysis of insulin binding in the presence of excess antibody indicated two distinct mechanisms of inhibition. One antibody (47-9) appeared to exert a direct steric blockade, presumably by reaction near the insulin binding site. Another antibody (83-14) decreased the binding affinity rather than the number of binding sites, which could reflect an allosteric action at a distance from the insulin binding site. Several different antibodies stimulated the binding of insulin, especially to solubilized receptor, by increasing the apparent binding affinity (M.A. Soos and K. Siddle, unpublished). These results indicate that the insulin receptor can exist in multiple conformations with different binding affinities, these states being readily interconvertible and independent of covalent modification, though influenced to some degree by the environment of the receptor.

The effect of antibodies on the rate of dissociation of ^{125}I-insulin previously bound to receptor was also investigated. Excess unlabelled insulin itself accelerates dissociation due to presumed negatively cooperative interactions which decrease binding affinity as occupancy increases (Gammeltoft, 1984). Antibody 47-9 mimicked this effect of insulin (M.A. Soos and K. Siddle, unpublished) as did some of the antibodies studied by Forsayeth et al, (1987a). In contrast antibody 83-14 slowed dissociation, and linearised the Scatchard plot for insulin binding, suggesting that it locked the receptor in a low affinity state without cooperative interactions.

Metabolic effects of antibodies were investigated in isolated human adipocytes (Taylor et al, 1987). Insulin stimulates glucose uptake and inhibits lipolysis in these cells. Antibody 47-9 blocked insulin binding and insulin action, indicating the independence of receptor activation from the changes underlying negatively cooperative affinity modulation. However, most antibodies tested mimicked insulin action on glucose uptake and lipolysis. This included antibodies reacting with both α and β subunits and antibodies which were inhibitory or non-inhibitory for insulin binding. It was concluded that reaction of antibodies at or near the insulin binding site was neither necessary nor sufficient for receptor activation. It appears rather that the insulin-like effects of antibodies depend on their bivalency and ability to cross-link receptors, and in agreement with this we have found monovalent Fab antibody fragments to be ineffective in mimicking insulin action (R. Taylor, M.A. Soos and K. Siddle, unpublished).

Further confirmation of this hypothesis was obtained by studying the effect of antibodies on the stimulation of tyrosine kinase activity of solubilized receptors from human placenta (R. O'Brien, M.A. Soos and K. Siddle, submitted for publication). With one exception the effects of antibodies on tyrosine kinase activity paralleled their metabolic effects on adipocytes. Thus 47-9 antagonised the effect of insulin, while most other antibodies themselves stimulated kinase activity. This stimulation was only apparent within a narrow concentration range, was not shown by monovalent antibody fragments, and was potentiated by a second, anti-(mouse IgG) antibody. Analysis of antibody-receptor complexes by centrifugation on sucrose density gradients suggested the presence of both intra- and inter-molecularly cross-linked complexes at optimally stimulating antibody concentrations, although increased autophosphorylation was associated preferentially with the former. We conclude that antibodies activate the receptor not by reaction at specific epitopes but by intramolecular cross-linking of extracellular domains, perturbing the steric relationship between subunits in a way which releases a constraint on kinase activity of the intracellular domains.

Monoclonal antibodies have been used in several ways to investigate the role of tyrosine kinase activity in mediating insulin action. We found a generally good correlation between the ability of monoclonal antibodies to stimulate kinase activity of solubilized receptor and to mimic metabolic effects of insulin in adipocytes (R.M. O'Brien, M.A. Soos and K. Siddle, submitted). However, one antibody (25-49) was identified which was potently insulin-like on fat cells, but failed significantly to activate the kinase and antagonised the stimulatory effect of insulin in line with its inhibition of insulin binding. Forsayeth et al, (1987b) described two similar antibodies. Such results may say more about the antibodies than the mechanism of insulin action. The ideal experiment in which kinase activation and metabolic stimulation are studied in parallel in the same cells is technically difficult to perform. In any case a maximal metabolic stimulation might be achieved by activation of only a small fraction of receptors whose increased autophosphorylation was undetectable.

More direct evidence in favor of the hypothesis that tyrosine kinase activation is an essential step in insulin action was provided by experiments in which antibodies reactive with the intracellular domain of the β subunit were injected into cells, either directly (Morgan et al, 1986) or by an osmotic technique (Morgan and Roth 1987). Antibodies which inhibited kinase activity blocked insulin action whereas non-inhibitory antibodies did not. These results are not conclusive of a role for kinase activity however, as the consequences of attaching a large antibody molecule to the receptor in situ may not be limited to inhibition of the kinase.

Experiments on cells transfected with normal or mutant receptor genes also strongly support an essential role for kinase activity. Receptors were engineered by site-directed mutagenesis in which a single lysine residue at the putative kinase active site was replaced by other amino acids (Ebina et al, 1987). These mutant receptors were devoid of kinase activity and would not mediate insulin-like effects of anti-receptor antibodies when expressed in rodent cells. Again these experiments must be interpreted with some caution given the restricted range of insulin-like effects studied, and the possibility that the consequences of the mutations introduced, however conservative, may not be restricted to abolition of kinase activity. Nevertheless, experiments with monoclonal antibodies provide some of the most convincing evidence so far available for the importance of tyrosine kinase activity in insulin action, although the nature of its involvement, in terms of the next steps on the signalling pathway, remain as unclear as ever.

The role of kinase activity and autophosphorylation in the endocytosis and down-regulation of receptors has also been studied by using monoclonal antibodies. In this case there is good evidence that antibody-induced down regulation is not dependent on receptor kinase activity (Roth et al, 1983b; Morgan et al, 1987; Forsayeth et al, 1987a). However, it is far from clear that receptors cross-linked by antibodies will be endocytosed and subsequently processed by the same pathway as those whose internalization is triggered by insulin binding.

CONCLUSIONS AND FUTURE PROSPECTS

The aim of this brief review has been to indicate the many different ways in which monoclonal antibodies have contributed to studies of insulin receptors, rather than to describe in detail the results of any one study. We anticipate that monoclonal antibodies will continue to provide valuable information, not least in studying the modulation of receptor activity which contributes to insulin resistance in diabetes and obesity, and in more detailed analysis of receptor structure. Antibodies prepared to predefined sites on the receptor, by using synthetic peptide antigens, are likely to be increasingly used as specific reagents, alongside the more randomly generated sets of antibodies already available.

Acknowledgements

The work in our laboratory is supported by the Wellcome Trust, the Medical Research Council, the Science and Engineering Research Council, the Bradlow Foundation and Serono Diagnostics Ltd. KS is a Wellcome Lecturer.

REFERENCES

Bernier, M., Laird, D. M., and Lane, M. D., 1987, Insulin-activated tyrosine phosphorylation of a 15-kilodalton protein in intact 3T3-L1 adipocytes, Proc.Natl.Acad.Sci.USA, 84:1844.

Bollag, G. E., Roth, R. A., Beaudoin, J., Mochly-Rosen, D., and Koshland, D. E., 1986, Protein kinase C directly phosphorylates the insulin receptor in vitro and reduces its protein-tyrosine kinase activity, Proc.Natl.Acad.Sci.USA, 83:5822.

Caro, J. F., Sinha, M. K., Raju, S. M., Ittoop, O., Pories, W. J., Flickinger, E. G., Meelheim, D., and Dohm, G. L., 1987, Insulin receptor kinase in human skeletal muscle from obese subjects with and without non insulin dependent diabetes, J.Clin.Invest., 79:1330.

Carpentier, J. L., Gorden, P., Robert, A., and Orci, L., 1986, Internalization of polypeptide hormones and receptor recycling, Experentia, 42:734.

Chou, C. K., Dull, T. J., Russell, D. S., Gherzi, R., Lebwohl, D., Ullrich, A., and Rosen, O. M., 1987, Human insulin receptors mutated at the ATP-binding site lack protein tyrosine kinase activity and fail to mediate postreceptor effects of insulin, J.Biol.Chem., 262:1842.

Cooper, D. R., Konda, T. S., Standaert, M. L., Davis, J. S., Pollet, R. J., and Farese, R. V., 1987, Insulin increases membrane and cytosolic protein kinase C activity in BC3H-1 myocytes, J.Biol.Chem., 262:3633.

Czech, M. P., 1985, The nature and regulation of the insulin receptor: structure and function, Annu.Rev.Physiol., 47:357.

Denton, R. M., 1986, Early events in insulin action, Adv.Cyclic.Nucl.Protei Phos.Res., 20:293.

De Pirro, R., Roth, R. A., Rossetti, L., and Goldfine, I. D., 1984, Characterization of the serum from a patient with insulin resistance and hypoglycaemia: evidence for multiple populations of insulin receptor antibodies with different receptor-binding and insulin-mimicking activities, Diabetes, 33:301.

Ebina, Y., Ellis, L., Jarnagin, K., Edery, M., Graf, L., Clauser, E., Ou, J., Masiarz, F., Kan, Y. W., Goldfine, I. D., Roth, R. A., and Rutter, W. J., 1985, The human insulin receptor cDNA: the structural basis for hormone-activated transmembrane signalling, Cell, 40:747.

Ebina, Y., Araki, E., Taira, M., Shimada, F., Mori, M., Craik, C. S., Siddle, K., Pierce, S. B., Roth, R. A., and Rutter, W. J., 1987, Replacement of lysine residue 1030 in the putative ATP-binding region of the insulin receptor abolishes insulin- and antibody-stimulated glucose uptake and receptor kinase activity, Proc.Natl. Acad.Sci.USA, 84:704.

Ellis, L., Clauser, E., Morgan, D. O., Edery, M., Roth, R. A., and Rutter, W. J., 1986, Replacement of insulin receptor tyrosine residues 1162 and 1163 compromises insulin-stimulated kinase activity and uptake of 2-deoxyglucose, Cell, 45:721.

Forsayeth, J. R., Montemurro, A., Maddux, B. A., De Pirro, R., and Goldfine, I. D., 1987a, Effect of monoclonal antibodies on human insulin receptor autophosphorylation, negative cooperativity, and down-regulation, J.Biol.Chem., 262:4134.

Forsayeth, J. R., Caro, J. F., Sinha, M. K., Maddux, B. A., and Goldfine, I. D., 1987b, Monoclonal antibodies to the human insulin receptor that activate glucose transport but not insulin receptor kinase activity, Proc.Natl.Acad.Sci.USA, 84:3448.

Gammeltoft, S., 1984, Insulin receptors: binding kinetics and structure-function relationship of insulin, Physiol.Rev., 64:1321.

Gammeltoft, S., and Van Obberghen, E., 1986, Protein kinase activity of the insulin receptor, Biochem J., 235:1.

Ganguly, S., Petruzzelli, L. M., Herrera, R., Stadtmauer, L., and Rosen, O. M., 1985, Insulin-dependent tyrosine protein kinase: cellular distribution and copurification with the insulin receptor, Curr.Top.Cell. Reg., 27:83.

Friedenberg, G. R., Henry, R. R., Klein, H. H., Reichart, D. R., and Olefsky, J. M., 1987, Decreased kinase activity of insulin receptors from adipocytes of non-insulin-dependent diabetic subjects, J.Clin. Invest., 79:240.

Haring, H. U., White, M. F., Machicao, F., Ermel, B., Schleicher, E., and Obermaier, B., 1987, Insulin rapidly stimulates phosphorylation of a 46-kDa membrane protein on tyrosine residues as well as phosphorylation of several soluble proteins in intact fat cells, Proc.Natl. Acad.Sci USA., 84:113.

Hedo, J. A., Collier, E., and Watkinson, A., 1987, Myristyl and palmityl acylation of the insulin receptor, J.Biol.Chem., 262:954.

Herrera, R., Petruzzelli, L., Thomas, N., Bramson, H. N., Kaiser, E. T., and Rosen, O. M., 1985, An antipeptide antibody that specifically inhibits insulin receptor autophosphorylation and protein kinase activity, Proc.Natl.Acad.Sci.USA., 82:7899.

Houslay, M. D., 1985, The insulin receptor and signal generation at the plasma membrane, in: "Molecular mechanisms of transmembrane signalling," P. Cohen and M.D. Houslay, eds., pp. 279-334, Elsevier, Amsterdam.

Houslay, M. D., Wakelam, M.J.O., and Pyne, N.J., 1986, The mediator is the message: is it part of the answer of insulin's action? Trends Biochem.Sci., 11:393.

Jacobs, S., and Cuatrecasas, P., 1981, Insulin receptor antibodies, Meth.Enzymol., 74:471.

Jacobs, S., and Cuatrecasas, P., 1983, Insulin receptors, Annu.Rev.Pharmacol.Toxicol., 23:461.

Jacobs, S., and Cuatrecasas, P., 1986, Phosphorylation of receptors for insulin and insulin-like growth factor 1, J.Biol.Chem., 261:934.

Jacobs, S., Kull, F. C., and Cuatrecasas, P., 1983, Monensin blocks the maturation of receptors for insulin and somatomedin C: identification of receptor precursors, Proc.Natl.Acad.Sci.USA., 80:1228.

Jacobs, S., Cook, S., Svoboda, M. E., and Van Wyk, J. J., 1986, Interaction of the monoclonal antibodies IR-1 and IR-3 with insulin and somatomedin-C receptors, Endocrinology, 118:223.

Jonas, H. A., and Harrison, L. C., 1986, Disulphide reduction alters the immunoreactivity and increases the affinity of insulin-like growth-factor-I receptors in human placenta, Biochem.J., 236:417.

Kahn, C. R., 1985, The molecular mechanism of insulin action, Annu.Rev. Med., 36:429.

Kahn, C. R., Baird, K. L., Flier, J. S., Grunfeld, C., Harmon, J. T., Harrison, L. C., Karlsson, F. A. Kasuga, M., King, G. L., Lang, U. C., Podskalny, J. M., and Van Obberghen, E., 1981, Insulin receptors, receptor antibodies and mechanisms of insulin action, Rec.Prog.Hormone Res., 37:477.

Kull, F. C., Jacobs, S., Su, Y. F., and Cuatrecasas, P., 1982, A monoclonal antibody to human insulin receptor, Biochem.Biophys.Res.Comm., 106:1019.

Kull, F. C., Jacobs, S., Su, Y. F., Svoboda, M. E., Van Wyk, J. J., and

Cuatrecasas, P., 1983, Monoclonal antibodies to receptors for insulin and somatomedin C, J.Biol.Chem., 258:6561.
LeBon, T. R., Jacobs, S., Cuatrecasas, P., Kathuria, S., and Fujita-Yamaguchi, Y., 1986, Purification of insulin-like growth factor I receptor from human placental membranes, J.Biol.Chem., 261:7685.
LeMarchand-Brustel, Y., Gremaux, T., Ballotti, R., and Van Obberghen, E., 1985, Insulin receptor tyrosine kinase is defective in skeletal muscle of insulin-resistant obese mice, Nature, 215:676.
Lerner, R. A., 1984, Antibodies of predetermined specificity in biology and medicine, Adv. Immunol., 36:1.
Magee, A. I., and Siddle, K., 1986, Human insulin receptor contains covalently bound palmitic acid, Biochem.Soc.Trans., 14:1103.
Morgan, D. O., and Roth, R. A., 1985, Plate binding assay for monoclonal anti-receptor antibodies, Endocrinology, 116:1224.
Morgan, D. O., and Roth, R. A., 1986a, Mapping surface structures of the human insulin receptor with monoclonal antibodies: localization of main immunogenic regions to the receptor kinase domain, Biochemistry, 25:1364.
Morgan, D. O., and Roth, R. A., 1986b, Identification of a monoclonal antibody which can distinguish between two distinct species of the type I receptor for insulin-like growth factor, Biochem.Biophys. Res.Comm., 138:1341.
Morgan, D. O., and Roth, R. A., 1987, Acute insulin action requires receptor kinase activity: introduction of an inhibitory antibody into mammalian cells blocks the rapid effects of insulin, Proc.Natl.Acad. Sci.USA, 84:41.
Morgan, D. O., Ho., L., Korn, L. J., and Roth, R. A., 1986, Insulin action is blocked by a monoclonal antibody that inhibits insulin receptor kinase, Proc.Natl.Acad.Sci.USA, 83:328.
Morgan, D. O., Ellis, L., Rutter, W. J., and Roth, R. A., 1987, Antibody-induced down regulation of a mutated insulin receptor lacking an intact cytoplasmic domain, Biochemistry, 26:2959.
Obermaier, B., Ermel, B., Kirsch, D., Mushack, J., Rattenhuber, E., Biemer, E., Machicao, F., and Haring, H. U., 1987, Catecholamines and tumor-promoting phorbol esters inhibit insulin receptor kinase and induce insulin resistance in isolated human adipocytes, Diabetologia, 30:93.
O'Brien, R. M., Soos, M. A., and Siddle, K., 1986, Immunoaffinity purification of insulin receptor by use of monoclonal antibodies, Biochem.Soc.Trans., 14:316.
Palfreyman, J. W., Aitcheson, T. C., and Taylor, P., 1984, Guidelines for the production of polypeptide-specific antisera using small synthetic oligopeptides as immunogens, J.Immunol.Meth., 75:383.
Pang, D. S., and Shafer, J. A., 1984, Evidence that insulin receptor from human placenta has a high affinity for only one molecule of insulin, J.Biol.Chem., 259:8589.
Perrotti, N., Accili, D., Marcus-Samuels, B., Rees-Jones, R. W., and Taylor, S. I., 1987, Insulin stimulates phosphorylation of a 120-kDa glycoprotein substrate (pp120) for the receptor-associated protein kinase in intact H-35 hepatoma cells, Proc.Natl.Acad.Sci.USA, 84:3137.
Richardson, P. J., and Luzio, J. P., 1986, Immunoaffinity purification of subcellular particles and organelles, Appl.Biochem.Biotechnol., 13:133.
Roth, R. A., and Beaudoin, J., 1986, A solid-phase competitive radio-immunoassay for the insulin receptor, Analyt.Biochem., 159:163.
Roth, R. A., and Beaudoin, J., 1987, Phosphorylation of purified insulin receptor by cAMP kinase, Diabetes, 36:123.
Roth, R. A., and Morgan, D. O., 1985, Monoclonal antibodies to the insulin receptor, Pharmacol.Therap. 28:1.
Roth, R. A., Cassell, D. J., Wong, K. Y., Maddux, B., and Goldfine, I. D., 1982, Monoclonal antibodies to the human insulin receptor block

insulin binding and inhibit insulin action, Proc.Natl.Acad.Sci.USA, 79:7312.

Roth, R. A., Maddux, B., Wong, K. Y., Styne, D., Van Vliet, G., Humbel, R. E., and Goldfine, I. D., 1983a, Interactions of a monoclonal antibody to the insulin receptor with receptors for insulin-like growth factor receptors, Endocrinology, 112:1865.

Roth, R. A., Maddux, B. A., Cassell, D. J., and Goldfine, I. D., 1983b, Regulation of the insulin receptor by a monoclonal anti-receptor antibody, J.Biol.Chem., 258:12094.

Roth, R. A., Morgan, D. O., Beaudoin, J., and Sara, V., 1986, Purification and characterization of the human brain insulin receptor, J.Biol. Chem., 261:3753.

Saltiel, A. R., 1987, Insulin generates an enzyme modulator from hepatic plasma membranes: regulation of adenosine 3',5'-monophosphate phosphodiesterase, pyruvate dehydrogenase, and adenylate cyclase, Endocrinology, 120:967.

Schechter, Y., Elias, D., Maron, R., and Cohen, I. R., 1984, Mouse antibodies to the insulin receptor developing spontaneously as anti-idiotypes, J.Biol.Chem., 259:6411.

Siddle, K., Soos, M. A., O'Brien, R. M., Ganderton, R. H., and Taylor, R., 1987, Monoclonal antibodies as probes of the structure and function of insulin receptors, Biochem.Soc.Trans., 15:47.

Soos, M. A., Siddle, K., Baron, M. D., Heward, J. M., Luzio, J. P., Bellatin, J., and Lennox, E. S., 1986, Monoclonal antibodies reacting with multiple epitopes on the human insulin receptor, Biochem.J., 235:199.

Stanley, K. K., Kocher, H. P., Luzio, J. P., Jackson, P., and Tschopp, J., 1985, The sequence and topology of human complement component C9, EMBO J., 4:375.

Taylor, R., Soos, M. A., Wells, A., Argyraki, M., and Siddle, K., 1987, Insulin-like and insulin-inhibitory effects of monoclonal antibodies for different epitopes on the human insulin receptor, Biochem.J., 242:123.

Tooze, S. A., and Stanley, K. K., 1986, Identification of two epitopes in the carboxyterminal 15 amino acids of the E1 glycoprotein of mouse hepatitis virus A59 by using hybrid proteins, J.Virol., 60:928.

Ullrich, A., Bell, J. R., Chen, E. Y., Herrera, R., Petruzzelli, L. M., Dull, T. J., Gray, A., Coussens, L., Liao, Y. C., Tsubokawa, M., Mason, A., Seeburg, P. H., Grunfeld, C., Rosen, O. M., and Ramachandran, J., 1985, Human insulin receptor and its relationship to the tyrosine kinase family of oncogenes, Nature, 313:756.

Ullrich, A., Gray, A., Tam, A. W., Yang-Feng, T., Tsubokawa, M., Collins, C., Henzel, W., Le Bon, T., Kathuria, S., Chen, E., Jacobs, S., Francke, U., Ramachandran, J., and Fujita-Yamaguchi, Y., 1986, Insulin-like growth factor I receptor primary structure, EMBO J., 5:2505.

Van Obberghen, E., and Kahn, C. R., 1981, Autoantibodies to insulin receptors, Mol.Cell.Endocrinol., 22:277.

Van Wyk, J. J., Graves, D. C., Casella, S. J., and Jacobs, S., 1985, Evidence from monoclonal antibody studies that insulin stimulates deoxyribonucleic acid synthesis through the type I somatomedin receptor, J.Clin.Endoc.Metab., 61:639.

Versphol, E. J., Roth, R. A., Vigneri, R., and Goldfine, I. D., 1984, Dual regulation of glycogen metabolism by insulin and insulin-like growth factors in human hepatoma cells (HEP-G2): analysis with an anti-receptor monoclonal antibody, J.Clin.Invest., 74:1436.

White, M. F., Stegmann, E. W., Dull, T. J., Ullrich, A., and Kahn, C. R., 1987, Characterization of an endogenous substrate of the insulin receptor in cultured cells, J.Biol.Chem., 262:9769.

Whittaker, J., Okamoto, A. K., Thys, R., Bell, G. I., Steiner, D. F., and

Hofmann, C. A., 1987, High level expression of human insulin receptor cDNA in mouse NIH 3T3 cells, Proc.Natl.Acad.Sci.USA, 84:5237.
Yu, K. T., and Czech, M. P., 1984, Tyrosine phosphorylation of the insulin receptor subunit activates the receptor-associated tyrosine kinase activity J.Biol.Chem., 259:5277.

MONOCLONAL ANTIBODIES FOR THE STUDY OF HUMAN GROWTH HORMONE FORMS AND VARIANTS

F. Frankenne, J. Closset, F. Gomez, G. Pirens,
J. Smal and G. Hennen

Endocrinologie Clinique et Expérimentale
Institut de Pathologie
C.H.U. B23, B-4000 Liège (Belgique)

INTRODUCTION

The molecular heterogeneity of human growth hormone (hGH) extracted from pituitary tissue has been extensively studied by Lewis et al, (1977, 1978, 1980) and Talamantes et al, (1981) who described most of the charge and size variants. These were considered as resulting from pre- and post-translational processing, affecting the final products originally coded by the GH-N gene. Indeed, the alternative splicing of the hGH pre-mRNA has been shown by Wallis (1980), elucidating the mechanism leading to the synthesis of the major 22K and 20K hGH pituitary forms, these proteins being then submitted to post-translational modifications.

Additional to the N gene, the GH genome locus contains a second gene, coded V for a form (or variant) of hGH which, until recently, was considered as non-expressed. Nevertheless, screening of the expression of this gene has not been systematic (Seeburg, 1982; Pavlakis et al, 1981).

Another area of uncertainty is the physiology of hGH in pregnancy. Indeed, assays employed to measure hGH in the serum of pregnant women were, up until recently, not sufficiently specific. They are significantly effected by the increasing serum concentration of hPL produced by the placental tissue in the developing pregnancy. The original cause for activation of somatomedin-C production and the related anabolic process observed in pregnant women remained largely unknown.

Our work was aimed at obtaining well characterized anti-hGH monoclonal antibodies to reassess the heterogeneity of hGH in tissues and serum, and, more specifically, to determine pituitary GH physiology during pregnancy in regard to the secretion of protein hormones of the same biochemical family, such as hPL. This approach produced results which led to the identification of a new growth hormone variant, now called human placental growth hormone (Hennen et al, 1985a; b; c).

Production and Characterization of K24 and 5B4 Monoclonal Antibodies

Hybridoma secreting anti-hGH monoclonal antibodies were obtained through the classical cell fusion technique between mouse myeloma cells and spleen cells from mice immunized against hGH (Gomez et al, 1985) as either

22K or 20 K hGH preparations extensively purified from human pituitaries (Closset et al, 1983).

Our selection criteria were not only a high specificity of the secreted MAbs, but also a high affinity for the binding of hGH as analyzed according to Scatchard (1949) during the screening step. Two MAbs, meeting these requirements, were finally selected. The affinity constants for the binding reaction to 22K hGH were 1.02×10^{11} L/M for 5B4 MAb and 5×10^{9} for K24 MAb respectively. Both were virtually unreactive towards hPL.

These antibodies were produced in bulk in ascitic fluid, and purified by salting out, chromatography on DEAE-Fractogel TSK 650 M (Merck) and Protein A Fractogel affinity chromatography.

The reactivity of these antibodies was assessed towards various forms of GHs either natural, chemically modified or that obtained by genetic engineering as well as towards peptides representing different portions of the GH peptide chain.

From Fig. 1, it appears that no human peptide hormone cross-reacted with the 5B4 MAb even at doses of 10μg/ml. Equine GH, unlike other GHs, exhibited a very weak reactivity. The lack of reactivity of hPL is of special interest since the anti-hGH MAbs were going to be used to measure GH in pregnant women serum. All highly purified hGH preparations (NIH, MRC, in House) appeared equipotent to the 22K hGH reference preparation. Compared to the monomeric 22K hGH, the absence of the 32-46 peptide which characterizes the 20K variant induces a loss of 30% in reactivity, which was also decreased by dimerization. Therefore, it appears that the GH epitope recognized by the 5B4 MAb is present in both the 22K and 20K variants and that it is not significantly affected by dimerization.

Carboxymethylation of the 22K hGH induces a loss of reactivity of only 17%, but the presence of N-terminal additional formyl-methionyl residue of the biosynthetic hGH produces a drastic drop in the potency which is only 1/500 when compared to that of native 22K hGH. The GH-peptides reproducing the N-terminal end of the native hGH, including the 1-5 peptide, were significantly recognized by the 5B4 MAb while the others were devoid of any reactivity. Since a modification of the N-terminal amino acid almost obliterated the GH immunoreactivity and since short hGH-N-terminal peptides were active, we concluded that the 5B4 recognizes an epitope located on the N-terminal part of the GH chain.

As shown in Fig. 2, the specificity of the K24 MAb towards human hormones other than GH is absolute since doses of 10 μg/ml were unreactive. This holds true for animal GHs, except for bGH which displayed a very low reactivity. Since hPL remained unreactive even at high concentrations, this antibody is also suitable for assaying GH in the serum of pregnant women.

The reactivity of the 20K hGH preparation with K24 MAb was very low: 2500 less than that of the 22K. Therefore, the K24 MAb should be directed either towards a 22K hGH epitope in which the 32-46 region is mainly involved or towards a structural epitope specific to the 22K variant. Moreover, that epitope appears to be hidden to a large extent by GH dimerization.

The additional N-terminal formyl-methionyl residue slightly increases the reactivity which is much enhanced by carboxymethylation. Hence, the integrity of the structure of 22K hGH is not required for the recognition of the epitope by the K24 MAb. Only one hGH related peptide, the 1-134 reacted weakly but significantly with the K24 MAb (Fig. 2).

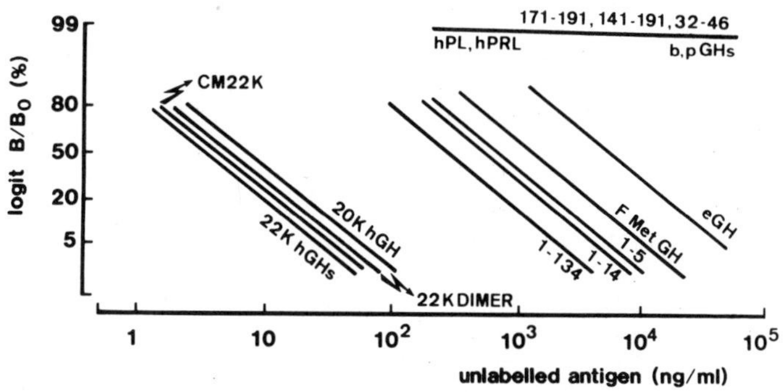

Fig. 1. Specificity of the 5B4-RIA. Response curves yielded by graded doses of several peptide hormones, GH variants, carboxy-methyl-22K hGH, Met-GH and GH-derived peptides.

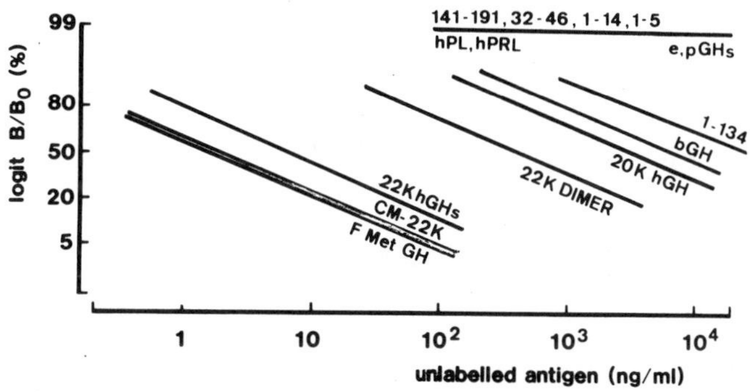

Fig. 2. Specificity of the K24-RIA. Response curves yielded by graded doses of several peptide hormones, GH variants, carboxy-methyl-22K hGH, Met-GH and GH-derived peptides.

From these results, it was concluded that the K24 MAb is directed towards an epitope which involves, at least partly, the 32-46 region of 22K hGH. Here again, the epitope depends upon the integrity of the peptide chain and is not affected by a significant structural change of the molecule, such as carboxymethylation.

SDS-PAGE, Gel IEF and Immunoblotting of Pituitary Extracts

An aliquot of the extract from a pool of normal pituitaries in 0.3 M KCl pH 5, routinely processed for producing therapeutic hGH, was desalted and lyophilized. Both assays, measuring GH in this extract yielded similar values, 219 ± 9 and 211 ± 11 ng hGH per g of extractable proteins for the 5B4- and K24-RIA respectively. This crude preparation was submitted to 5-25% acrylamide gradient SDS-PAGE (Millard et al, 1969) or to gel electrofocusing. In the latter case, Immobiline (LKB) gels containing 10% polyacrylamide were cast between two sheets of Gel Bond (FMC Corp. USA) according to the manufacturer's recommendations, but with the hydrophilic side of both sheets facing the gel, in order to allow easy peeling. The Immobiline gels were electroblotted for one hour at 150 volts onto nitrocellulose sheets (Millipore) following the method of Towbin et al, (1979) and using a home-made chamber, while a Nova-blot transfer device (LKB) was used for other gels, using 25 mM and 300 mM Tris in 20% methanol, pH 10.4

as anode buffers and 40 mM 6-N-hexanoic acid as cathode solution and setting the current at 0.8 A/cm^2 for one hour. The enzymatic immuno-staining was performed using the HRP Immunblot Assay kit from BioRad (USA). The nitrocellulose sheets were incubated with the first antibody in sealed PVC bags at 20°C for sixteen hours. Purified preparations of both MAbs, were used at a dose of 10 µg/ml in 1 ml of 25 mM Tris, 192 mM glycine, 20% methanol, containing 1% BSA.

The immunoblots recorded after SDS-PAGE of the extract from a normal pituitary pool using either the 5B4 or the K24 MAbs are shown in Fig. 3. The K24 pattern (Fig. 3a) essentially revealed the 22K hGH. The 20K variant and the 24K forms are not detected by the K24 MAb, while the 44K dimeric hGH band is only very weakly stained. The pattern yielded by the 5B4 MAb (Fig. 3b) corresponds to the classical SDS-PAGE pattern of purified pituitary hGH, displaying the 44K, the 24K, the 22K and the 20K forms (Lewis et al, 1980). This fits perfectly with the immunoreactivities shown by these GH forms (Fig. 1 and 2). The integrity of the GH peptide chain is required for K24 MAb recognition and is illustrated by the lack of reactivity of the 24K form, which only differs from the 22K by a nick between residues 139 and 140 (Lewis et al, 1977).

Fig. 4 shows the immunoblots of the polyacrylamide-IEF of normal pituitary extract. The patterns yielded by 5B4 MAb displays both the 20K and 22K variants together with a set of deamidated forms and a more basic component whose I.P. is 6.1. All variants except the 20K reacted with the K24 MAb and thus bear both epitopes recognized by the respective anti-bodies.

Developments of GH Assays. Clinical Use.

RIA methods were preferred since we intended to carry out an inventory of the GH forms and variants in various circumstances.

GH labelling and RIA conditions were previously described as well as the related data processing (Gomez et al, 1984). As seen in Table 1, the 5B4 and K24 RIA sensitivity were 0.5 and 2 µIU/ml respectively in terms of the hGH MRC 66/217 International standard. Both MAb RIAs were applied to samples collected from normal subjects for determination of GH baseline levels as well as to samples drawn during GH secretory dynamic tests. Fig. 5 illustrates the parallel evolution of the recorded GH values from samples related to an insulin test. Both systems yielded values in good corre-lation ($P<0.01$) when applied either to control samples for determination of basal level ($r=0.958$, $N=53$) or to samples from dynamic tests ($r=0.94$, $N=18$). Nevertheless, the K24 RIA yielded figures which were significantly lower than those from the 5B4 RIA ($P<0.01$, $N=53$). The mean basal serum levels calculated from 53 control samples were comprised between 6.3 and 9.5 µIU/ml ($P<0.01$) for the 5B4 RIA and between 3.5 and 6.13 µIU/ml ($P<0.01$) for the K24 RIA.

Since both systems yielded similar data when measuring GH from a pituitary extract, their discrepancies appear restricted to serum assays. The interpretation of this difference is not obvious, even if one can hypothesize on GH binding factors (Baumann et al, 1986) having a greater affect on the K24 MAb data because of the lower Ka of that antibody compared with that of the 5B4 MAb. Alternatively, a preferential degradation in serum of the K24 epitope could be considered. It must be stressed that both RIAs are highly specific, and that they remain un-affected by hPL and hPRL.

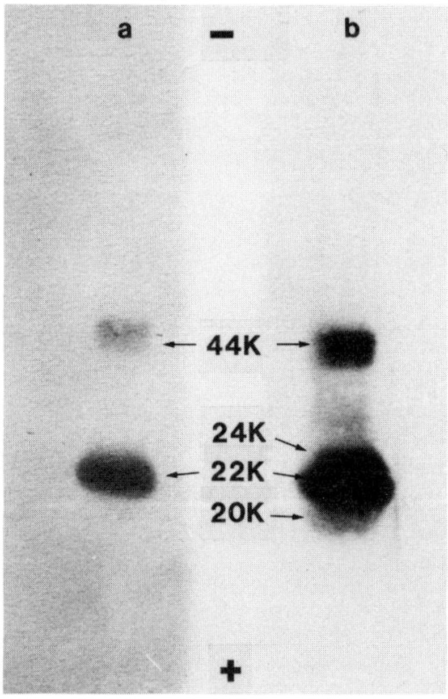

Fig. 3. Immunoblots of GH extracted from pituitaries and submitted to SDS-PAGE electrophoresis. a) reactivity to the anti-GH K24-MAb; b) reactivity to the anti-GH 5B4-MAb.

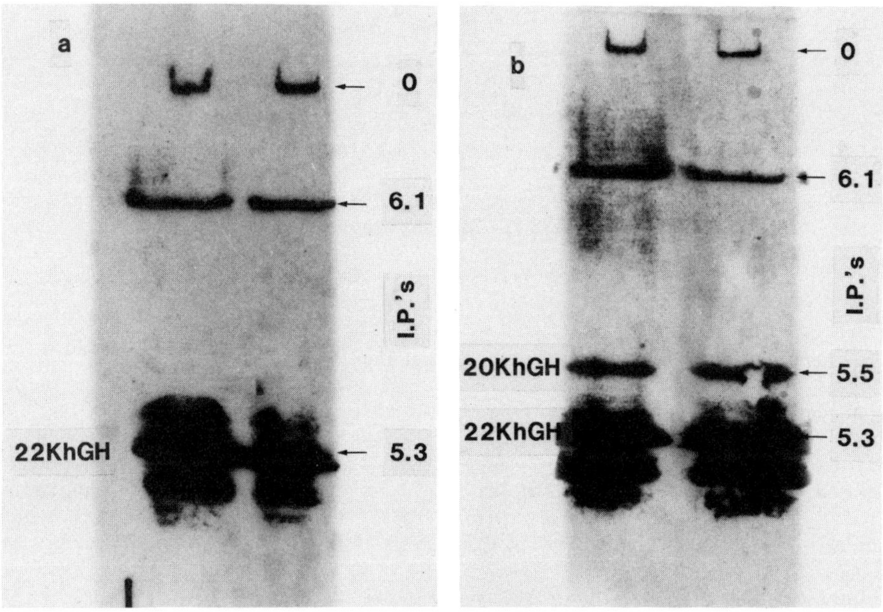

Fig. 4. Immunoblots of GH extracted from pituitaries and submitted to isoelectrofocusing; a) reactivity to the anti-GH K24-MAb; b) reactivity to the anti-GH 5B4-MAb.

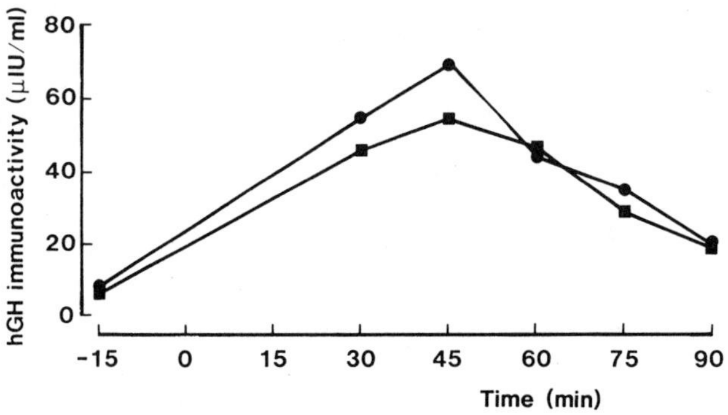

Fig. 5. Time-course evolution of GH serum levels after a single insulin injection, as measured using both the 5B4-RIA (■) and the K24-RIA (●).

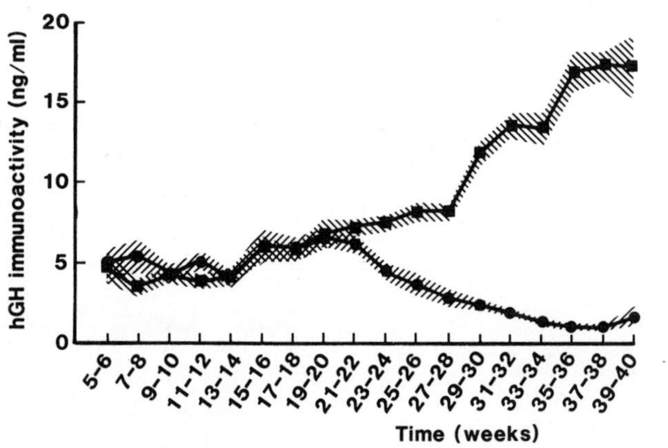

Fig. 6. Evolution of serum GH immunoactivities in pregnant women throughout pregnancy. Values yielded by the 5B4 MAb-RIA (■) and by the K24 MAb-RIA (●).

Physiology of Growth Hormones During Pregnancy

Serum samples were obtained from 900 women at various times of pregnancy. Paired mother and cord serum as well as amniotic fluid were obtained at parturition. Early post-partum studies were made in 5 women by serial sampling before and after delivery, as diagnosed by ultrasonography monitoring. Fig. 6 shows the evolution of GH immunoreactivities in pregnant women serum throughout pregnancy.

Until week 20-22, both assay systems yielded similar data with GH values within the normal range of non-pregnant control subjects. From that period up to term the level of pituitary 22K hHG, specifically measured by the K24 RIA, dropped progressively to become undetectable at week 36 and after. On the contrary, during the same period, the 5B4 RIA yielded steadily increasing values up to week 37-38, thus accounting for an antigen distinct from pituitary hGH. The antigen responsible for this typical immunoreactivity was tentatively called human placental GH (hPGH).

Table 1. Characteristics of the GH-RIA using the K24 MAb, Compared to those of the 5B4 MAb-RIA.

GH-assay	Affinity constant (Ka)	ED50 (µIU/ml)	Sensitivity (µIU/ml)	Slope	Cross-reactivities (%) hPL	hPRL
K24-RIA	$0.5\ 10^{10}M^{-1}$	15.1 ± 3.04 (N=7)	2 µIU/ml	-0.93	0.01	0.01
5B4-RIA	$1.02\ 10^{11}\ M^{-1}$	11.4 ± 2.2 (N=55)	0.5 µIU/ml	-1.03	0.01	0.01

The apparent stabilization of hPGH levels at 17 ng/ml (in terms of 22K hGH immunoreactivity) during the last four weeks of pregnancy is not significant, since individual variations were evident in that period.

Fig. 7 gives the evolution of the values recorded with both RIAs in early post-partum. Immediately after delivery, circulating hPGH, as measured by the 5B4 RIA, dropped sharply to reach, within one hour, very low levels (0 to 3 ng/ml in terms of 22K hGH immunoreactivity). It remained constant thereafter. The values assayed using the K24 RIA, accounting essentially for pituitary GH, remained at the same low level as seen in late pregnancy.

Low immunoreactivities were recorded with both RIAs in 11 amniotic fluid samples: 3.2 ± 3 µIU/ml (5B4-RIA) and 3 ± 2.8 µIU/ml (K24-RIA). Hence these low levels correspond to pituitary GH.

Comparable high concentrations of GH were recorded by assaying cord serum using the 5B4 as well as the K24 RIA (Fig. 3a). This pattern differed from that yielded by maternal samples (Fig. 3b) which displayed the typical immunoreactivity of hPGH, recorded in the 5B4 but not in the K24 assay. Therefore, it is confirmed that high levels of pituitary GH are present at term in fetal serum, which contains no significant amount of hPGH.

The GH immunoreactivity recorded in a placental extract accounted for the presence of hPGH since only the 5B4 RIA elicited a positive response. Measured using that assay, the tissue concentration amounted to 0.2 µg/g. This is ten times higher than the mean hPGH level recorded in maternal serum in late pregnancy.

Purification of hPGH From Placental Tissue

Table 2 gives the purification steps and the recovery throughout the purification scheme. By DEAE cellulose chromatography, a clearcut separation of hPL and hPGH was achieved, hPGH being essentially non-adsorbed at 0.09 M NH_4 HCO_3 while hPL desorbtion required a higher ionic strength. In parallel experiments, pituitary 20K and 22K hGH behaved as hPL, giving evidence for the more basic character of hPGH when compared to its pituitary counterparts or hPL.

For affinity chromatography, 5B4 MAb (20 mg) was coupled to 10 ml of CNBr-activated Sepharose (Pharmacia). THe DEAE chromatography fraction containing hPGH (2 g) was chromatographed on this support and the adsorbed fraction was eluted by 3 M KSCN. After desalting, the fraction was lyophilized. An apparent 100 fold enrichment in hPGH was obtained with 1% recovery. Nevertheless, these figures are likely to be underestimated since 5B4 MAb was present in that fraction, as described below, and probably interfered in hPGH measurement.

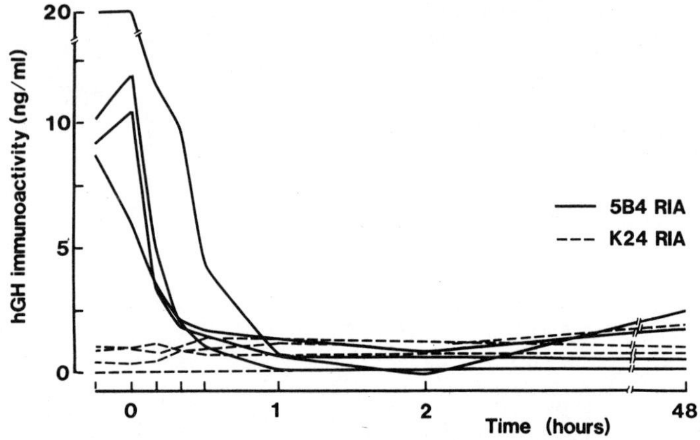

Fig. 7. Post-partum evolution of serum GH immunoactivities in five pregnant women. Values yielded by the 5B4 MAb-RIA (——) and by the K24 MAb-RIA (---). Delivery is coded zero time.

Table 2. Partial Purification of hPGH from Placenta.

Fraction	Total protein (g)	hPL (mg)	hPGH+ (μg)	Purification factor	Overall recovery (%)
Crude extract	50	150	200	1	100
25–55 % $(NH_4)_2SO_4$ pp	24	125	150	1.5	75
DE I	10	0.4	100	2.5	50
DE II	2.3	90	3		
Immunoaffinity	0.0008	-	0.8	250	0.4

+Quantitatively, hPGH is expressed in terms of the immunoactivity recorded in the 5B4 MAb-RIA, using 22K hGH as standard.

By polyacrylamide gel electrophoresis, the desorbed material displayed a multiband pattern (Fig. 8), with additional components migrating to positions close to that of 22K GH separated from material identified as the monoclonal antibody itself.

In order to ascertain whether hPGH was present in the fraction desorbed from the immunosorbant, the proteins separated by PAGE were blotted onto nitro-cellulose sheets and immunostained using either the 5B4 MAb or the K24 MAb as first antibody. After staining the mouse IgGs, bands corresponding to blotted MAb appear after incubation with both antibodies, whereas two components of 22K and ± 25K MW, respectively, appeared only when the incubation was made with 5B4 MAb (Fig. 9). We conclude therefore that the material desorbed from the affinity column is composed of MAb leaked from the affinity support and of two proteins reacting with 5B4 MAb and not with K24 MAb. These two proteins, presenting with molecular weights of 22K and 25K, obviously correspond to hPGH.

hPGH components were eluted together from a preparative PAGE. No more than 100 ng (in terms of 22K hGH immunoreactivity) of pure material could thus be obtained.

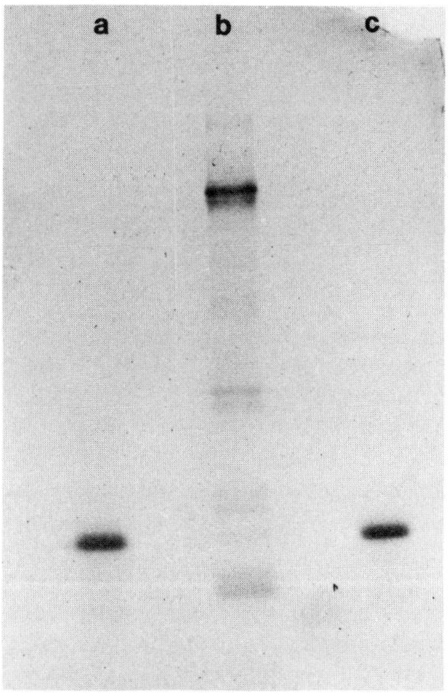

Fig. 8. SDS, 5-25% gradient PAGE of the adsorbed fraction from immunoaffinity chromatography in hPGH purification (track b). Tracks a and c: pituitary hGH (22K). Coomassie blue staining.

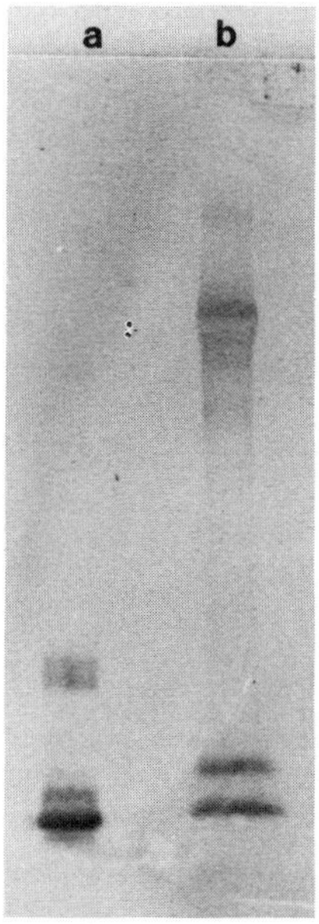

Fig. 9. Immunoblot of the adsorbed fraction from the immunoaffinity chromatography, in hPGH purification (track b) compared to pituitary hGH (track a). The anti GH 5B4 MAb was used as specific antibody.

Binding of hPGH to GH Receptors

The GH radioligand receptor assay employed in this study was described previously (Closset et al, 1983). As shown in Fig. 10, purified hPGH obtained through preparative PAGE behaved as a complete agonist to the pituitary 22K hGH in binding to hepatic GH receptors from pregnant rabbit liver. Moreover, it appeared more potent than the pituitary 22K variant on the basis of a similar 5B4 immunoreactivity.

DISCUSSIONS AND CONCLUSIONS

This work was designed to study the heterogeneity of hGH found in serum and in pituitary extracts, using two high affinity and highly specific anti-hGH MAbs coded 5B4 and K24. In view of the reactivities of these antibodies towards several native and modified GHs as well as related peptides, it was concluded that the 5B4 MAb recognizes a N-terminal epitope of GH, while the K24 MAb recognizes an internally located epitope, absent in the 20K variant.

From immunoblots performed on PAGE and electrofocusing of pituitary extracts, it appeared that the 5B4 N-terminal epitope is present in all known GH pituitary variants, while the internal epitope is clearly identified in 22K hGH and its charge-isomers.

Assays of serum GH from control subjects using two RIAs based on either MAb yielded well correlated values. Yet, the K24-RIA mean GH level (4.8 ± 1.32 µIU/ml) was significantly lower than that calculated from 5B4-RIA data (7.9 ± 1.6 µIU/ml). Since both RIAs assayed similar values when applied to a pituitary pool extract, the interference of a serum factor has therefore to be considered. Additionally, these MAb based RIAs were shown to be unaffected by hPL; they could therefore be used to study GH physiology in pregnancy.

Evolution of serum GH in a large group of pregnant women was studied using these highly specific GH RIAs based on two MAbs recognizing distinct GH epitopes. From the recorded data, it was concluded that pituitary 22K GH progressively disappears during the second half of pregnancy, while a placental GH variant (hPGH) appears in the circulation and rises in concentration up to term. After delivery hPGH disappears from maternal serum within one hour. Amniotic fluid was shown to contain low concentrations, and cord serum high levels, of GH from pituitary origin, while no hPGH was detected in these media. hPGH appears thus selectively secreted in the maternal compartment.

hPGH has been partially purified from placental extracts. According to its chromatographic behavior, it appears more basic than pituitary 22K and 20K hGHs, and presents with a size dimorphism, being composed of two entities of 22K and 25K M.V., respectively.

Pure hPGH, obtained on a very small scale by preparative electophoresis, was shown to bind to GH hepatic receptors with an apparent high potency as compared to pituitary GH. hPGH is thus expected to behave as a GH agonist sharing most of the biological properties of GH.

Besides the GH-N gene, responsible for the expression of the pituitary GH variants, the human genome bears a second GH gene, the GH-V gene, which codes for a protein differing from the 22K hGH by 13 amino acid positions and whose characteristics should be similar to those of hPGH (Seeburg, 1982; Pavlakis et al, 1981). We have recently reported on the placental expression of the GH-V gene (Frankenne et al, 1987), which therefore most likely codes for hPGH.

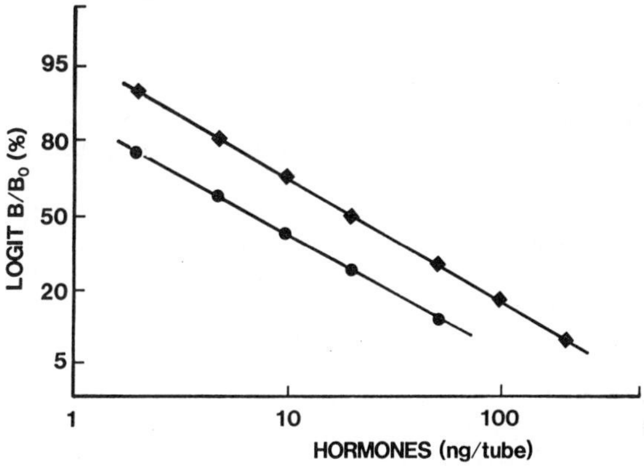

Fig. 10. Binding of hPGH (O) and 22K hGH (◆) on GH hepatic receptors. Quantitatively, hPGH is expressed in terms of the immunoactivity recorded in the 5B4 MAb-RIA, using 22K hGH as reference preparation.

One of the original tasks of this work was to determine which factor was responsible for the high circulating somatomedin-C activity in pregnant women during the second half of pregnancy. Our results demonstrated that pituitary GH can be ruled out while hPL and the placental GH variant had to be considered.

Preliminary results from a study on the relationships between maternal serum levels of pituitary GH, placental GH, hCS and IGF-I were recently reported (Caufriez et al, 1987a; 1987b). From partial correlation analysis (N=94), it was concluded that IGF-I is positively correlated with hPGH ($P<0.01$), and negatively with pituitary GH ($P<0.05$), while there is no significant correlation with hPL. Moreover, the individual evolution of these parameters throughout pregnancy, as determined from serial hormone assays in 11 subjects, showed that the IGF-I level evolution paralleled only that of hPGH. Thus, in the second half of pregnancy, hPGH could be considered as responsible for the enhancement of maternal metabolism which would be under the specific control of the placenta. At term, high somatomedin-C level could negatively control pituitary GH secretion.

The known placental lactogens from other mammalian species, such as oPL and bPL, exhibit a high degree of both GH and PRL activities, while hPL essentially displays a PRL activity. Indeed, the mammotrophic and lactotrophic potencies of hPL and hPRL are of the same order while hPL displays a somatotrophic activity of 0.2% that of hGH.

An evolutionary process should thus be considered, through which the placental PRL and GH activities would have been dissociated in the human species and distributed between hPL and hPGH.

Acknowlegements

This work was supported by grants from "Exécutif de la Région Wallonne" and from "Fonds National de la Recherche Scientifique de Belgique".

REFERENCES

Baumann, G., Stolar, M. W., Amburn, K., Barsano, C. P., Devries, B. C., 1986, A specific growth hormone-binding protein in human plasma: initial characterization, J.Clin.Endocrinol.Metab., 62:134.

Caufriez, A., Frankenne, F., Hennen, G., Cantraine, F., Goldstein, J., Copinshi, G., 1987, Close relationship between maternal somatomedin-C and placental hGH levels during normal human pregnancy. 69th Annual Meeting of the Endocrine Society, Indianapolis (Abstract 672).

Caufriez, A., Frankenne, F., Hennen, G., Golstein, J., Cantraine, F., Englert, Y., Copinshi, G., 1987, Possible mechanisms of somatomedin-C elevation during normal human pregnancy. First European Congress of Endocrinology, Copenhagen (Abstract 21).

Closset, J., Smal, J., Gomez, F., Hennen, G., 1983, Purification of the 22000 and 20000 mol.wt. forms of human somatotropin and characterization of their binding to liver and mammary binding sites, Biochem.J., 214:885.

Frankenne, F., Rentier-Delrue, F., Scippo, M. L., Martial, J., Hennen, G., 1987, Expression of the growth hormone variant gene in human placenta, J.Clin.Endocrinol.Metab., 64:635.

Gomez, F., Pirens, G., Schaus, C., Closset, J., Hennen, G., 1984, A highly sensitive radioimmunoassay for human growth hormone using a monoclonal antibody, J.Immunoassay, 5:145.

Hennen, G., Frankenne, F., Closset, J., Gomez, F., Pirens, G., Smal, J., Beckers, A., El Khayat, N., Lambotte, R., 1985, Monoclonal antibody to growth hormone: the discovery of a new variant, human placental growth hormone, in: "International Symposium on Monoclonal Antibodies: Basic Principles, Experimental and Clinical Applications in Endocrinology. ARES Serono Symposia, Florence," G. Forti, M. Serio and M.B. Lipsett, eds., Raven Press, New York, 30:29.

Hennen, G., Frankenne, F., Pirens, G., Gomez, F., Closset, J., Schaus, C., El Khayat, N., 1985, New chorionic GH-like antigen revealed by monoclonal antibody radioimmunoassays, The Lancet, Feb.16:399.

Hennen, G., Frankenne, F., Pirens, G., Gomez, F., El Khayat, N., Closset, J., Schaus, C., 1985, A chorionic GH-like antigen: increasing levels during second half of pregnancy with pituitary GH suppression as revealed by monoclonal antibody radioimmunoassays, Int.J. Fert., 30:27.

Lewis, U. J., Dunn, J. T., Bonewald, L. F., Seavey, B. K., and Vanderlaan, W. P., 1978, A naturally occurring structural variant of human growth hormone. J.Biol.Chem., 253:2679.

Lewis, U. J., Peterson, S. M., Bonewald, L. F., Seavey, B. K., Vanderlaan, W. P., 1977, An interchain disulfide dimer of human growth hormone, J.Biol.Chem., 252:3697.

Lewis, U. J., Singh, R. N., Tutwiler, G. F., Sigel, M. B., Vanderlaan, E. F., Vanderlaan, W. P., 1980, Human growth hormone: a complex of proteins, Recent Prog.Horm.Res., 36:477.

Millard, S. A., Kubose, A., Gal, E. M., 1969, Brain lipoyl dehydrogenase, Purification, properties and inhibitors, J.Biol.Chem., 244:2511.

Pavlakis, G. N., Hizuka, N., Gorden, P. H., Seeburg, P. H., Hamber, D. H., 1981, Expression of two human growth hormone genes in monkey cells infected by simian virus 40 recombinants, Proc.Natl.Acad.Sci.USA., 78:7398.

Scatchard, G., 1949, The attraction of proteins for small molecules and ions, Ann.NY Acad.Sci., 51:660.

Seeburg, P. H., 1982, The human growth hormone gene family: nucleotide sequences show recent divergence and predict a new polypeptide hormone, DNA, 1:239.

Talamantes, F., Lopez, J., Lewis, U. J., Wilson, C. B., 1981, Multiple forms of growth hormone: detection in medium from cultured pituitary adenoma explants, Acta Endocrinol.(Copenhagen), 98:8.

Towbin, H., Stachelin, T., Gordon, J., 1979, Electrophoretic transfer of proteins from polyacrylamide gel to nitrocellulose sheets: procedure and some applications, Proc.Natl.Acad.Sci. USA, 76:4350.

Wallis, M., 1980, Growth hormone: deletions in the protein and intron in the gene, Nature, 284:512.

SECTION C
CLINICAL AND VETERINARY APPLICATIONS OF MONOCLONAL ANTIBODIES

MONOCLONAL ANTIBODIES AGAINST RABIES VIRUS

W.H. Wunner, B. Dietzschold, C.E. Rupprecht and H. Koprowski

Wistar Institute of Anatomy and Biology
36th Street at Spruce
Philadelphia, PA 19104

INTRODUCTION

Monoclonal anti-virion antibodies have played a central role in the study of rabies and rabies virus antigens over the past decade. Their effectiveness in discriminating antigenic differences among rabies virus strains in recent epidemiological surveys and in the identification of viruses causing rabies in humans and animals has been particularly significant. Studies of the antigenic structure of viral components and mechanisms of rabies virus neutralization have been greatly facilitated by the use of monoclonal antibodies. Monoclonal antibodies have also provided an opportunity to demonstrate that vaccination against rabies may be possible in the absence of viral antigen through a series of networks that depend on idiotype-anti-idiotype recognition to induce an immune response.

Rabies is primarily a zoonotic disease caused by viruses belonging to the Lyssavirus genus of the family Rhabdoviridae which includes the rabies viruses and the rabies-related viruses. These viruses are transmitted in saliva from one warm-blooded animal to another and from animals to human. Viruses which have been identified as close relatives of the rabies virus and have caused rabies-like symptoms in animals and humans currently include Lagos bat, Mokola, and Duvenhage viruses. All were originally isolated from mammals in Africa (Boulger and Porterfield, 1958; Shope et al, 1970; Kemp et al, 1972; Tignor et al, 1977). Although as a group the rabies-related viruses appear to have a more limited geographic distribution and host range than rabies viruses, their importance in maintaining enzootic rabies is still a matter of considerable speculation, and therefore the antigenic relationship of these viruses to rabies virus has continued to be a focus of investigation in rabies research. In this regard, the viruses which have recently been isolated from rabid bats in Europe that resemble, but are not identical to, the African Duvenhage virus have caught our attention. These viruses have stimulated speculation about their origin and mode of transmission (Schneider et al, 1985). A particular concern has been whether vaccines prepared with rabies virus strains are as protective against these rabies-related viruses as against wildlife rabies viruses (Dietzschold et al, 1987c). A more detailed analysis of the antigenic relationship between the various viruses of the Lyssavirus groups is needed before these practical concerns can be adequately addressed. Prior to the application of hybridoma technology to the study of antigenic relationships, differentiation between rabies viruses rested largely on tests involving polyclonal antibodies of anti-rabies strain-specific sera.

In 1977, Koprowski proposed that antigenic relationships among rabies viruses would be more reliably defined by using monoclonal antibodies with different specificities for various rabies virus strains. The antibodies could also be used to determine to what extent the few strains of "fixed" virus were cross-reactive with "street" strains. "Fixed" viruses are those viruses which have been adapted to grow in laboratory animals or tissue culture systems with reproducible and characteristic biological behavior. They are the strains of rabies virus that are used for vaccines. "Street" virus strains are those found in wildlife. It was not known to what extent humans in a given area were protected by a fixed virus vaccine after exposure to a street rabies virus from wildlife. The use of monoclonal antibodies was suggested for analysis of rabies virus isolates after a previous successful analysis of influenza virus antigenic variants (Koprowski et al, 1977; Gerhard et al, 1978). Wiktor and Koprowski (1978) described the first production of monoclonal antibodies to rabies virus from hybridomas obtained by fusion of mouse myeloma cells with spleen cells from BALB/c mice immunized with rabies virus vaccine. In an earlier review of the hybridoma antibodies derived from such mice immunized with different strains of rabies or rabies-related viruses, Koprowski and Wiktor (1980) defined the properties and classification of monoclonal antibodies against rabies virus. This review focuses on the application of specific monoclonal antibodies in further studies of rabies and rabies virus antigens.

SPECIFICITIES OF MONOCLONAL ANTIBODIES AGAINST RABIES VIRUS

Monoclonal antibodies produced at the Wistar Institute were derived from fusions of P3 x 63 Ag8 mouse myeloma cells with splenocytes obtained from BALB/c mice that were immunized with an intraperitoneal (i.p.) inoculation of purified β-propiolactone (BPL) - inactivated virus vaccines (Wiktor and Koprowski, 1978) or with purified viral nucleocapsids prepared as described (Flamand et al, 1980a). For the various fusions, mice were immunized with vaccines prepared from the ERA, CVS, or Kelev strains of rabies virus or the rabies-related Mokola or Lagos bat viruses, and with nucleocapsids isolated from the rabies virus ERA strain. Mice received an intravenous (i.v.) booster inoculation of the same immunogen 4-8 weeks after primary immunization. Monoclonal antibodies produced at the Centers for Disease Control (CDC) in Atlanta which are illustrated in this communication were derived from a single mouse inoculated i.p. and in the left and right footpads with a suckling mouse brain suspension of the ERA vaccine strain of rabies virus. The i.v. booster inoculation was prepared from gamma-irradiated, ERA virus-infected mouse neuroblastoma (MNA) cell culture supernate and given 8 weeks later (Smith et al, 1984).

The specificity of antibodies produced by these hybridomas has been determined by

i radioimmunoassay (RIA) using rabies virus-infected cells, purified virions, viral glycoprotein or nucleocapsids isolated from different fixed rabies and rabies-related viruses, and by enzyme immunofiltration assay (EIA) using rabies virus-infected cells,

ii indirect immunofluorescence (IF) with live or acetone- or Formalin-fixed virus-infected cells,

iii virus neutralization assayed in mice or by plaque reduction, iv immunolysis in the presence of complement, and

v immunoprecipitation (Wiktor et al, 1980; Flamand et al, 1980a, 1980b, Smith et al, 1984).

The monoclonal antibodies were directed against one of the structural proteins of rabies virus (Koprowski and Wiktor, 1980); the viral antigens to which most antibodies were directed included the surface glycoprotein (G), and two nucleocapsid proteins (N and NS). Hybridomas produced from mice that had been immunized with purified nucleocapsids, as expected, secreted antibody directed against the nucleocapsid proteins of the virus. Monoclonal antibodies reportedly directed against the matrix (M) protein (Wiktor and Koprowski, 1978) do not appear to be as specific for M protein (Dietzschold, unpublished); no monoclonal antibody with activity directed against the virion-associated transcriptase (L protein) has been detected as yet.

Monoclonal Antibodies Against the Glycoprotein

Monoclonal antibodies directed against the viral glycoprotein antigen bind to rabies virus-infected unfixed cells and lyse them in the presence of complement. The indirect fluorescent antibody staining pattern which is observed with anti-glycoprotein monoclonal antibody on the membranes of unfixed cells is one of intense but diffuse fluorescence (Wiktor and Koprowski, 1978). The antibodies also bind specifically to the purified glycoprotein in an RIA. In the initial screening of hybridomas for antibody activity by RIA, anti-glycoprotein antibodies were identified that reacted with purified virions from fixed rabies and rabies-related viruses. Some of these anti-glycoprotein antibodies cross-reacted with all of the rabies and rabies-related viruses, some antibodies reacted with all viruses of the rabies group and cross-reacted with one, two, or three viruses of the rabies-related group, some showed cross-specificity with different combinations of viruses within the rabies group, and still others showed a single specificity for CVS, Kelev or Mokola viruses. The original hybridoma clones from which these antibodies were derived could be classified into 14 distinct groups based on their antibody specificities in the RIA (Flamand et al, 1980b).

Not all of the glycoprotein-specific antibodies that react positively in the RIA are capable of neutralizing the viruses to which they appear to bind. For example, of 19 monoclonal antibodies which showed binding in the RIA to the CVS strain of rabies virus in the initial evaluation, 12 neutralized the virus (Flamand et al, 1980b). In the course of studying antigenic differences among rabies virus strains, larger panels containing 41-44 anti-glycoprotein monoclonal antibodies (all with virus-neutralizing activity) have been used to discriminate between virus isolates. The test panel of virus-neutralizing anti-glycoprotein monoclonal antibodies has been subdivided into groups of antibodies according to their reactivity with major antigenic sites I, II, and III on the rabies virus glycoprotein (Lafon et al, 1983, 1984).

Monoclonal Antibodies Against The Nucleocapsid (N and NS) Proteins

Monoclonal antibodies directed against the nucleocapsid antigens bind to intracytoplasmic inclusions of acetone - or formalin-fixed virus-infected cells or tissue, and they show no virus-neutralizing or immunolytic activity. Indirect immunofluorescent detection of nucleocapsid antigen reveals it to be in distinct foci rather than evenly distributed throughout the cell culture or tissue (Wiktor and Koprowski, 1978; Reid et al, 1983). Anti-nucleocapsid antibodies also bind specifically to purified nucleocapsids in an RIA. While this technique works well with nucleocapsids from fixed viruses which replicate efficiently in tissue culture, it does not work well for detection of street viruses because frequently the replication of street viruses in tissue culture is insufficient and the level of detection or production of nucleocapsids for use in an RIA is so limited. Assays of the specificities of anti-nucleocapsid antibodies for street viruses are therefore performed using an IF test on impression smears of cells from mouse brains infected with street virus (Wiktor et al, 1980). The results may be directly compared with impressions of brain tissue infected with the standard fixed laboratory strains. The intensity of fluorescence is usually scored on a scale of 1-4.

The anti-nucleocapsid monoclonal antibodies have been particularly useful in distinguishing the rabies from rabies-related viruses and in distinguishing individual viruses within each group. Presently, a smaller panel of anti-nucleocapsid monoclonal antibodies produced at the Wistar Institute identifies all Lyssaviruses and distinguishes between the rabies and rabies-related viruses. The panel consists of one antibody (502-3) which cross-reacts with all strains of fixed rabies virus and rabies-related street viruses but not any other virus, one (422-5) that reacts only with all three rabies-related viruses [except the Duvenhage-like strains isolated from bats in Europe (Dietzschold et al, 1987a)], and one (103-7) that reacts only with fixed or street rabies viruses. By adding other strain-specific antibodies to this basic panel, it is possible to make distinctions between strains of virus within each of the serotypes, such as between CVS-11 and Flury HEP strains or between Mokola virus on the one hand and Lagos bat and Duvenhage viruses on the other (Wiktor et al, 1980). It should not be forgotten, however, that monoclonal antibodies against the glycoprotein of virus particles as previously mentioned can also be used in neutralization tests to discriminate between rabies and rabies-related viruses (Wiktor et al, 1984).

Larger panels of 36 and 37 anti-nucleocapsid monoclonal antibodies are necessary to discriminate viruses that are distinguished by fewer antigenic differences. The expanded panels are generally used for a secondary screening of virus isolates in epidemiological studies. They have been particularly effective in the differential diagnosis of antigenically similar viruses which nevertheless reveal one or two unique antigenic features that distinguish the strains by their host species and geographic origins. Two of the antibodies in the large panel of anti-nucleocapsid antibodies are each directed against independent antigenic sites on the NS protein of the nucleocapsid complex, the remaining 34-35 hybridoma antibodies react specifically with the N protein and are subdivided into groups according to their reactivity with antigenic sites I, II, and III on the N protein (Lafon and Wiktor, 1985). One further distinction between the properties of Wistar panel of anti-NS and anti-N protein antibodies, which can be useful in discriminating epitopes on the two antigens, is that the anti-NS protein antibodies detect intracytoplasmic inclusions in rabies virus-infected cells only after fixation with formalin and not after fixation with acetone (Wiktor et al, 1980).

ANTIGENIC ANALYSIS OF RABIES VIRUSES

Fixed Virus Strains

The standard laboratory strains of fixed rabies viruses are often used for vaccine production (Wiktor et al, 1980). Recently, the antigenic relationships between the glycoproteins and the nucleocapsid protein components of several laboratory strains maintained at the Wistar Institute which include the PM, CVS-24, CVS-11, ERA, Flury LEP, FLury HEP, and Kelev strains described by Wiktor et al, (1980) and the PVP and Pasteur strains of rabies virus provided by the Pasteur Institute were evaluated with the antigen-specific panels of monclonal antibodies as described elsewhere (Dietzschold et al, 1987a). Comparative antigenic analysis of these 9 fixed rabies virus strains with the anti-glycoprotein monoclonal antibodies (Fig. 1) has revealed different patterns of virus-neutralizing activity indicating that the glycoproteins of the various fixed virus strains display considerable antigenic diversity. Based on the neutralization results, it was estimated that 50-85% of the determinants on the glycoprotein were shared among the fixed laboratory (and vaccine) strains (Dietzschold et al, 1987a). The analysis also showed that by the use of anti-glycoprotein monoclonal antibodies in a virus neutralization or antibody binding test it is possible to identify laboratory strains of rabies virus and distinguish fixed (vaccine) rabies virus strains from street viruses in cases of suspected vaccine-induced rabies.

In contrast to the diversity of determinants on the glycoprotein of the fixed rabies virus strains, the antigenic specificity of the N and NS protein determinants detected by the panel of anti-nucleocapsid monoclonal antibodies is highly conserved in all of the viruses except the Flury HEP strain (Fig. 2). This is indicative of the fact that the nucleocapsid antigens of laboratory and vaccine strains provide the type specificity that is necessary to distinguish rabies from rabies-related virus strains.

Street Virus Strains

Recently, several comparative antigenic analysis studies have been conducted in which "street" virus isolates collected from various wildlife species and human cases of rabies associated with different rabies enzootic areas were examined with anti-nucleocapsid and anti-glycoprotein monoclonal antibodies (Charlton et al,1982; Sureau et al, 1983; Smith et al, 1984; Webster et al, 1985; Smith et al, 1986; Rupprecht et al, 1987; Dietzschold et al, 1987a). The purpose of these studies has been to define antigenic relationships between viruses isolated from the same or different host species and geographic regions and to identify distinctive antigenic characteristics relating to the transmission of rabies among wildlife species. In all, more than 1,000 street rabies virus isolates originating in different regions of Canada, North and South America, Europe, Africa, and Asia have been analyzed.

Geographic Distribution of Antigenically Distinct Street Strains Defined by Reactivity of Viral Isolates with Anti-Nucleocapsid Antibodies

A regularly observed characteristic of street viruses which has emerged from these studies is the identical or nearly identical nature of strains which originate from the same geographic area. The first study to clearly illustrate this point included 204 strains of street rabies viruses collected from Europe (France), Africa, and Asia (Sureau et al, 1983). On the basis of the reactivity of the viral isolates with a panel of 20 anti-nucleocapsid monoclonal antibodies, the viruses from France (15 strains), North and Central Africa (64 strains), South Africa (9 strains), and India (10 strains) formed groups which could be differentiated from each other; within each group the viruses were nearly identical irrespective of species origin.

In contrast to these rather homogeneous groupings, the viruses which were collected in Thailand (78 strains), Madagascar (12 strains) and Iran (16 strains) were more diversified and could be divided into 3, 4, and 5 groups, respectively. One of the groups from Madagascar and one from Iran were indistinguishable from the strains from India and the PV-11 vaccine strain (Sureau et al, 1983).

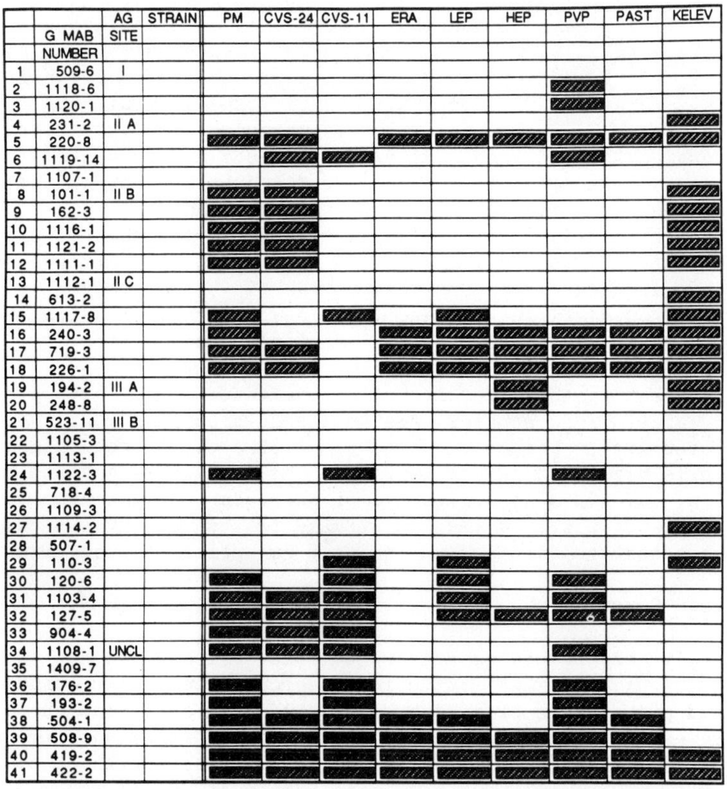

Fig. 1. Comparative antigenic analysis of fixed rabies strains with anti-glycoprotein monoclonal antibodies. Open boxes, positive virus neutralization (> 2 $\log_{10}$ reduction in titer; neutralization index of 100 or higher); shaded boxes, no virus neutralization. (Dietzschold et al, 1987a).

Several studies of North American rabies virus isolates have recently been conducted using viral isolates which were collected from various geographic regions in the United States (Smith et al, 1984; 1986; Dietzschold et al, 1987a) and Canada (Webster et al, 1985; 1986). For some of these studies only panels of anti-nucleocapsid antibodies were used for the comparative antigenic analysis. One of the studies by Smith and co-workers using the CDC panel of antibodies (Smith et al, 1984) focused on the antigenic characteristics of viral isolates associated with the current mid-Atlantic epizootic of raccoon rabies in the United States which began in 1980. The most important question concerning the mid-Atlantic raccoon rabies outbreak which seemed possible to answer with the use of monoclonal antibodies was whether the new rabies epizootic could have originated in the enzootic area of raccoon rabies in the southeastern United States. After gathering isolates from raccoons and other mammals in the mid-Atlantic area, from raccoons in the enzootic area of the southeastern United States, and from animals in other areas of the United States and Canada where rabies is enzootic in skunks and foxes, a comparative antigenic analysis using anti-nucleocapsid monoclonal antibodies showed that the isolates could be grouped according to geographic area and that strains could define "ecocycles" of transmitted rabies. For example, the anti-nucleocapsid antibody reactivity patterns with all 8 rabies virus isolates from raccoons identified with the mid-Atlantic outbreak were identical and indistinguishable from those isolates examined from other terrestrial

		SITE	PM	CVS-24	CVS-11	ERA	LEP	HEP	PVP	PAST	KELEV
		#	3650	3673	3627	1402-2	3661	3640	3661	3666	3654
	# isolates										
	MAb #										
1	701-5	NS I									
2	721-2	NS II									
3	377-7	N I		■							
4	808-2	N II									
5	715-3										
6	801-1										
7	803-6										
8	806-1										
9	120-2										
10	209-1										
11	229-1										
12	590-2										
13	364-1										
14	805-3							■			
15	237-3	N III						■			
16	816-1							■			
17	502-2										
18	515-3										
19	206-1										
20	222-9							■			
21	103-7										
22	804-9							■			
23	703-8							■			
24	239-1							■			
25	104-4							■			
26	111-2										
27	389-1							■			
28	111-1										
29	422-5	UNCL									
30	1403-4					■					
31	102-27										
32	802-2										
33	807-5										
34	817-5							■			
35	822-7										
36	714-3										
37	818-5										

Fig. 2. Comparative antigenic analysis of fixed rabies virus strains with anti-nucleocapsid monoclonal antibodies. Open boxes, positive immunofluorescence of viral antigen in acetone-fixed virus-infected cells; shaded boxes, no fluorescence.

mammals which include 3 skunks, 1 groundhog, and 1 cow from the same area (Smith et al, 1984). The fact that the antigenic character of the raccoon rabies isolates in the mid-Atlantic area was the same among all viral isolates examined regardless of host species in the area suggests that the one strain of virus was responsible for the terrestrial rabies in the mid-Atlantic region (Smith et al, 1984). Furthermore, since the antigenic character was also identical to that of isolates from the enzootic area of the southeastern United States it was suggested that the mid-Atlantic raccoon epizootic originated from animals which were transported from the southeastern area and released into the mid-Atlantic area (Nettles et al, 1979). The potential use of monoclonal antibodies as diagnostic reagents in the investigation of rabies transmissions is clearly demonstrated in this case. It is perhaps more significant to find that in a survey of 231 rabies virus isolates collected from terrestrial animals in six areas of the United States, the "mid-Atlantic and southeastern" group of rabies virus strains was one of four antigenically distinct groups discerned (Smith et al, 1986). The other 3 groups were identified as the "north eastern United States and Canada" group which is characterized by the virus strain of the predominant host, the red fox; the "south central United States" group which is identified by a typical skunk virus isolate; and the "north central United States" group which is represented by a second antigenically distinct prototype strain of skunk virus.

In a separate survey of rabies virus isolates from a variety of terrestrial mammals and bats in Canada, several groups of antigenically distinct isolates were similarly identified using a combined panel of 43 anti-nucleocapsid antibodies from the Wistar and CDC laboratories (Webster et al, 1986). For example, identical isolates from wildlife animals (mainly skunk) in the "south mid-central" region that runs along the United States- Canada border spanning the Manitoba to Alberta Provinces were recognized as one antigenic group (Group I). A second group (Group II) of isolates, which is also a collection of rabies viruses from skunks in a restricted area in Alberta close to the first group, represents viruses with a slight antigenic variation (a difference of one anti-nucleocapsid antibody binding site) that could be recognized by the panel of anti-nucleocapsid monoclonal antibodies. It may be that the group II viruses signify the emergence of a separate "ecocycle" of rabies transmission in the skunk population, and if this turns out to be the case (on the basis of testing a larger sample pool), it will be extremely interesting from a molecular epidemiological view point. It suggests that a new "ecocycle" has been established by an antigenic variant strain bearing a single amino acid change (in the nucleocapsid antigen), but a detailed molecular analysis of the nucleoprotein and glycoprotein genes from the Group I and Group II viruses would have to confirm this hypotehsis.

It is curious that a similar phenomenon may also have been detected in establishing the distribution pattern of two antigenically distinct groups of rabies viruses within the regions of Ontario, Quebec, and the Northwest Territories (Webster et al, 1986). The antigenic group which represents the largest proportion of isolates (51 of 65) collected from these regions, called the "Canadian Arctic" group, spans the vast area associated with these regions. The rest of the viral isolates come from a relatively restricted region in the vicinity of the "south-eastern Georgian Bay Region" (a group distinction) of Ontario, close to an area which contains rabies viruses belonging to the Canadian Arctic group. This smaller group of identical isolates (14 of 65) represents an antigenically distinct virus group which can be distinguished from the Canadian Arctic isolates by a single epitope on the viral nucleocapsid protein. These data suggest that as a result of a slight antigenic variation, an independent cycle of rabies transmission may be established.

Rabies virus isolates from bats in Canada were also classified in distinctive groups based on their reactivity patterns with the combined panel of 43 anti-nucleocapsid monoclonal antibodies (Webster et al, 1986). One group was made up exclusively of isolates from the big brown bat (*Eptesicus fuscus*), but the isolates, which were entirely from southern Ontario, were not uniform in their antigenic profiles. A second, third, and fourth group of bat isolates represented antigenically distinct viruses despite having come from a variety of bat species and a wide geographic area.

Antigenically Distinct Street Viruses Associated with Different Wildlife Species

In another comparative study of more than 300 isolates collected from a variety of wildlife species throughout North America, viruses from the major host species were classified by antigenic type on the basis of their reactivity against a panel of anti-glycoprotein monoclonal antibodies as well as a panel of anti-nucleocapsid monoclonal antibodies (Dietzschold et al, 1987a). To illustrate some of the characteristic features of the antigenic profiles of these viral isolates, the patterns of reactivity between the individual viral isolates and the two monoclonal antibody panels are shown in Figs. 3 and 4. First, it can be seen that the street isolates of different species exhibit considerably different patterns of reactivity with the monoclonal antibodies. The constancy in the antigenic reactivity pattern of the viral isolates from the raccoon (*Procyon lotor*), is contrasted by the more heterogeneous isolates from some of the other terrestrial hosts and bats. In the analysis of street isolates from terrestrial mammals in this particular study, two antigenically distinct strains of skunk viruses were identified in the midwestern United States - a geographically distinct area not included in the surveys conducted by Smith et al, (1986). Second, it can also be seen that the viral isolates exhibit geographic compartmentalization, as described above. The four antigenically distinct bat virus isolates, for example, were each isolated in different geographic areas in the United States. Two of the predominant patterns detected among big brown bat (*Eptesicus fuscus*) isolates were found in eastern (Type I) and northern or western (Type IV) North American populations. The virus strains isolated from these rather sedentary colonial bats were clearly differentiated from the strains isolated from the ever migrating red bats (*Lasiurus borealis*) (Type II), regardless of their geographic location. The fourth group of isolates was from the Mexican free-tailed bat (*Tadarida brasiliensis*) collected from the southwestern United States (Type VI). All the North American insectivorous bat strains were distinct from those isolated from South American bats, including vampire bats (Dietzschold et al, 1987a). Third, it can be seen with most of the examples, except the raccoon (Type R-I) and fox (Type F-I), that the assignment of a street virus isolate to a specific antigenic group may be made by testing either the viral nucleocapsid or glycoprotein antigens against their respective monoclonal antibody panels. However, to reduce the ambiguity of an assignment, it might be best to test the antigenicity of an isolate with both antibody panels.

The advantage of using two different antigen-specific monoclonal anti-body panels instead of one to characterize a collection of rabies virus isolates is illustrated in Figs. 5 and 6 as described by Dietzschold et al, (1987a). The isolates from South America included 11 street strains from Argentina, 9 from Brazil, 2 from Guyana, 1 from Trinidad, and 4 from Colombia. Of the 11 street viruses isolated in Argentina, 3 were from cattle, 1 was from a human, 5 were from dogs, 1 was from a goat, and 1 was from a vampire bat. These viruses were propagated on BHK-21 cells and then analyzed with the panels of anti-glycoprotein and anti-nucleocapsid monoclonal antibodies.

On the basis of the antigenic analysis of the various Argentinian strains with the anti-nucleocapsid antibodies, three distinct patterns of antigenic reactivity emerged separating the bovine isolates from the human, dog, and goat isolates, and also from the vampire bat isolate (Fig. 5). In this analysis, the bovine isolates (Group I) and the fixed (ERA) strain of rabies virus were indistinguishable. Further subgrouping of these viruses was achieved after testing the virus-neutralizing activity of the panel of anti-glycoprotein antibodies with the different viruses (Fig. 6). In the neutralization analysis, 2 of the bovine isolates and 1 of the dog isolates were distinguishable from the other isolates in their respective groups; each revealed an additional neutralizing epitope, the 504-1 epitope, on the glycoprotein. The bovine isolates (Group I) and human, dog, and goat isolates (Groups II) also revealed reactivity patterns with anti-glycoprotein antibodies which were not only similar to each other but were similar to patterns observed with the fixed ERA (Fig. 6) and Pasteur virus (Fig. 2) strains and isolates obtained from foxes in western Europe (data not shown). The ERA strain of rabies virus, however, could be distinguished from the virus isolates from Argentina Groups I and II, as well as the Pasteur virus and European fox virus, by its neutralization with monoclonal antibody 127-5. Even with this single epitope difference on the glycoprotein, it was tempting to speculate that the Argentina isolates from Groups I and II, the Pasteur virus and the European fox isolate were

G MAb NUMBER	AG. SITE	BAT N=38 TYPE BT-I	BAT N=35 TYPE BT-II	BAT N=31 TYPE BT-IV	BAT N=16 TYPE BT-VI	RACCOON N=11 TYPE R-I	FOX N=5 TYPE F-I	ARCT. FOX N=2 TYPE F-II	SKUNK N=6 TYPE S-I	SKUNK N=5 TYPE S-II
509-6	I	■	■	■	■					
231-2	II A	■	■	■	■	■				
220-8		■	■	■	■	■		■	■	■
1119-14										
1107-1										
101-1	II B									
162-3		■					■			
1116-1		■								
1121-2		■								
1111-1		■								
1112-1	II C									
613-2			■				■			
1117-8		■	■	■	■	■	■	■	■	■
240-3		■	■	■	■	■		■	■	■
719-3		■	■	■	■	■		■	■	■
226-1		■	■	■	■	■	■	■	■	■
194-2	IIIA		■				■			
248-8		■	■							
523-11	III B	■	■				■			
1105-3			■							
1113-1		■	■		■					
1122-3										
718-4										
1109-3										
1114-2							■			■
507-1		■	■							
110-3		■	■	■	■	■	■			■
120-6			■	■	■	■		■	■	
1103-4		■	■		■	■		■		
127-5		■	■	■	■	■		■	■	■
1118-6								■		
1120-1			■					■		
904-4			■	■	■	■		■	■	■
1108-1	UNCL						■			
176-2		■		■	■	■		■	■	
193-2			■		■			■		
504-1		■	■	■	■	■		■	■	■
508-9		■	■	■	■	■	■	■	■	■
419-2		■	■	■	■	■	■	■	■	■
422-2		■	■	■	■	■	■	■	■	■

Fig. 3. Antigenic analysis of rabies street strains and rabies-related viruses from North American wildlife with the panel of anti-glycoprotein monoclonal antibodies. N, number isolates per group. Open boxes, virus neutralization index 100 or higher; shaded boxes, virus neutralization index lower than 100. (Dietzschold et al, 1987a).

derived from a common ERA vaccine strain, since ERA had been used as a vaccine in both Argentina and Europe (Dietzschold et al, 1987a). An alternative possibility is that the Group I and Group II viruses, which are very similar to the western European fox isolates, may have been introduced by animals transported from Europe to South America. However, the large number of epitope changes on the nucleocapsid antigens of the Argentina isolates must be considered when comparing them with the ERA nucleocapsid proteins. It is possible that the antigenic differences on the nucleocapsid could have arisen through adaptation processes in the different hosts. This is not a phenomenon that can be easily analyzed, but further study of virus evolution may lead to an understanding of the possible mechanisms responsible for acquired antigenic diversity.

The third antigenic group (Group III) of the Argentina viruses is represented by a single isolate from a vampire bat which differs considerably from the viruses in Groups I and II, but has an identical antigenic pattern of reactivity with the anti-glycoprotein and anti-nucleocapsid antibodies as the DR 19 virus which was isolated from a vampire bat (*Desmodus rotundus*) in Brazil (Wiktor et al, 1980). The close antigenic relationship among isolates from vampire bats is an interesting one, and it suggests that this virus may cycle independently in the vampire bat. Occasionally, however, the virus strain appears in terrestrial animals, particularly cattle. When this occurs, the virus infection which was transmitted reaches an apparent dead end.

NC MAb NUMBER	AG SITE	BAT N=38 TYPE BT-I	BAT N=35 TYPE BT-II	BAT N=31 TYPE BT-IV	BAT N=16 TYPE BT-VI	RACCOON N=11 TYPE R-I	FOX N=5 TYPE F-I	ARCT. FOX N=2 TYPE F-II	SKUNK N=6 TYPE S-I	SKUNK N=5 TYPE S-II
701-5	NS I	■	■	■	■	■	■			■
721-2	NS II									
377-7	N I				■					
808-2	N II									
715-3										
801-1										
803-6										
806-1										
120-2										
209-1										
229-1										
237-3	N III		■	■						■
239-1			■	■						■
590-2										
364-1								■		
805-3				■						
816-1				■						
502-2										
515-3										
206-1										
222-9				■						
103-7										■
804-9				■						
703-8				■						
104-4				■						
111-2				■						
389-1				■						
111-1				■				■		
422-5	UNCL.	■	■	■	■	■	■	■	■	■
102-27		■	■	■	■	Variable	■	■		
802-2	9									
807-5										
817-5				■						
822-7				■						
714-3										
818-5										

Fig. 4. Antigenic analysis of rabies virus strains isolated from North American wildlife with panel of anti-nucleocapsid monoclonal antibodies. Open boxes, positive fluorescent staining; shaded boxes, no staining. (Dietzschold et al, 1987a).

ANTIGENIC RELATIONSHIP BETWEEN RABIES AND RABIES-RELATED VIRUSES

The distinction between rabies and rabies-related viruses, based on their antigenic relationships, had already been determined by serological tests at the time that the hybridoma technology was introduced (Tignor and Shope, 1972; Shope, 1975; Tignor et al, 1977). When monoclonal antibodies (derived mostly from fixed rabies virus determinants) were applied to the study of antigenic relationships among all Lyssaviruses, they confirmed the distinction between rabies-related viruses (Flamand et al, 1980a, 1980b; Wiktor et al, 1980; Schneider et al, 1985).

Despite the low level of antigenic homology between the nucleocapsid antigens and glycoproteins of fixed rabies virus strains and the rabies-related Mokola, Lagos bat, or Duvenhage (both the African and the related European) viruses, the expanded panels of anti-nucleocapsid and anti-glycoprotein antibodies against the fixed strains have managed to differentiate up to 5 antigenic types of Mokola virus from Zimbabwe and 5 antigenic types of Duvenhage viruses (including African and European strains) (Dietzschold et al, 1987a). Of the 37 anti-nucleocapsid antibodies used in the panel, 9 cross-reacted with all rabies-related virus strains; antibody 502-2 was found to react with all the Lyssaviruses, antibody 103-7 reacted only with rabies viruses, and antibody 422-5 reacted only with the Mokola and Lagos bat virus strains. Monoclonal antibody 422-5, which was originally considered to react only with rabies-related viruses, including the African Duvenhage strain (Wiktor et al, 1980), is now regarded as less specific for the African Duvenhage strain and nonspecific for the European bat strains that are similar but not identical to the African Duvenhage isolates. (Dietzschold et al, 1987a). Despite the

					ARGENTINA					BRASIL					GUY		TRI	COLOMBIA		
		AG		DR	B	H	D	G	VB	B	B	BT	B	B	BT	B	B	H	D	B
	# isolates	SITE	ERA	1 9	n=3	n=1	n=5	n=1	n=1	n=4	n=2	n=1	n=1	n=1	n=1	n=1	n=1	n=2	n=1	n=1
	MAb #				I	II	II	II	III	a-d	e-f	g	h	i						
1	701-5	NS I		■					■	■	■	■			■		■			
2	721-2	NS II																		
3	377-7	N I		■					■		■	■		■						
4	808-2	N II																		
5	715-3																			
6	801-1																			
7	803-6																			
8	806-1					■	■	■												
9	120-2																			
10	209-1																			
11	229-1																			
12	590-2																			
13	364-1																			
14	805-3																			
15	237-3	N III																		
16	816-1					■	■	■												
17	502-2																			
18	515-3																			
19	206-1																			
20	222-9																			
21	103-7																			
22	804-9																			
23	703-8																			
24	239-1																			
25	104-4																			
26	111-2					■	■	■												
27	389-1					■	■	■												
28	111-1					■	■	■												
29	422-5	UNCL	■	■	■	■	■	■	■	■	■	■	■	■	■	■	■	■	■	■
30	102-27										■	■								
31	802-2								■		■				■					
32	807-5																			
33	817-5																			
34	822-7																			
35	714-3																			
36	818-5																			

Fig. 5. Antigenic analysis of rabies strains isolated from cattle (B), human (H), dog (D), goat (G), bat (unclassified, BT), and vampire bat (VB) in South America using the panel of anti-nucleocapsid monoclonal antibodies. Open boxes, positive fluorescence by indirect immunofluorescence test; shaded boxes, no fluorescence. (Dietzschold et al, 1987a).

usefulness of available monoclonal antibodies, (albeit produced from fixed virus strains as diagnostic reagents), it is still possible that monoclonal antibodies capable of cross-reacting with the major epitopes of viral proteins from different street strains or detecting minor antigenic heterogeneity in street strains may not yet have been isolated. Future research work will undoubtedly produce additional monoclonal antibodies specifically against a variety of street viruses isolated from wildlife species.

THE ROLE OF MONOCLONAL ANTIBODIES IN THE STUDY OF RABIES VIRUS ANTIGENIC STRUCTURE

To begin to understand the nature of antigenic variation among rabies viruses, some basic knowledge of the structure of antigenic determinants of the viral proteins is required. The ability of monoclonal antibodies to discriminate antigenic differences, even to the level of single amino acid differences, makes them ideal reagents for investigating changes in the primary structure of individual antigenic determinants (epitopes) that occur during antigenic variation. The process of identifying and mapping virus-neutralizing determinants on the rabies virus glycoprotein was initiated by selecting neutralization-resistant (escape) mutants from cloned parental ERA and CVS-11 rabies virus strains in the presence of an excess amount of virus-neutralizing monoclonal antibody (Wiktor and

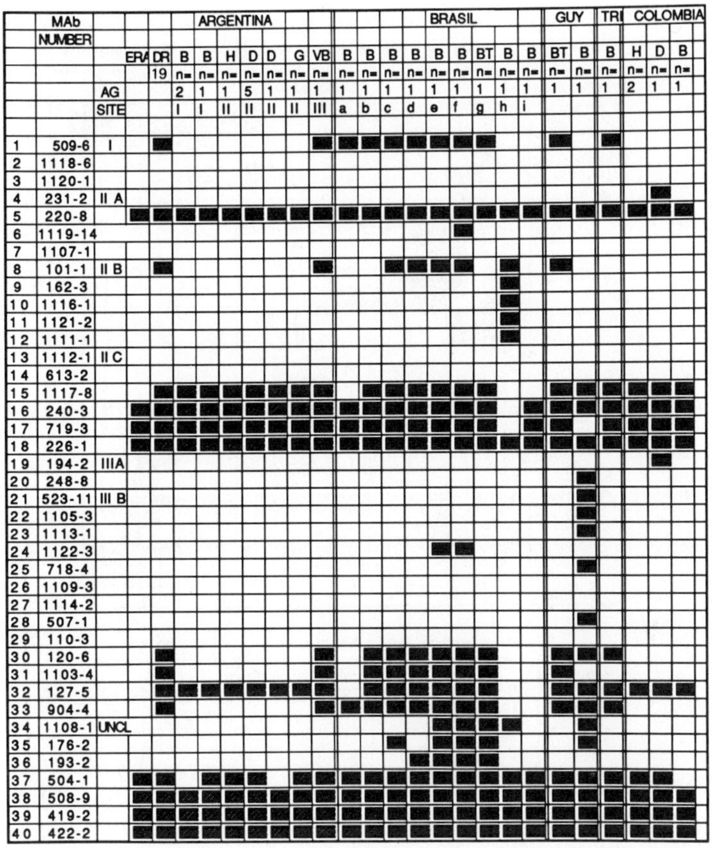

Fig. 6. Antigenic analysis of rabies street strains isolated from cattle (B), human (H), dog (D), goat (G), bat (unclassified, BT), and vampire bat (VB), and the DR19 isolated from vampire bat in South America compared with fixed rabies ERA strain, using the anti-glycoprotein monoclonal antibodies. Open boxes, virus neutralization index of 100 or higher; shaded boxes, virus neutralization index lower than 100. (Dietzschold et al, 1987a).

Koprowski, 1980). From 52 escape mutants of ERA virus, 17 unique epitopes were identified, and from 90 mutants of CVS-11 virus, 13 epitopes were found (Lafon et al, 1983; 1984). By grouping the neutralization-resistant variant (RV) viruses into three antigenically distinct sets, on the basis of the individual patterns of resistance to neutralization by the panel of anti-glycoprotein monoclonal antibodies used for the selection of mutant viruses (Fig. 7), it was possible to delineate several major functional antigenic sites (unique regions of one or several overlapping epitopes) on the rabies virus glycoprotein. Three of the functionally defined sites were common to both the ERA and CVS-11 strains, and two independently by defined virus-neutralizing antibody binding sites were also delineated on the ERA strain virus glycoprotein.

The strategy employed to locate the major antigenic sites on the rabies virus glycoprotein was similar to that developed for the analysis of antigenic structure of the influenza virus hemaglutinin (Caton et al, 1982). The complete nucleotide sequence of the glycoprotein gene was determined for the various 'monoclonal' variants of the ERA and CVS-11 strains of rabies virus which delineate the functionally defined antigenic

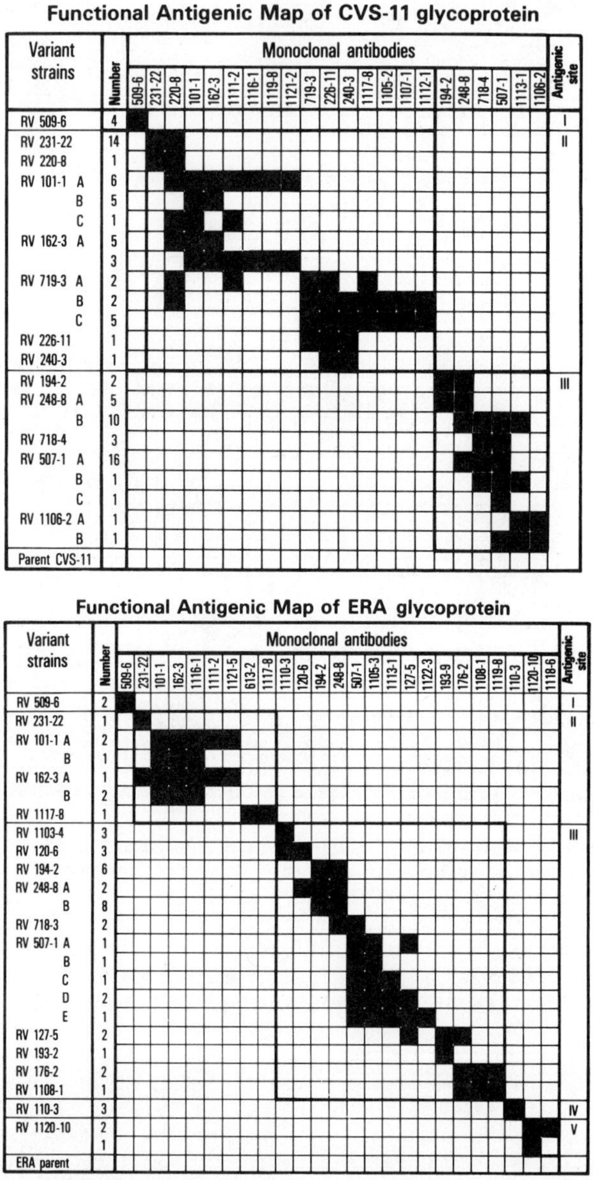

Fig. 7. Functional epitope maps of CVS-11 and ERA strain rabies virus glycoproteins. Neutralization-resistant variants (RV) were selected using monoclonal antibodies, and then tested for susceptibility (□) or resistance (■) to neutralization by antibodies in the panel. (Lafon et al, 1984).

sites on the glycoprotein. In the glycoprotein gene of each of the "monoclonal" variants, a single nucleotide change (point mutation) was found which could be translated into a single amino acid substitution. The locations of the various amino acid changes in the primary structure thus identify the antigenic site(s). The changes that define distinct sites may be scattered along the primary sequence of the glycoprotein, as in antigenic sites I and II (Smith and Wunner, unpublished), or may be closely linked in the primary structure, as in antigenic site III (see Fig. 8). Those which appear to be widely separated in the primary sequence can lie very close to each other in the three-dimensional or secondary structure of the glycoprotein molecule. As shown in Fig. 8, the single amino acid changes associated with antigenic site III were found at four different positions within a short stretch of amino acids in the glycoprotein. These were identified in a total of eight independently selected "monoclonal" mutants all mapping to the same functional antigenic site (Wunner et al, 1984; 1985; Smith and Wunner, in preparation). Three of the eight mutants (ERA RV 194-2, CVS RV 194-2, ERA RV 248-8) showed a substitution at position 333; three mutants (ERA RV 507-1, ERA RV 127-5, ERA RV 718-4) revealed a change at position 336; one (CVS RV 523-11) showed a change at position 338; one (ERA RV 120-6) showed a change at position 357. The location of the map positions occupied by these amino acid changes in relation to the rest of the linear sequence suggests that at least part of the structure recognized by the antibodies (excepting residue 357) in antigenic site III is represented by a continuous stretch of amino acids forming overlapping epitopes. That is, when a change of one amino acid occurs within the continuous determinant of the antigenic site, the binding of one or more antibodies could be simultaneously affected. It seems likely, therefore, that the loss of a single epitope in antigenic variation may have major consequences for virus neutralization if several overlapping epitopes that bind other antibodies occupy positions close to each other, even in the three-dimensional structure.

STUDIES ON THE MECHANISM OF NEUTRALIZATION OF RABIES VIRUS BY ANTIBODY

Virus neutralization by antibody results in a loss of viral infectivity which can be measured in animals or cells in culture. The mechanism(s) by which antibodies neutralize infectivity of rabies virus has been unclear and monoclonal antibodies against rabies virus have been used to investigate the phenomenon of virus neutralization. It has been well established that rabies virus-neutralizing antibodies (VNA) are directed exclusively against the surface glycoprotein of the virus (Cox et al, 1977; Wunner et al, 1983). Yet, the rabies virus glycoprotein is also the effector molecule for virus attachment to cell surface receptors (Wunner et al, 1984) and pH-dependent fusion of cell membranes (Reagan and Wunner, 1984). To determine the mechanism of virus neutralization by antibody, much will depend on whether the cellular binding site and portion on the glycoprotein responsible for cell-cell fusion is distinct from the neutralization (antibody binding) sites. Recently, it was shown that when the virus was treated with VNA, before adsorption, or when cell-adsorbed virus was exposed to VNA, virus replication was prevented (Dietzschold et al, 1985). Also, anti-idiotypic antibodies prepared (as described below) against anti-glycoprotein monoclonal antibodies which strongly neutralized virus infectivity were ineffective at blocking virus attachment to the cell surface receptor (Reagan and Wunner, unpublished). In both instances, the observations indicated that the cellular binding site on the rabies virus glycoprotein is distinct from the neutralization sites.

The biological activities of the rabies virus glycoprotein-specific monoclonal VNA differ considerably with respect to their ability to neutralize virus depending on whether neutralization occurs before adsorption of virus to cells or after virus has been adsorbed to cells (Dietzschold et al, 1987b). The ability of VNA to neutralize virus after virus adsorption to cells correlated well with ability to inhibit acid-catalyzed fusion of rabies virus-infected cells. In a recently conducted series of experiments, two biologically distinctive monoclonal VNA were used to analyze the mechanism(s) of rabies virus neutralization. The two antibodies which were selected from the Wistar panel of anti-glycoprotein

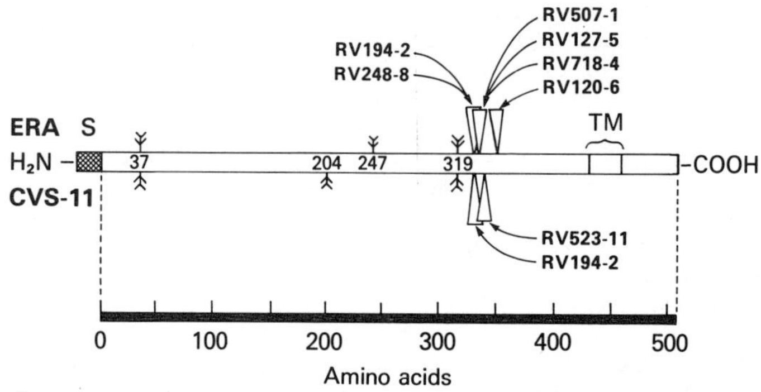

Fig. 8. Map of antigenic site III of the rabies virus glycoprotein identifying the location of amino acids that represent functional epitopes. The diagram shows the location of point mutations (darts) in antigenically altered glycoproteins of rabies neutralization-resistant variant (RV) viruses of ERA (above) and CVS-11 (below) strains. The RV number refers to the virus-neutralizing monoclonal antibody used for selection of corresponding variant virus. Other identifiable features of the rabies virus glycoprotein include the signal peptide (S) which is cleaved from the mature glycoprotein molecule, the transmembrane (TM) domain, and location of asparagine residues which serve as potential carbohydrate acceptance sites (Y). (Wunner and Dietzschold, 1987).

monoclonal antibodies recognized topographically distinct antigenic sites on the glycoprotein. To determine whether rabies virus neutralization is caused by inhibition of virus binding to specific cell surface receptors or inhibition of the intraendosomal acid-catalyzed fusion step following virus uptake into susceptible cells, the following experiments were performed (Dietzschold et al, 1987b).

The two monoclonal VNA (509-6 and 528-2), having similar virus-neutralizing activities but vastly different capabilities to neutralize cell-adsorbed virus and to inhibit acid-catalyzed fusion, were compared firstly for their ability to prevent virus attachment to susceptible cells. At dilutions of 1:50 and 1:250, the antibodies were individually mixed with radiolabeled rabies virus for 1 hour at 37°C. After this time, the antibody-treated virus which was completely neutralized (as measured by infectivity assay in cells in culture) was allowed to adsorb to BHK-21 cells for 2 hours at 4°C, and the amount of cell-associated labeled virus was determined. After 2 hours of incubation, 40% of radiolabeled virus treated with antibody 509-6 and 66% of virus treated with antibody 528-2 was adsorbed to cells. This indicated that despite virus neutralization, a large proportion of virus still attached to cells. Secondly, since virus adsorbed to cells may have been removed by antibody as part of the neutralization, it was necessary to determine whether these two antibodies were capable of removing adsorbed virus from the cell surface. The study showed that antibody 509-6 released up to 30% of adsorbed virus from cells after 1 hour at 4°C and antibody 528-2 released up to 18%; only 11% of adsorbed virus was released from cells not exposed to antibody. Finally, there remains a portion of the virus adsorbed to cells which is capable of entering the cell by endocytosis. In general, rabies virus particles which are taken up by cells presumably fused, with the membranes of acidic endocytic vacuoles (Marsh, 1984).

To investigate whether monoclonal VNA affect the endocytosis of rabies virus following virus attachment to the cell surface, temperature shift experiments were performed with rabies virus-infected BHK-21 cells in the presence and absence of the 509-6 and 528-2 antibodies. It was found that when virus was added to cells at 4°C and the cells were incubated with the monoclonal antibodies before shifting the temperature to 37°C, virus uptake was not delayed after the temperature shift, compared with virus that had penetrated the cells in the absence of antibody. These results suggested that the antibodies did not impede the internalization of adsorbed virus by the cell; approximately the same amount of virus was taken up by the cell in the presence or absence of antibody. Rather, these results suggested that the antibodies may have altered the course of infection by the internalized virus by inhibiting the pH-dependent fusion reaction between the membranes of the virus particles and the endocytic vacuoles, thus preventing the intraendosomal uncoating of viral nucleocapsids and initiation of replication (Marsh, 1984). Instead, the unfused virus particles were presumably transferred to lysosomes and inactivated by hydrolysis and degraded to acid-soluble material (Gollins and Porterfield, 1986). It was demonstrated, using as an assay the amount of acid-soluble radioactive material, that both radiolabeled virus bound to unlabeled antibody and radiolabeled antibody bound to unlabeled virus were internalized and degraded via the endocytic and lysosomal pathways to acid-soluble products (Dietzschold et al, 1987b). The data imply that the antibody-mediated neutralization of rabies virus is most likely the result of several different mechanisms, including inhibition of virus attachment, release of cell-bound virus, and inhibition of intraendosomal fusion. All of these mechanisms and their consequences must be considered if and when monoclonal antibodies are selected for therapeutic use in post-exposure treatment of rabies.

VIRAL ANTIGEN-INDEPENDENT IMMUNIZATION AGAINST RABIES USING ANTI-IDIOTYPE ANTIBODIES OF VIRUS-NEUTRALIZING MONOCLONAL ANTIBODIES

Anti-idiotypic (id) antibodies bearing images of the antigenic (ag) (and immunogenic) determinants on the rabies virus glycoprotein can be generated as a subset of molecules in the idiotype network of the immune system (Jerne, 1974; Urbain et al, 1983). Anti-idiotypic antibodies (Ab2 molecules) are raised against the idiotopes expressed on an antigen (in this case, the rabies virus glycoprotein) specific antibody (Ab1 molecule). The Ab2 molecules give rise to a diverse population of anti-anti-idiotypic antibodies (Ab3 molecules) which as a population include molecules that recognize idiotopes of Ab2 but not the antigen and may or may not share idiotopes with Ab1 (id^+ag^- or id^-ag^-) and molecules which may or may not share an idiotope with Ab1 and exhibit antigen specificity (id^+ag^+ or id^-ag^+). This last subset contains idiotopes that have a similar antigenic binding specificity as the variable region determinant(s) of the Ab1 molecule.

Immunization with anti-idiotypic antibodies has been regarded as a potentially useful antigen-independent method of developing protection against rabies particularly in the newborn animal or human. To study the effect of anti-idiotypic antibodies on the immune system of animals, we prepared polyclonal anti-idiotypic antibodies against five anti-glycoprotein monoclonal antibodies (αG MAb) of defined specificity (Reagan et al, 1983). These αG MAB (Ab1 molecules) were directed against non-overlapping epitopes in the functionally defined antigenic sites I, II, and III of the rabies virus glycoprotein, and all but one of the αG MAb, 1104-2, strongly neutralized virus infectivity. None of the anti-idiotypic antibodies revealed major cross-reactive idiotypes among the original Ab1 molecules. Although each anti-idiotypic antibody preparation reacted only with its homologous Ab1; the proportion of anti-idiotypic antibodies directed against framework versus antigen-combining site determinants on the Ab1 molecules varied with the different polyclonal Ab2 sera. Two of the anti-idiotypic antibody preparations, against monoclonal antibodies 101-1 and 719-3, had either a minor population or no population of antigen-combining site specificities (Reagan et al, 1983). The three remaining anti-idiotypic antibody preparations (against monoclonal antibodies 509-6, 507-1, and 1104-2) showed measurable inhibition (20-50%) of binding between homologous Ab1 and rabies virus glycoprotein, and these were used for the induction of Ab3 molecules with virus-neutralizing activity in mice.

When outbred animals were inoculated with individual preparation of anti-idiotypic antibodies of 509-6, 507-1, and 1104-2 (50μg per mouse) and the sera were tested for a population of Ab3 molecules that not only could bind to Ab2 but also could neutralize virus, two of the preparations (509-6 and 1104-2) elicited specific but low levels of VNA (Reagan et al, 1983). A significant observation was that the specificity of the Ab3 molecules induced by both the 509-6 and 1104-2 Ab2 preparations was targeted to the 509-6 epitope. This was demonstrated by the fact that only viruses containing the 509-6 epitope (and not the RV 509-6 mutant virus) were neutralized by the Ab3 polyclonal sera (Reagan et al, 1983). In the case of the Ab3 molecules induced by the Ab2 1104-2 preparation, the fact that the specificity was not equivalent to the original non-neutralizing Ab1 (αG MAb 1104-2) indicates that induction of multispecific antibodies by an anti-idiotypic antibody can occur, as observed by others (Krieger et al, 1983).

To investigate the significance of this low but specific VNA response in the absence of viral antigen exposure, mice were immunized with a mixture of the three anti-idiotypic antibody preparations (509-6, 507-1, and 1104-2) and then injected 3 weeks later with a titrated amount (0.04 IU) of rabies virus vaccine (Reagan et al, 1985). When VNA titers were again measured immediately before and 7 days after the vaccine was given, the mice treated with the anti-idiotypic antibodies showed a booster response to the rabies virus vaccine rather than the primary immune response produced by the control animals which received only phosphate-buffered saline (PBS) before immunization. The booster response to the anti-idiotypic antibodies produced VNA titers (160-320) approaching those of animals pretreated with 0.1 IU of the vaccine (320-640), suggesting that the anti-idiotypic antibodies were capable of priming the immune system.

Because the titers of VNA produced after anti-idiotypic antibody priming followed by a boost with the vaccine approached protective levels similar to the response to vaccine priming and vaccine boost, the significance of this anti-idiotypic antibody plus vaccine regimen on survival of mice against an intracerebral rabies virus challenge was investigated. Groups of 10 outbred mice were inoculated with PBS, rabies virus vaccine (0.04 IU), or anti-idiotypic antibodies (individually and in a mixture). Two weeks later, all animals except some of the PBS controls received an i.p. injection of rabies virus vaccine (0.04 IU) and 1 week later all received 10 LD_{50} (50% lethal doses) of rabies virus intracerebrally (Reagan et al, 1985). Survival of mice up to 3 weeks postchallenge is shown in Table 1. From these data, it is apparent that the intracerebral challenge was sufficient to kill all nonimmune (PBS control) animals and that the amount of vaccine given by itself (without priming) did not provide better than 20% protection, while all animals pretreated with vaccine and boosted with vaccine were protected. The data also indicate that the 50-μg dosages of anti-idiotypic antibodies given individually did not provide any significant degree of protection, although 50 μg of each anti-idiotypic antibody in a mixture did increase survival. Thus, it may be possible to achieve significant protection with a mixture of paratope-specific anti-idiotypic antibodies. In a separate experiment, however, using inoculations of 10-fold dilutions of the 509-6 anti-idiotypic antibody preparation per mouse from 50 μg to 0.005 μg, an optimal effect on survival

Table 1. Effect of Preimmunization with Anti-Idiotypic Antibodies Plus Vaccine on Survival against Intracerebral Challenge with Rabies Virus.

Preimmunization	Amount	Vaccine	Survival (%)
PBS	-	-	0
PBS	-	+	20
Vaccine	0.04 IU	+	100
Ab2 509-6	50 μg	+	40
Ab2 507-1	50 μg	+	30
Ab2 1104-2	50 μg	+	50
Ab2 (mixture)	50 μg each	+	78 (n = 9)

ICR mice (in groups of 10) were preimmunized by one i.p. injection (0.5 ml) with Ab2, or rabies virus vaccine (0.04 IU) in PBS, 2 weeks prior to the i.p. injection of vaccine (0.04 IU) in PBS. One week later, animals received 10 LD_{50} of rabies virus intracerebrally (Reagan et al., 1985).

was noted at 5 μg (Reagan et al, 1985). This suggests, as noted by others (references in Reagan, 1985), that manipulation of specific immunity by individual paratope-specific anti-idiotypic antibodies is dose dependent and that perhaps a more refined methodology will be required for modulating more effective immune responses.

CONCLUDING REMARKS

To date, the uses of monoclonal antibodies against rabies virus antigens have been numerous and varied. There has been significant progress in rabies research resulting from their comprehensive application - from the original use of site-specific antibodies for the antigenic analysis of virus strains and identification of antigenic determinants on viral antigens, to the use of specifically defined idiotype-bearing antibodies for production of polyclonal anti-idiotypic antibodies. Moreover, the use of these reagents has proved effective in a wide spectrum of investigative disciplines. In the study of rabies, monoclonal antibodies are being used by virologists, immunologists, cell biologists, and molecular biologists as they attempt to resolve particular problems in their areas of rabies research.

In light of our present knowledge of rabies and rabies virus antigens, it seems fair to suppose that monoclonal antibodies against rabies viruses will play an integral part in research efforts of the future. The major usefulness of monoclonal antibodies will be in the area of rabies epidemiology and identification of wildlife strain determinants. An important issue in ensuing investigations of enzootic rabies will be whether monoclonal antibodies directed against the different antigens of street strains will identify a broadly cross-reactive epitope among antigens of rabies and rabies-related virus isolates. Once a cross-reactive epitope is identified, a monoclonal anti-idiotypic antibody which defines the specific cross-reactive epitope will be prepared and evaluated as a vaccine primer. Since Ab2 molecules have been demonstrated as having a priming effect on the anti-rabies immune system, one of the most interesting practical uses of anti-idiotypic antibodies to be studied with this system will be the priming of neonates for induction of immunity. Neonatal priming with monoclonal or polyclonal anti-idiotypic antibody could substitute for the requirement for rabies virus antigen in very early life and potentially induce substantial amounts of anti-rabies virus-neutralizing antibody upon subsequent administration of vaccine in later stages of the young animal's life (Stein, 1985).

Another potential use of the anti-rabies virus-neutralizing monoclonal antibodies will be for post-exposure treatment of human rabies. Production of human-mouse hybrid antibodies with the mouse variable region determinant in a human immunoglobulin framework would seem to be the ideal type of chimeric antibody for treatment of human rabies, rather than use mouse monoclonal antibodies which may cause undesirable side effects in the immunocompetent individual. Other research efforts involving monoclonal antibodies will focus on developing probes for detection of cell-surface receptors that bind rabies virus and T cell receptors that recognize viral antigen determinants. In all these areas of research, questions will arise that may well be answered in part with the use of monoclonal antibodies. There can be no doubt that these reagents will provide an indispensible tool for the future.

Acknowledgements

The authors' research is supported by Grants AI-18562, AI-18883, and AI-09706 from the National Institutes of Health.

REFERENCES

Boulger, L. R., and Porterfield, J. S., 1958, Isolation of a virus from Nigerian fruit bats, Trans.Royal Soc.Trop.Med.Hyg., 52:421.

Caton, A. J., Brownlee, G. G., Yewdell, J. W., and Gerhard, W., 1982, The antigenic structure of the influenza virus A/PR/8/34 hemaglutinin (H1 subtype), Cell, 31:417.

Charlton, K. M., Casey, G. A., Boucher, D. W., and Wiktor, T. J., 1982, Antigenic variants of rabies virus, Comp.Immun.Microbiol.Infect. Dis., 5:113.

Cox, J. H., Dietzschold, B., and Schneider, L. G., 1977, Rabies virus glycoprotein. II. Biological and serological characterization, Infect.Immun., 16:743.

Dietzschold, B., Rupprecht, C. E., Tollis, M., Lafon, M., Mattei, J., Wiktor, T. J., and Koprowski, H., 1987a, Antigenic diversity of the glycoprotein and nucleocapsid proteins of rabies and rabies-related viruses: implications for epidemiology and control of rabies in: "Research Towards Rabies Prevention," G.M. Baer, R.E. Shope, and W.H. Wunner, eds., Rev.Inf.Dis., in press.

Dietzschold, B., Tollis, M., Lafon, M. Wunner, W. H., and Koprowski, H., 1987b, Mechanisms of rabies virus neutralization by glycoprotein-specific monoclonal antibodies, Virology, in press.

Dietzschold, B., Tollis, M., Rupprecht, C. E. Celis, E., and Koprowski, H., 1987c, Antigenic variation in rabies and rabies-related viruses: cross-protection, independent of glycoprotein-mediated virus-neutralizing antibody, J.Inf.Dis., in press.

Dietzschold, B., Wiktor, T. J., Trojanowski, J. Q., Macfarlan, R. I., Wunner, W. H., Torres-Anjel, M. J., and Koprowski, H., 1985, Differences in cell-to-cell spread of pathogenic and apathogenic rabies virus in vivo and in vitro, J.Virol., 56:12.

Flamand, A., Wiktor, T. J., and Koprowski, H., 1980a, Use of hybridoma monoclonal antibodies in the detection of antigenic differences between rabies and rabies-related virus proteins. I. The nucleocapsid protein, J.Gen.Virol., 48:97.

Flamand, A., Wiktor, T. J., and Koprowski, H., 1980b, Use of hybridoma monoclonal antibodies in the detection of antigenic differences between rabies and rabies-related virus proteins. II. The glycoprotein, J.Gen.Virol., 48:105.

Gerhard, W., Croce, C. M., Lopes, D., and Koprowski, H., 1978, Repertoire of antiviral antibodies expressed by somatic cell hybrids, Proc. Natl.Acad.Sci.USA, 75:1510.

Gollins, S. W., and Porterfield, J. S., 1986, A new mechanism for neutralization of enveloped viruses by antiviral antibody, Nature, 321:244.

Jerne, N. K., 1974, Towards a network theory of the immune system, Ann. Immunol., (Paris), 125C:373.

Kemp, G. E., Causey, O. R., Moore, D. L., Odelola, A., and Fabiyi, A., 1972, Further studies on IbAn 27377, a new rabies-related etiologic agent of zoonosis in Nigeria, Am.J.Trop.Med.Hyg., 21:356.

Koprowski, H., Gerhard, W., and Croce, C. M., 1977, Production of antibodies against influenza virus by somatic cell hybrids between mouse myeloma and primed spleen cells, Proc.Natl.Acad.Sci.USA., 74:2985.

Koprowski, H., and Wiktor, T. J., 1980, Monoclonal antibodies against rabies virus in: "Monoclonal Antibodies," R.H. Kennett, T.J. McKearn, K.B. Bechtol, eds., Plenum Publishing Corp., New York.

Krieger, N. J., Pesce, A. J., and Michael, J. G., 1983, Induction of multispecific antibodies to bovine serum albumin after production of anti-idiotypic antibodies to an albumin-specific monoclonal antibody, Ann.N.Y.Acad.Sci., 418:305.

Lafon, M., Ideler, J., and Wunner, W. H., 1984, Investigation of the antigenic structure of rabies virus glycoprotein by monoclonal antibodies, Devel.Biol.Standard., 57:219.

Lafon, M., and Wiktor, T. J., 1985, Antigenic sites on the ERA rabies virus nucleoprotein and non-structural protein, J.Gen.Virol., 66:2125.

Lafon, M., Wiktor, T. J., and Macfarlan, R. I., 1983, Antigenic sites on the CVS rabies virus glycoprotein: analysis with monoclonal antibodies, J.Gen.Virol., 64:843.

Marsh, M., 1984, The entry of enveloped viruses into cells by endocytosis, Review article, Biochem.J.., 218:1.

Nettles, V. F., Shaddock, J. H., Sikes, R. K., Reyes, C. R., 1979, Rabies in translocated raccoons, Am.J.Public Health., 69:601.

Reagan, K. J., 1985, Modulation of immunity to rabies virus induced by anti-idiotypic antibodies in: "Current Topics in Microbiology and Immunology," vol. 119, H. Koprowski and F. Melchers, eds., Springer-Verlag, Berlin, Heidelberg.

Reagan, K. J., and Wunner, W. H., 1984, Early interactions of rabies virus with cell surface receptors in: "Nonsegmented Negative Strand Viruses," D.H.L. Bishop, R.W. Compans, eds., Academic Press, New York.

Reagan, K. J., Wunner, W. H., Wiktor, T. J., and Koprowski, H., 1983, Anti-idiotypic antibodies induce neutralizing antibodies to rabies virus glycoprotein, J.Virol., 48:660.

Reagan, K. J. Wunner, W. H., and Koprowski, H., 1985, Viral antigen-independent immunization: induction by anti-idiotype antibodies of neutralizing antibodies to defined epitopes in rabies virus glycoprotein in: "High Technology Route to Virus Vaccines," G.R. Dreesman, J.G. Bronson, R.C. Kennedy, eds., Am.Soc.Microbiol., in press.

Reid, R. L., Hall, N. H., Smith, J. S., and Baer, G. M., 1983, Increased immunofluorescent staining of rabies-infected, formalin-fixed brain tissue after pepsin and trypsin digestion, J.Clin.Micro., 18:968.

Rupprecht, C. E., Glickman, L. T., Spencer, P. A., and Wiktor, T. J., 1987, Epidemiology of rabies virus variants: differentiation using monoclonal antibodies and discriminant analysis, Am.J.Epidemiol., 126:298.

Schneider, L. G., Barnard, B. J. H., and Schneider, H. P., 1985, Application of monoclonal antibodies for epidemiological investigations and oral vaccination studies in: "Rabies In the Tropics," E. Kuwert, C. Merieux, H. Koprowski, K. Bogel, eds, Springer-Verlag, Berlin.

Shope, R. E., Murphy, F. A., Harrison, A. K., Causey, O. R., Kemp, G. E., Simpson, D. I. H., and Moore, D. L., 1970, Two African viruses serologically and morphologically related to rabies virus, J.Virol., 6:690.

Shope, R. E., 1975, Rabies virus antigenic relationships in: "The Natural History of Rabies," Vol. I, G. M. Baer, ed., Academic Press Inc., New York.

Smith, J. S., Reid-Sander, F. L., Roumillat, L. F., Trimarchi, C., Clark, K., Baer, G. M., and Winkler, W. G., 1986, Demonstration of antigenic variation among rabies virus isolates by using monoclonal antibodies to nucleocapsid proteins, J.Clin.Microbiol., 24:573.

Smith, J. S., Summer, J. W., and Roumillat, L. F., 1984, Enzyme immunoassay for rabies antibody in hybridoma culture fluids and its application to differentiation of street and laboratory strains of rabies virus, J.Clin.Microbiol., 19:267.

Stein, K. E., 1985, Network regulation of the immune response to bacterial polysaccharide antigens in: "Current Topics in Microbiology and Immunology," Vol. 119, H. Koprowski and F. Melchers, eds., Springer-Verlag, Berlin-Heidelberg.

Sureau, P., Rollin, P., and Wiktor, T. J., 1983, Epidemiologic analysis of antigenic variations of street virus: detection by monoclonal antibodies, Am.J.Epidemiol., 117:605.

Tignor, G. H., and Shope, R. E., 1972, Vaccination and challenge of mice with viruses of the rabies serogroup, J.Inf.Dis., 125:322.

Tignor, G. H., Murphy, F. A., Clark, H. F., Shope, R. E., Madore, P., Bauer, S. P. Buckley, S. M., and Meredith, C. D., 1977, Duvenhage virus: morphological, biochemical, histopathological and antigenic relationships to the rabies serogroup, J.Gen.Virol., 37:595.

Urbain, J., Francotte, M., Frenssen, J. D., Hairnaux, J., Leo, O., Moser, M., Slaoni, M., Urbain-Vansanten, G., Van Acker, A., and Wikler, M., 1983, From clonal selection to immune networks: induction of silent idiotypes, Ann.N.Y.Acad.Sci., 418:1.

Webster, W. A., Casey, G. A., and Charlton, K. M., 1986, Major antigenic groups of rabies virus in Canada determined by anti-nucleocapsid monoclonal antibodies, Comp.Immun.Microbiol.Infect.Dis., 9:59.

Webster, W. A., Casey, G. A., Charlton, K. M., and Wiktor, T. J., 1985, Antigenic variants of rabies virus isolates from eastern, central, and northern Canada, Can.J.Comp.Med., 45:186.

Wiktor, T. J., Flamand, A., and Koprowski, H., 1980, Use of monoclonal antibodies in diagnosis of rabies virus infection and differentiation of rabies and rabies-related viruses, J.Virol.Methods, 1:33.

Wiktor, T. J., and Koprowski, H., 1978, Monoclonal antibodies against rabies virus produced by somatic cell hybridization: detection of antigenic variants, Proc.Natl.Acad.Sci.USA., 75:3938.

Wiktor, T. J., and Koprowski, H., 1980, antigenic variations of rabies virus, J.Exp.Med., 152:99.

Wiktor, T. J., Macfarlan, R. I., Foggin, C. M., and Koprowski, H., 1984, Antigenic analysis of rabies and Mokola virus from Zimbabwe using monoclonal antibodies, Devel.Biol.Standard., 57:199.

Wunner, W. H., and Dietzschold, B., 1987, Rabies virus infection: Genetic mutations and the impact on viral pathogenicity and immunity in: "Contributions to Microbiology and Immunology," J. M. Cruse and R.E. Lewis, eds., S. Karger, Basel.

Wunner, W. H., Dietzschold, B., Macfarlan, R. I., Smith, C. L., Golub, E., and Wiktor, T. J., 1985, Localization of immunogenic domains of the rabies virus glycoprotein, Annls. Virol.Inst.Pasteur, Paris 136E:353.

Wunner, W. H., Dietzschold, B., Curtis, P., and Wiktor, T. J., 1983, Rabies subunit vaccines, Review article, J.Gen.Virol., 64:1649.

Wunner, W. H., Smith, C. L., Lafon, M., Ideler, J., and Wiktor, T. J., 1984, Comparative nucleotide sequence analysis of the glycoprotein gene of antigenically altered rabies viruses in: "Nonsegmented Negative Strand Viruses," D.H.L.Bishop, R. W. Compans, eds., Academic Press, San Diego.

THE DIAGNOSIS, PREVENTION AND TREATMENT OF HUMAN CYTOMEGALOVIRUS INFECTIONS USING HUMAN AND MURINE MONOCLONAL ANTIBODIES

David Emanuel, Joanne Peppard, Jihed Chehimi,
Ulrich Hammerling and Richard O'Reilly

Bone Marrow Transplant Service, Department of Pediatrics
Memorial Hospital and the Sloan-Kettering Institute
Memorial Sloan-Kettering Cancer Center
New York City, New York

INTRODUCTION

These investigations were supported by PHS Grant Number CA-23766 and CA-00984, awarded by the National Cancer Institute, DHHS. Dr. Emanuel is the recipient of Clinical Investigator Award K08-CA00984 from the National Cancer Institute.

Human Cytomegalovirus (CMV), the largest of the human Herpes viruses, is an important human pathogen producing a wide spectrum of clinical syndromes in immunocompromised patients (Weller, 1981). The virus is a major cause of morbidity and is the commonest infectious cause of death in patients receiving allogeneic bone marrow transplants (Meyers et al, 1986).

IMMUNOPROPHYLAXIS OF CYTOMEGALOVIRUS INFECTIONS

Approaches to the prevention of Cytomegalovirus infections in recipients of bone marrow and renal transplants have included the use of CMV seronegative blood products and passive immunoprophylaxis with CMV immune, plasma or globulins (Winston et al, 1982; Meyers et al, 1983; Condie and O'Reilly 1984; Bowden et al, 1986; Snydman et al, 1984; Winston et al, 1987). Most of these studies have reported that passive immunization with intravenous immune globulin can modify the severity of CMV infection and prevent interstitial pneumonia after bone marrow transplantation. However the assumption that passive immunoprophylaxis and the use of screened blood products are indicated for the majority of transplant patients remains to be definitively determined by carefully controlled prospective clinical trials. The role of passive immunoprophylaxis for CMV seropositive marrow transplant recipients or for the seronegative recipient receiving a seropositive marrow is being investigated in a number of ongoing clinical studies. Thus the effectiveness in general of passive immunoprophylaxis for CMV infections remains controversial.

Comparison of data collected in previous CMV immunoprophylaxis studies is difficult as there are many variables in these studies. Primarily the quantity and quality of products administered has been markedly different. This was because the availability of immune globulin preparations containing high titers of CMV-specific antibodies was limited. Also the criteria for selection of patients for these studies has varied. A number of studies have only enrolled CMV-seronegative recipients on the grounds that (i) CMV seropositive patients have a significantly lower risk of interstitial pneumonia, or (ii) CMV infection is due to reactivation of latent virus in seropositive recipients and hence it is unlikely that passive immunoprophylaxis would be worthwhile in these patients (Winston et al, 1982; Meyers et al, 1983; Bowden et al, 1986; and Winston et al, 1987).

In addition virological surveillance and diagnostic techniques in these studies have been markedly variable, again making comparisons difficult (Winston et al, 1982; Condie and O'Reilly, 1984; and Winston et al, 1987).

Recently, however a number of new intravenous immune globulin products have become available for use in the USA and Europe. By comparing the in vitro characteristics of specific lots of 2 immune globulins that had been used successfully in previous clinical trials, with similar data from randomly selected lots of three commercially available immune globulins, we hoped to define some baseline criteria for the selection of an immune globulin for use in a large prospective CMV immunoprophylaxis study planned for our patients. Each lot of product was assayed for CMV-specific immunoreactive antibody by ELISA and for the titer of CMV neutralizing antibodies as determined by a standard virus plaque reduction assay (Middeldorp et al, 1984; Wentworth and French, 1970). As we proceeded with the evaluation we unexpectedly noted no direct or predictable correlation between the amount of CMV-IgG by ELISA (Immunoreactive antibody), and the titer of neutralizing antibodies (Functional antibody). In addition there were notable differences between individual lots of the same product (Tables 1 and 2). These data suggest that beside there being no direct relationship between the quantity of immunoreactive antibodies and functional antibodies in any of the products tested, it is in addition incorrect to assume that these immune globulins are standardized products. This may be of relevance to the interpretation of previous clinical data and to the design of future CMV immunoprophylaxis studies.

If the assumption is made that passive immunoprophylaxis can prevent or reduce the severity of CMV-related disease in some allogeneic transplant recipients, the nature of the clinically protective antibodies remains unknown. Analyses both in vitro and in vivo of the CMV antibody profiles of different lots of immune globulins currently being used in clinical trial at Memorial Hospital, are currently being conducted and correlations made with clinical outcome. These studies may further define what particular antibodies may be of clinical relevance in these products.

We have recently developed a panel of CMV-specific human monoclonal antibodies with in vitro neutralizing activity that are shortly to be introduced in Phase 1 trial in recipients of marrow transplants at Memorial Hospital (Tables 3 and 4). The availability of highly specific, functionally active human monoclonal antibodies to CMV may obviate some of the inherent problems of using immune globulins for CMV immunoprophylaxis, e.g. supply, standardization, cost, specificity. We have selected 2 human monoclonals with broad cross-reactivity to both laboratory and clinical strains of CMV, both by immunofluorescence and virus neutralization, for use in the initial Phase 1 studies. It is likely that more than a single monoclonal will be required to cover the marked genetic heterogeneity of human CMV, as we have found that an individual antibody will not recognize or neutralize up to 20% of clinical isolates. We anticipate that a panel of 3 or more antibodies will be required to both, a) interact with CMV at different stages of the infection cycle and, b) to recognize as wide a variety of CMV strains as possible.

Table 1. HCMV Antibody Reactivity of Different Intravenous Human Immune Globulins

Immune Globulin	Elisa Titer[1]	Neutralization Titer[1]	µg/ml of IgG required for 50% virus plaque reduction
Gamimune N[(2)]	1:3,200	1:500	100
Gammagard[(3)]	1:6,400	1:1,000	50
Minnesota[(4)]	1:25,600	1:1,000	50
Sandoglobulin[(5)]	1:1,600	1:50	600
Biotest	1:25,600	1:1,000	110

(1) Results represent the average of 3 different experiments and a median of lots tested.

Table 2. Relationship Between the CMV-Specific Immunoreactive and Functional Activity of Different Lots of 5 Intravenous Immune Globulins, as Measured by ELISA and Virus Neutralization

		TITER OF CMV-IgG BY ELISA					
		800	1600	3200	6400	12800	25600
	0		S,S				
	50		S				
TITER OF CMV	100				GN		
NEUTRALIZING	200	S	S	GD,GD GD			
ANTIBODIES	500			GN,GN			
	1000			GD		GD,GD GD,GD	M,M, B
	2000			GN		GD,GD GD	
	4000			GD			

S=Sandoglobulin
GN=Gamimune-N
GD=Gammagard
M=Minnesota Hyperimmune Globulin
B=Biotest Hyperimmune Globulin

Table 3. Immunofluorescence Pattern and Polypeptides Recognized by Four Human Monoclonal Antibodies to Cytomegalovirus

Monoclonal Designation	Molecular Weight of Polypeptide Recognized	Immunofluorescence Pattern
312 A	75K	Nuclear
D7.6.20.28	56K Major 110K Minor	Nuclear and Cytoplasmic
D7.6.2.96	56K Major 110K Minor	Nuclear and Cytoplasmic
48.23.36	58K	Cytoplasmic

Table 4. Characteristics of Four Human Monoclonal Antibodies to CMV

HYBRID CELL LINE	IMMUNOGLOBULIN SUBCLASS	ELISA TITER (1 mg/ml purified antibody)	NEUTRALIZATION (Lab and clinical strains)	TYPE OF CMV PROTEIN DETECTED	REACTIVITY WITH MEMBRANE ANTIGEN
D7.6.20.28	IgG_2	$1:10^6$	YES	LATE	YES
D7.6.2.96	IgG_2	$1:10^5$	YES	LATE	YES
48.23.36	IgM	$1:10^6$	YES	LATE	YES
312 A	IgG_1	$1:10^4$	NO	LATE	NO

IMMUNODIAGNOSIS OF CMV INFECTIONS

Approximately 50 proteins specified by the viral genome are expressed in the CMV infected cell (Stinski 1984). After infection of the permissive cell, CMV polypeptides are expressed in a sequential cascade and according to their time of maximal synthesis are classified as immediate-early, early or late proteins (Stinski, 1984). The immediate early proteins are the first to appear and are detectable within 1-2 hours of infection. The early proteins require the de-novo synthesis of immediate-early proteins and begin to appear within 3-6 hours of infection prior to the synthesis of the viral DNA. The late proteins are produced after viral DNA synthesis begins and are mainly viral structural proteins, reaching maximal levels of expression 3-6 days after infection of the cell. Since 1982, a series of both human and murine monoclonal antibodies have been produced, directed to a wide spectrum of CMV-specific antigens (Pereira et al, 1982; Goldstein et al, 1982; Rogers and Mundin, 1985; Emanuel et al, 1984), which have facilitated the characterization of the CMV glycoproteins and those epitopes mediating virus neutralizations (Rasmussan et al, 1984; Rasmussen et al, 1985; Britt, 1984).

Some of the antibodies described above have been utilized for the rapid detection of CMV and the diagnosis of CMV infections in immunocompromised patients (Swenson and Kaplan 1985; Griffiths et al, 1984; Hackman et al, 1985), including a pool of murine and human monoclonal antibodies developed in our laboratory and used for the rapid immunodiagnosis of CMV pneumonia in immunocompromised patients (Emanuel et al, 1986). (Table 5). These antibodies have facilitated the early diagnosis and thereby the successful treatment of CMV infections, primarily CMV pneumonia. With the advent of new and potentially useful anti-viral therapies, early diagnosis has become more important, and in early studies, appears to be critical to the successful management of CMV-related disease.

Immunodetection of CMV Using Monoclonal Antibodies

We recently described the use of CMV-specific monoclonal antibodies for the rapid immunodiagnosis of CMV pneumonia by the detection of CMV-specific antigens in alveolar macrophages (Emuanuel et al, 1986) using the scheme outlined in Fig. 1. In this study we showed that the presence of CMV-specific proteins in the alveolar macrophages from patients with interstitial pneumonia always correlated with the histopathological diagnosis of CMV pneumonia and provided a rapid, highly sensitive and specific means of making a diagnosis of CMV pneumonia. Since this study was published, we have extended our investigations to include another 89 patients with interstitial pneumonia of unknown etiology. CMV-specific antigens were detected in 31 of these patients (35%). We detected CMV in all samples where it was detected by any other method (cytology, culture) or procedure (open lung biopsy), either at the time of lavage or in the subsequent 2 months, i.e., it had a sensitivity of 100%.

In addition to indirect immunofluorescent staining of alveolar macrophages, we also co-cultured lavage fluid with human embryonic fibroblasts (MRC-5) for 48-72 hours and then performed indirect immunofluorescent staining of these fibroblasts with a pool of 3 antibodies directed to CMV early antigens (Table 5). We have performed these procedures in parallel for the past year and have noted the following:

1. Where the staining of alveolar macrophages is positive for CMV, the co-cultured fibroblasts are always positive for CMV early proteins by indirect immunofluorescence using the pool of antibodies to early CMV proteins ("Early Pool").

Table 5. CMV Monoclonal Antibodies

Antibody	Class	Fluorescence Pattern	Antigen Detected
154-1-14-1	G_2bk	Cytoplasmic	Late
154-10-20-3	G_2bk	Cytoplasmic	Late
154-20-2-7	G_2bk	Nuclear+Cytoplasmic	Late
158-2-19-2	G_1k	Nuclear+Cytoplasmic	Early
179-17-2-1	G_2bk	Nuclear+Cytoplasmic	Early
205-39-17-1	G_1k	Nuclear+Cytoplasmic	Early
312-A-91-4 (human)	G_2bk	Nuclear	Late

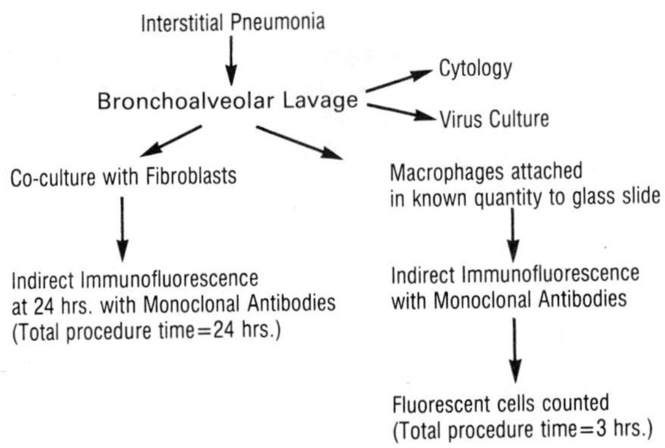

Fig. 1. Immunodetection of cytomegalovirus in patients with interstitial pneumonia.

2. We have had 10 cases in the past 12 months where CMV early antigens were detected in fibroblasts co-cultured with lavage material, but where the alveolar macrophages were negative for CMV proteins.

In 9 of these cases (90%), CMV was subsequently cultured from lavage fluid after a period of 2-8 weeks. All these 10 patients either improved or recovered from their interstitial pneumonias with specific "non-CMV" related treatment, e.g., with therapy for pneumocystis pneumonia or fungal pneumonia. In those patients who had autopsies performed (4 patients) no evidence of CMV pneumonia was detected by histopathology or immunopathological staining. However, where CMV was detected in alveolar macrophages, six of these patients died rapidly of CMV pneumonia proven at autopsy.

We are thus postulating the following:

a) Infection of the alveolar macrophage is possibly a prerequisite for development of overt CMV pneumonia in the severely immunocompromised patient. Where the alveolar macrophage can be shown to be infected with CMV by monoclonal antibodies, CMV is probably the primary etiology of the pneumonia.

b) The mere presence of CMV in alveolar lavage fluid obtained from the severely immunocompromised patient with interstitial pneumonia is not in itself enough evidence for CMV to be implicated as the primary etiology.

Recently we have been staining alveolar macrophages with separate pools of CMV-specific early and late antibodies. Once we had established that a mixed pool of early and late antibodies was a useful means of making a diagnosis of CMV pneumonia, we then investigated the type of CMV antigen expressed in the alveolar macrophage in patients with CMV pneumonia, i.e., CMV early or late proteins. This was done to prospectively evaluate

i whether the alveolar macrophage is in fact "permissive" for CMV, i.e., whether the virus can reproduce in the cell, and

ii the relevance of early versus late CMV proteins for the diagnosis of CMV pneumonia.

We have prospectively processed 48 bronchoalveolar lavage specimens from 44 patients in a 6 month period. 33 of these patients were recipients of allogeneic bone marrow transplants and 11 patients had AIDS. All patients had clinical and radiological evidence of interstitial pneumonia of unknown etiology. To date we have only detected CMV early proteins in alveolar macrophages in marrow transplant recipients with CMV pneumonitis. CMV late proteins have been detected in only 2 patients with AIDS with CMV pneumonitis. This unexpected finding may mean:

a) the alveolar macrophage is not permissive for human CMV, except in the patients with AIDS.

b) the alveolar macrophage in the patients with AIDS may in an undetermined way, be dysfunctional, thereby allowing for the replication of CMV therein.

These preliminary data need to be confirmed by continued careful prospective analysis using both immunological and molecular biological methodologies. Although it has long been assumed that the primary target cell for CMV in the lung is the alveolar macrophage (Craighead, 1971) and that this cell is intrinsically permissive of CMV replication (Drew et al, 1979), these assumptions may be incorrect. Little is known about the basic mechanisms of lung injury in CMV pneumonitis. Recent studies indicating that there was still a 90% mortality rate in patients with CMV pneumonitis treated with DHPG, despite evidence of a consistent 99.9% reduction in virus titer in the lungs of these patients, suggests that there are probably other mediating factors in the pathogenesis of the disease (Shepp et al, 1985).

TREATMENT OF CMV PNEUMONIA

Although there has been some encouraging progress in the treatment of CMV retinitis and colitis in the immunocompromised patient using Ganciclovir (DHPG), attempts to treat CMV pneumonia has not been success-ful, using Ganciclovir or a variety of other anti-viral and immunomodulatory agents alone or in combination (Shepp et al, and Felsenstein et al, 1985). Despite evidence of excellent in vivo anti-viral activity in 10 patients with CMV pneumonia treated with DHPG in Seattle, 9 of the 10 patients nevertheless still died of the disease (Shepp et al, 1985).

Recently, the Bone Marrow Transplant Unit at the Medical College of Wisconsin reported a 40% survival rate (4 of 10 patients) in patients with CMV pneumonia treated with DHPG combined with intravenous immune globulin (Bratanow et al, 1986). This is markedly better than the historical norm of a 90% mortality rate in this group of patients.

We have now treated 10 patients with CMV pneumonia following allogeneic bone marrow transplantation with a combination of DHPG and high dose intravenous immune globulin, from January - May 1987. The diagnosis of CMV pneumonia was made by bronchoalveolar lavage in 8 patients and by open lung biopsy in 2 patients. The virus was isolated by culture in all 10 patients. The diagnosis was made by detection of CMV-specific antigens in tissue (alveolar macrophages, lung) using CMV-specific monoclonal antibodies as described. CMV intra-nuclear inclusions were noted in 2 of the 10 patients (20%). Treatment consisted of DHPG at a dose of 2.5 mg/kg/dose three times a day for 14-21 days. Combined with intravenous immune globulin (Gammagard, Hyland-Travenol) at a dose of 500 mg/kg/dose daily for 14-21 days. Patients were then treated with a "maintenance schedule" of DHPG 5 mg/kg/day five times a week for 4 weeks, combined with Gammagard 500 mg/kg/dose twice a week for 4 weeks.

Eight of the 10 patients (80%) are currently alive without recurrence of disease, either off therapy or in the latter stages of maintenance treatment. There were two deaths in this group of patients. One patient died of Gram negative sepsis with respiratory distress syndrome following an initial successful course of treatment for CMV pneumonia. No autopsy was obtained so it remains unclear as to whether there was evidence of recurrence of CMV disease. The second patient died of leukemic relapse 5 months after treatment without clinical or radiological evidence of CMV pneumonia.

Although this pilot study is far from complete and these data are preliminary, we are encouraged by this dramatic change in the outcome for our patients with CMV pneumonia, treated with this combination of drugs. A careful prospective analysis of clinical, radiological, pulmonary function, immunological and virological studies performed on these patients is currently in progress. We thus report these very preliminary data in the hope that other bone marrow transplant groups will investigate this approach to the treatment of CMV pneumonia. We postulate that the improved outcome for our patients with CMV pneumonia using DHPG combined with immune globulin may be due to the following factors:

1. Early diagnosis by CMV immunodetection using monoclonal antibodies. We now perform bronchoalveolar lavage at the earliest possible time following a clinical and radiological diagnosis of interstitial pneumonia. Using the immunodetection methods outlined we are able to institute treatment at an earlier stage of disease than previously possible.

2. A synergistic effect of anti-viral agent (DHPG) with immune globulin possibly mediated through the augmentation of intrinsic cellular immune mechanisms of the host, i.e., antibody-dependent cellular cytotoxicity (ADCC) or through other undefined immunomodulatory effects of the immune globulin. These areas are currently under investigation in vitro and in vivo.

Three of the 4 human monoclonal antibodies to CMV previously described, (Table 3), appear to mediate ADCC and may also be useful as adjuncts to the treatment of CMV pneumonia. This will be investigated in future prospective clinical trials.

REFERENCES

Bowden, R. A., Sayers, M., Fluornoy, N., et al, 1986, Cytomegalovirus immune globulin and seronegative blood products to prevent primary cytomegalovirus infection after marrow transplantation, N.Engl.J.Med., 314:1006.

Britt, W., 1984, Neutralizing antibodies detect disulphide linked glycoprotein complex within the envelope of human cytomegalovirus, J.Virology, 135:369.

Bratanow, N., Ash, Turner, P., Smith R., Chitambar C., Hansen, R., Casper, J., 1986, Successful treatment of serious cytomegalovirus disease with 9-[1,3-Dihydroxy-2-propoxymethyl]guanine in Bone Marrow Transplant Patients, Blood, 68:280A.

Condie, R. N., and O'Reilly, R. J., 1984, Prevention of cytomegalovirus infection by prophylaxis with an intravenous hyperimmune native unmodified cytomegalovirus globulin: Randomized trial in bone marrow transplant recipients, Am.J.Med., 76:(3A)134.

Craighead, J. E., 1971, Pulmonary cytomegalovirus infections in the adult, Am,J.Path., 63:487.

Drew, L., Mintz, L., Hoo, R., Finley T., 1979, Growth of Herpes simplex and

cytomegalovirus in cultured human alveolar macrophages, Am.Rev.Resp. Dis., 119:287.
Emanuel, D., Gold, J., Colacino, J., Lopez C., Hammerling U. A ., 1984, A human monoclonal antibody to cytomegalovirus, J.Immun., 133:2202.
Emanuel, D., Peppard, J., Stover, D., Gold, J., Armstrong, D., Hammerling, U., 1986, Rapid immunodiagnosis of cytomegalovirus pneumonia by bronchoalveolar lavage using human and murine monoclonal antibodies, Ann.Int.Med., 104:476.
Felsentein, D., D'Amico, D., Hirsch, M., Neumeyer, D., Cederberg, D., Miranda, P., Schooley, R., 1985, Treatment of cytomegalovirus retinitis with 9-[2-hydroxy-1-(hydroxymethyl)ethoxymethyl]guanine, Ann.Int.Med., 103:377.
Goldstein, L., McDougall, J., Hackman, R., Meyers, J., Thomas, E. D., Nowinski, R., 1982, Monoclonal antibodies to cytomegalovirus. Rapid identification of clinical isolates and preliminary use in diagnosis of cytomegalovirus pneumonia, Infect.Immun., 38:273.
Griffiths, P., Panjwani, D., Stirk, R., Ball, M., Ganczarowski, M., Blacklock, H., Prentice, H. G., 1984, Rapid diagnosis of cytomegalovirus infection in immunocompromised patients by detection of early antigen fluorescent foci, Lancet, 1:1242.
Hackman, R., Myerson, D., Meyers, J., Shulman, H., Sale, G., Goldstein, L., Rastetter, M., Fluornoy, N., Thomas E. D., 1985, Rapid diagnosis of cytomegalovirus pneumonia by tissue immunofluorescence with a murine monoclonal antibody, J.Inf.Dis., 151:325.
Middeldorp, J., Jongsma, J., Ter Haar A., Schirai, J., 1984, The detection of immunoglobulin M and G Antibodies against cytomegalovirus early and late antigens by enzyme linked immunosorbent assay, J.Clin.Micro., 20:763.
Meyers, J. D., Fluornoy, N., Thomas, E. D., 1986, Risk factors for cytomegalovirus infection after human marrow transplantation, J.Inf.Dis., 153:478.
Meyers, J. D., Lesczynski, J., Zaia, J. A., et al., 1983, Prevention of cytomegalovirus infection by cytomegalovirus immune globulin after marrow transplantation, Ann.Int.Med., 98:442.
Pereira, L., Hoffman, M., Gallo D., 1982, Monoclonal antibodies to human cytomegalovirus: Three surface membrane proteins with unique immunological and electrophoretic properties specify cross-reactive determinants, Infect.Immun., 36:924.
Rasmussen, L., Nelson, R. M., Kelsall, D. C., Merigan, T., 1984, Murine monoclonal antibodies to a single protein neutralizes the infectivity of human cytomegalovirus, Proc.Nat.Acad.Sci. USA., 81:876.
Rasmussen, L., Mullenax, J., Nelson, R., Merigan, T., 1985, Viral polypeptides detected by complement dependent neutralizing murine monoclonal antibodies to human cytomegalovirus, J.Virol., 55:274.
Rogers, B. C., Mundin, J., Sissons, J., 1985, Monoclonal antibodies recognizing early and late antigens of human cytomegalovirus: heterogeneity of polypeptides recognized between virus isolates, J.Gen.Virol., 66:2045.
Shepp, D. H., Dandliker, P., de Miranda, P., Burnette, T., Cederberg, D., Kirk, L., Meyers, J., 1985, Activity of 9-[2-hydroxy-1-(hydroxymethyl) ethoxymethyl]guanine in the Treatment of cytomegalovirus pneumonia, Ann.Int.Med., 103:368.
Snydman, D. R., McIver, J., Leszczynski, J., et al., 1984, A pilot trial of a novel CMV immune globulin in renal transplant recipients, Transplantation, Nov. 38(5):553.
Stinski, M., 1984, The proteins of human cytomegalovirus, in: "CMV; Pathogenesis and prevention of human infection, p. 49-62, Alan Liss, New York.
Swenson, P. D., Kaplan, M.H., 1985, Rapid detection of cytomegalovirus in cell culture by indirect immunoperoxidase staining with monoclonal antibody to an early nuclear antigen, J.Clin.Micro., 21:669.

Weller, T. H., 1981, Clinical spectrum of cytomegalovirus infections, *in*: "The Human Herpesviruses," p. 20-30, A.J. Nahmias, W.R. Dowalle, R. Schinaze, eds., Elsevier, New York.

Wentworth, B., and French, L., 1970, Plaque assay of cytomegalovirus strain of human origin, *Proc.Soc.Exp.Biol.Med.*, 135:253.

Winston, D. J., Pollard, R. B., Ho, W. G., et al, 1982, cytomegalovirus immune plasma in bone marrow transplant recipients, *Ann.Int.Med.*, 97:11.

Winston, D. J., Ho, W., Lin, C. H., et al, 1987, Intravenous immune globulin for the prevention of cytomegalovirus infection and interstitial pneumonia after marrow transplantation, *Ann.Int.Med.*, 106:12.

GENERATION AND APPLICATIONS OF HUMAN MONOCLONAL ANTIBODIES TO HERPES VIRUSES

P.A. Bradshaw*, S. Perkins, E.T. Lennette**
J. Rowe and S.K.H. Foung

Department of Pathology, Stanford University Medical Center
Stanford, California 94305, USA
*Stanford Medical School Blood Center, Stanford University
Blood Bank, 800 Welch Road, Palo Alto, California 94304, USA
**Virolab, Incorporated, 1204 Tenth Street, Berkeley
California, 94710, USA

INTRODUCTION

Human cytomegalovirus (CMV) and varicella zoster virus (VZV) are two herpes viruses associated with severe infections in immunosuppressed or immunodeficient patients. For example, bone marrow transplant recipients are especially susceptible to CMV infections and juvenile leukemia patients are vulnerable to VZV infections (Gershon, 1980; Meyers et al, 1986). Three approaches are used to control these infections: antiviral drugs, active immunization with attenuated viruses, and passive immunization with hyperimmune immunoglobulin preparations. Two nucleoside analogs, 9-(1,3-dihydroxy-2-propoxymethyl)guanine (DHPG) and acyclovir (ACV), have recently been studied for treatment of CMV infections. Although DHPG (Koretz, et al, 1986) and, in some cases, ACV (Balfour, et al, 1982), have some efficacy in treating CMV-infected patients, their use is limited by toxic effects and neither is effective against CMV pneumonia (Wade, et al, 1982; Myers, 1985). Antiviral drug therapy against severe VZV infections, however, is more promising. Both ACV and vidarabine, for example, are effective in reducing the severity of VZV infections in immunosuppressed patients (Vildé, et al, 1986).

Immunization with attenuated virus is the second approach to control primary VZV and CMV infections (Gershon, 1985; Plotkin, 1985; Takahashi, et al, 1986). For VZV, immunization with attenuated varicella vaccine provides protection in the majority of cases and when subsequent infections do occur, the infections are subclinical or mild. However, in some studies, immunity appears not to persist lifelong in immunocompromised patients, necessitating repeat injections of virus. In contrast, trials with CMV vaccine show that vaccination is not effective in preventing disease but there may be moderate reduction in the occurrence of severe CMV disease. The effectiveness of immunization in patients, particularly when immunosuppressed, is variable. Of concern are the long term effects of vaccination, i.e. viral latency and reinfection, and viral oncogenicity, especially with CMV.

Passive immunization by prophylactic administration of intravenous hyperimmune CMV globulin (Condie & O'Reilly, 1984) or zoster immune globulin (Paryani, et al, 1984) has been effective in reducing the morbidity and mortality of primary CMV or VZV infections. Immunization with varicella zoster immune globulin within 48-96 hours after exposure reduces both the incidence and severity of disease in susceptible, immunocompromised patients (Centers for Disease Control, 1984). For both diseases, immune globulin must be administered before the infection develops. However, supplies of both hyperimmune globulin preparations are limited and expensive. Human monoclonal antibodies (HMabs) against herpes viruses are potentially useful in the control of such infections among susceptible individuals. Specific HMabs produced in quantity may alleviate shortages of hyperimmune globulins. Furthermore, their use would avoid the potential transmission of other infectious agents through preparations from blood products. Since few patients experience only mild adverse reactions to hyperimmune globulin treatment, it is anticipated that the use of HMabs should be well tolerated. We have developed a strategy to produce stable antigen-specific HMabs with the ultimate goal of replacing hyperimmune globulin with HMabs. Eventually these HMabs may be used to develop subunit vaccines for CMV and VZV.

PRODUCTION OF HUMAN MONOCLONAL ANTIBODIES WITH THE SBC-H20 CELL LINE

Two approaches have been sued to produce HMabs: transformation of human B cells with Epstein-Barr virus (EBV) or fusion of human B cells with mouse or human myeloma and human lymphoblastoid cell lines (Steinitz et al, 1977; Croce et al, 1980; Nowinski et al, 1980; Olsson and Kaplan, 1980). While a number of antigen-specific HMabs have been produced by transforming B-lymphocytes with EBV, a common problem with this approach has been the stability of immunoglobulin (Ig) secretion. Frequently, the desired EBV clones, isolated with great effort, will stop secreting Ig within two to four months. Analogous to the generation of murine Mabs, considerable effort has been devoted by many laboratories to derive suitable fusion partners to generate human hybridomas. Three types of cell lines have been explored extensively: mouse myelomas, EBV lymphoblastoid cell lines, and human myelomas. While the fusion efficiency between human lymphocytes and murine myelomas is high, the hybridomas that are generated are usually unstable. A preferential loss of human chromosomes leads to a rapid loss of Ig secretion. Usually, poor fusion efficiency limits hybridoma production between normal human B-lymphocytes and human B cell lines.

To overcome these problems, we have developed a fusion partner, SBC-H20, which fuses efficiently to stimulated normal human B-lymphocytes and produces stable human immunoglobulin secreting hybridomas. The SBC-H20 cell line was derived from a fusion between a mouse myeloma line, SP2/08AZ, and human peripheral B-lymphocytes isolated from a normal donor. After the establishment of 6-thioguanine resistance, a non-Ig-secreting subclone was isolated and designated SBC-H20. SBC-H20 is sensitive to medium containing hypoxanthine-aminopterin-thymidine (HAT) or azaserine-hypoxanthine (AH) (Foung et al, 1982). Furthermore, SBC-H20 is relatively ouabain resistant and will grow at 1 μM ouabain. SBC-H20 expresses human MHC class I antigens as defined by indirect immunofluorescence binding of W6/32, a murine Mab to an HLA class I framework determinant (Brodsky et al, 1979). The SBC-H20 cell line has been successfully used for producing human hybridomas secreting antigen-specific HMabs since 1983.

Our strategy to produce antigen-specific HMabs involves either fusion of in vivo stimulated B cells directly to SBC-H20 or an initial expansion of antigen-specific B cells with Epstein-Barr virus followed by fusion to SBC-H20 (Foung et al, 1984, 1985a, 1985b; Foung and Lennette, 1986).

Lymphocytes from spleen or peripheral blood are isolated by Ficoll-Hypaque density gradient centrifugation and B-lymphocytes are separated from T-lymphocytes by a standard rosetting technique with sheep erythrocytes. To expand the enriched B-lymphocyte population, culture supernatant from the marmoset cell line, B95-8 (Sly, et al, 1976), is used to activate cells. Transformed cells are plated in microtiter wells and cultured for 2 weeks. Proliferating cells are identified by light microscopy and supernatants from those cells are collected and assayed for the production of relevant antibodies by enzyme-linked immunoassay. Cells selected by this procedure are expanded, or cloned and expanded, prior to fusion with SBC-H20 cells. EBV-transformed cells are relatively ouabain-sensitive and generally will not survive at 1-2 x 10^{-7}M ouabain.

Fusion between selected lymphocytes and SBC-H20 cells is accomplished with either polyethylene glycol (PEG) or electrofusion, as indicated in the pertinent section. After identification of the antibody-producing hybridomas, those cells are cloned by soft agar or limiting dilution to ensure monoclonality. A subclone is selected from each parent hybridoma for antibody production and characterization. The specific reactivities of the HMabs are defined in comparison to positive and negative sera or HMab, as well as isotype matched irrelevant HMab control.

ANTIBODIES TO VARICELLA ZOSTER VIRUS

Adults with acute varicella infections have elevated levels of specific antibody producing B-lymphocytes. Sufficient cell numbers can be obtained at appropriate times to generate productive hybridomas by PEG fusion. We have produced two IgG_1 HMabs to VZV (Foung et al, 1985a, 1985b) from B-lymphocytes obtained six days after the onset of acute varicella in an otherwise healthy individual. Equal cell numbers of B-lymphocytes from the donor and SBC-H20 fusion partners were fused using PEG as described in detail by Foung, et al, (1985b). Proliferation of hybridomas was observed in 37 of 58 seeded wells after three weeks from the onset of fusion. Supernatants from wells containing hybridomas were screened for VZV-specific IgG antibodies by a solid phase radioimmunoassay (Arvin and Koropchak, 1980). Approximately 30 percent of the supernatants contained VZV-specific antibodies. Two of these hybridoma populations were selected and cloned for further characterization. Antibody 1-A2 has kappa light chains and 2-1D5 has lambda light chains.

Specific immunofluorescence is detectable at approximately 1 µg immunoglobulin/ml of HMab 1-A2 and HMab 2-1D5 (Foung, et al, 1985a, 1985b). Both antibodies show membrane fluorescence with antibody 1-A2 showing strong cytoplasmic staining as well. Nonrelevant human monoclonal antibodies or human serum without VZV-specific antibody show no binding to VZV-infected fibroblasts. These antibodies are able to fix complement as demonstrated by anti-complement immunofluorescent assays.

The antibodies were tested for viral neutralization against VZV with a standard plaque-reducing assay using human melanoma cells (Grose, et al, 1979). Three VZV isolates and one herpes simplex virus-1 (HSV-1) were used to infect the cells in the assay. Both antibodies neutralize VZV activity at 1-5 µg IgG/ml and fail to neutralize HSV-1 activity at much higher concentrations (Table 1).

The VZV antigens identified by 1-A2 have been determined by a solid phase immunoisolation technique (Tamura, et al, 1984) and by standard Western blot methods. Under reducing conditions antibody 1-A2 identifies three proteins of 97,000, 95,000 and 60,000 Daltons. The 2-1D5 antibody

Table 1. Neutralizing Activity of Human Monoclonal Antibodies to VZV

Antibody	Virus[a]	Inoculum (pfu/ml)	Neutralizing activity[b] of immunoglobulin (μg/ml)
1-A2	VZV-32	196	1.0
	VZV-BB	160	5.0
	VZV-CH	88	4.0
	HSV-1	180	>1,000
2-ID5	VZV-BB	160	5.0
	VZV-CH	88	2.0
	HSV-1	180	>1,000

a VZV–32 is a laboratory strain; VZV–BB and VZV–CH are clinical isolates from patients with varicella; HSV–1 is a laboratory strain of Herpes Simplex 1 (MacIntyre).

b Immunoglobulin-neutralizing activity was calculated as the lowest concentration of antibody that produced 50% plaque reduction compared with the mean number of plaques per well with challenge virus alone.

identifies a 66,000 molecular weight protein under reducing conditions with the solid phase immunoisolation technique (Foung, et al, 1985a).

ANTIBODIES TO CYTOMEGALOVIRUS

In contrast to our experience with VZV, obtaining sufficient numbers of CMV antibody-producing B-lymphocytes for hybridoma production has been difficult. Current methods of expanding the antigen-specific B cells have been unsuccessful for a variety of reasons. To overcome these problems, we have implemented electric-field-induced cell fusion (Zimmermann, 1986) to produce hybridomas. The process is based on the observation that close membrane contact between two cells can be established in an alternating electric field. Application of a second direct current can induce breakdown of the cell membrane, creating a cytoplasmic bridge leading to cell fusion. Utilizing these two properties, electric field-induced cell fusion, or electrofusion, is an alternative approach to PEG-induced fusion in hybridoma formation. The advantages of this process are an increase in fusion efficiency and a concomitant reduction in the number of B cells necessary for adequate fusion frequency.

Preliminary experiments evaluated electrofusion compared with PEG-induced fusion. A relative measure of fusion efficiency, the number of hybridoma-containing wells per 10^5 input B cells, was used to compare the two methods. In comparison with PEG-induced fusion, electrofusion generated more hybridomas. Electrofusion efficiency ranged from 10^{-3} to 10^{-4} while PEG-induced fusion efficiency ranged from 10^{-4} to 10^{-5}. The electrofusion efficiency was highest with the lower numbers of input B cells.

Antigen-specific B cells were obtained from two CMV seropositive individuals:

(i) spleen cells from a patient with Hodgkin's disease who required splenectomy and
(ii) peripheral blood from a normal blood donor.

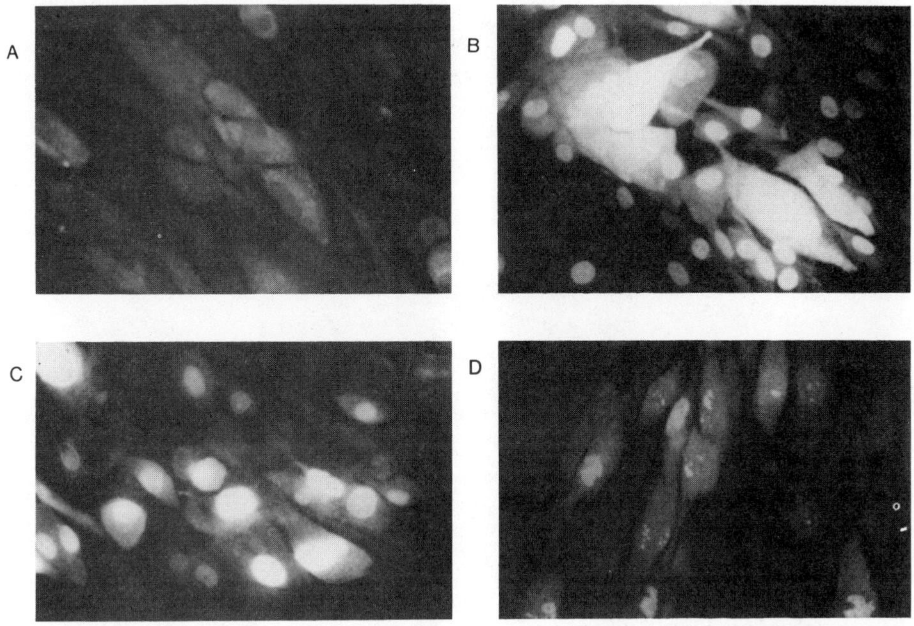

Fig. 1. Indirect immunofluorescence assay of HMab binding to CMV infected fibroblasts. Fibroblasts of the FS-4 cell line were infected with strain AD-169 virus and fixed at 4 days post-infection. The following primary antibodies were used. A. an irrelevant HMab of the IgG_1 class, B. HMab Z01, C. HMab Z02, D. HMab Z10. The secondary antibody was fluorescein-conjugated goat anti-human IgG Fc fragment.

These cells were EBV activated and fused to SBC-H20 by the electrofusion protocol. Multiple wells containing hybrids secreting CMV-specific antibody were produced and three hybridomas were chosen for further investigation. These were designated Z01, Z02, and Z10. The Z01 and Z02 antibodies were derived from splenic B cells. The Z10 antibody was derived from peripheral B cells. All three antibodies are IgG_1 with kappa light chains (Table 2). All three hybridoma cell lines have continued to secrete relevant Mabs for over 12 months.

The anti-CMV specificity of all three antibodies is confirmed by indirect immunofluorescent assays (IFA) (Table 2 and Fig. 1). Antibody binding at 2 days post-infection is low, albeit detectable, and limited to the nucleus for all three antibodies. By 4 days post-infection antibody binding increases to that shown in Figs. 1B, C, D. The fluorescence pattern is distinctive with each antibody. On day 4, nuclear staining by antibody Z01 increases in intensity compared with day 2 post-infection, and cytoplasmic staining, undetectable at 2 days, is intense, especially in the megaloblastic cells. Antibody Z02 nuclear staining is also more intense than at day 2, often with a very bright perinuclear ring. At 4 days post-infection antibody Z02 also shows diffuse cytoplasmic staining. The punctate staining of antibody Z10 increases between 2 and 4 days, but is restricted to irregularly shaped areas within the nucleus. None of the three antibodies react with cells infected with other herpes viruses.

Since a critical aspect of the protective humoral immune response to CMV is the development of neutralizing antibodies, each antibody has been evaluated using a standard plaque reduction assay and a microneutralization

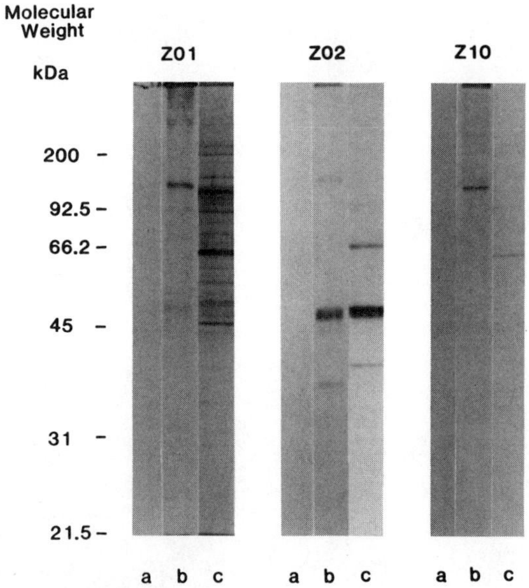

Fig. 2. ^{35}S-methionine labeled CMV proteins immunoprecipitated by human monoclonal antibodies. Lane *a* in each panel shows the result of immunoprecipitation from uninfected labeled FS-4 fibroblast cells. Lanes *b* and *c* show antigens immunoprecipitated from CMV AD-169 infected cells. Lanes *a* and *b* are antigens electrophoresed under nonreducing conditions while lane *c* shows the effect of 1.5% dithiothreitol on the electrophoretic pattern. Antigens were separated on 10% polyacrylamide gels and identified by autoradiography. Results with isotype matched irrelevant HMabs on CMV infected and uninfected cells were negative and identical to lane *a* of each panel.

assay procedure. In the presence of complement, all three HMabs neutralize CMV in both assay systems (Table 2). Using the microneutralization assay and an endpoint of 50% reduction in the number of infected cells compared to virus infected cells without antibody, neutralizing activity is observed at 0.2 μg/ml or less of specific immunoglobulin.

CMV antigens recognized by these HMabs have been determined by immunoprecipitation of ^{35}S-methionine labeled material from infected cells (Fig. 2). Under nonreducing conditions, Z01 identifies a major protein at 115,000 Da. After reduction, multiple polypeptide bands appear with the most intensely labeled proteins migrating at about 100,000, 60,000 and 42,000 Da. Mab Z02 immunoprecipitates a major protein which migrates with an apparent molecular weight of 48,000 Da under nonreducing conditions. Other polypeptides are seen at 37,000 and 130,000 Da. Under reducing conditions, the major protein band and the lower molecular weight protein migrate with apparent molecular weights of 50,000 Da and 38,000 Da. One explanation for this behavior is that the reduction of intrachain disulfide bonds produces a conformational change which alters the migration properties of the proteins (Scheid and Choppin, 1977; Kreis and Lodish, 1986). A protein band appears at 65,000 Da under reducing conditions. Under non reducing conditions, Z10 precipitates a CMV antigen with an apparent molecular weight near 105,000 Da. In the presence of dithiothreitol, the 105,000 Da protein is replaced with a 60,000 Da band. Immunoprecipitations with nonspecific antibodies under the same conditions used for Z01, Z02 and Z10 are negative.

Table 2. Characterization of Human Monoclonal Antibodies to CMV

Antibody	Heavy Chain	Light Chain	Ig[a]	IFA[b]	ACIF[c]	Microneutralization[d] of AD-169 (μg/ml)
Z01	IgG_1	kappa	1	+	+	+ (<0.12)
Z02	IgG_1	kappa	25	+	+	+ (<0.22)
Z10	IgG_1	kappa	4	+	+	+ (<0.20)
Anti-VZV	IgG_1	lambda	10	-	-	-

a Ig - Concentration (μg/ml) of immunoglobulin in spent supernatant.

b IFA - Indirect immunofluorescence on CMV-infected fibroblasts.

c ACIF - Anticomplement indirect immunofluorescence on CMV-infected fibroblasts.

d Microneutralization - The number of infected cells on each cover slip is counted by immunofluorescence microscopy. The highest serum dilution yielding 50% or more infected cell reduction compared to virus control was considered as the neutralizing titer.

APPLICATIONS FOR THESE ANTIBODIES

These antibodies to CMV and VZV are potentially useful in three ways. First, they will have clinical applications as substitutes for hyperimmune sera in passive immunization of patients susceptible to severe CMV or VZV infections. Therapeutically, there are several possible advantages to HMabs. The inherent specificities of the antibodies permit administration of immunoglobulins of known activity which contain minimal extraneous foreign proteins. The supply of these should be essentially unlimited; their quality consistent and predictable. HMabs should circumvent the difficulties observed with therapeutic use of murine monoclonal antibodies. When mouse antibodies are administered to humans, an anti-murine response frequently occurs (Miller, et al, 1983), leading to rapid inactivation of the murine antibodies and to induction of serum sickness. In addition, the carbohydrate portions of immunoglobulins, which are species-specific, are essential for much of the biological activity of antibodies (Nose and Wigzell, 1983). Often, in other antigenic systems, such as the Major Histocompatibility Complex antigens, the majority of murine antibodies are directed toward monomorphic instead of polymorphic determinants. And for some important antigens such as Rh red cell antigens, rodents have failed to mount an antibody response.

Second, these antibodies provide tools for investigating the human immune response to disease. The sites of latency, the molecular events involved in reactivation, as well as the specific proteins that stimulate cell-mediated or humoral immune responses are not well established. The mechanisms whereby antibodies to CMV can modulate infection include neutralization of extracellular virus, complement-mediated lysis of CMV-infected cells, and cell-mediated immunity by antibody-dependent cellular cytotoxicity. The relative importance and the interrelationships among these mechanisms in recovery from primary or reactivation disease are not known. With antibodies to specific viral antigens, we will be able to further our understanding of the immune response to CMV and VZV.

Third, these neutralizing antibodies provide a means to identify viral antigens suitable for use as subunit vaccines which avoid the risks of live vaccines. Murine monoclonal antibodies have been used to localize specific genes on the CMV genome and to clone those gene fragments (Pande et al, 1984; Mocarski et al, 1985). Now neutralizing HMabs can be used in the same way to clone gene fragments for viral antigens required for infection. These can be used to produce polypeptides suitable for subunit vaccines.

Acknowledgement

We thank Judy Halmos for technical assistance and Judy Campbell in the preparation of this manuscript. This work was supported in part by grants AI22557 and HL33811 from the National Institutes of Health.

REFERENCES

Arvin, A. M., and Koropchak, C. M., 1980, Immunoglobulins, M and G to varicella-zoster virus measured by solid-phase radioimmunoassay: Antibody responses to varicella and herpes zoster infections, J.Clin.Microbiol., 12:367.

Balfour, H. H., Bean, B., Mitchell, C. D., Sachs, G. W., Boen, J. R., and Edelman, C. K., 1982, Acyclovir in immunocompromised patients with cytomegalovirus disease: A controlled trial at one institution, Am.J.Med., 73:241.

Brodsky, F. M., Parham, P., Barnstable, C. J., Crumpton, M. J., and Bodmer, W. F., 1979, Monoclonal antibodies for analysis of the HLA system, Immunol.Rev., 47:3.

Centers for Disease Control, 1984, Varicella-zoster immune globulin for the prevention of chicken pox: Recommendations of the immunization practices advisory committee, Ann.Intern.Med., 100:859.

Condie, R. M., and O'Reilly, R. J., 1984, Prevention of cytomegalovirus infection by prophylaxis with an intravenous, hyperimmune, native, unmodified cytomegalovirus globulin, Am.J.Med., 76:134.

Croce, C. M., Linnenbach, A., Hall, W., Steplewski, Z., and Koprowski, H., 1980, Production of human hybridomas secreting antibodies to measles virus, Nature, 288:488.

Foung, S. K. H., and Lennette, E. T., 1986, A strategy to produce human monoclonal antibodies to herpes viruses in: "Human Herpes virus Infections: Pathogenesis, Diagnosis, and Treatment," C. Lopez and B. Roizman, eds, Raven Press, New York.

Foung, S, K. H., Perkins, S., Arvin, A., Lifson, J., Mohagheghpour, N., Fishwild, D., Grumet, F. C., and Engleman, E. G., 1985a, Production of human monoclonal antibodies using a human-mouse fusion partner in: "Human Hybridomas and Monoclonal Antibodies," E.G.Engleman, S.K.Foung, J.W.Larrick, and A. Raubitschek, eds., Plenum Press, New York.

Foung, S. K. H., Perkins, S., Koropchak, C., Fishwild, D., Wittek, A. E., Engleman, E. G., Grumet, F. C., and Arvin, A. M., 1985b, Human monoclonal antibodies neutralizing varicella-zoster virus, J.Infect.Dis., 152:280.

Foung, S. K. H., Perkins, S., Raubitschek, A., Larrick, J., Lizak, G., Fishwild, D., Engleman E. G., and Grumet, F. C., 1984, Rescue of human monoclonal antibody production from an EBV-transformed B cell line by fusion to a human-mouse hybridoma, J.Immunol.Meth., 70:83.

Foung, S. K. H., Sasaki, D., Grumet, F. C., and Engleman, E. G., 1982, Production of functional human T-T hybridomas in selection medium lacking aminopterin and thymidine, Proc.Natl.Acad.Sci., U.S.A., 79:7484.

Gershon, A. A., 1985, Live attenuated varicella vaccine in: "The Herpes viruses, vol. 4," B. Roizman and C. Lopez, eds., Plenum Press, New York.

Gershon, A., 1980, Live attenuated varicella-zoster vaccine, Rev.Infect. Dis., 2:393.

Grose, C., Edmond, B. J., and Brunell, P. A., 1979, Complement enhanced neutralizing antibody response to varicella-zoster virus, J.Infect.Dis., 139:432.

Koretz, S. H., and members of the Collaborative DHPG Study Group, 1986, Treatment of serious cytomegalovirus infections with 9-(1,3-dihydroxy-2-propoxymethyl) guanine in patients with AIDS and other immunodeficiencies, N.Engl.J.Med., 314:801.

Kreis, T. E., and Lodish, H. F., 1986, Oligomerization is essential for transport of vesicular stomatitis viral glycoprotein to the cell surface, Cell, 46:929.

Meyers, J. D., 1985, Cytomegalovirus infection after organ allografting: prospects for immunoprophylaxis in: "The Herpes viruses, vol. 4," B. Roizman and C. Lopez, eds., Plenum Press, New York.

Meyers, J. D., Flournoy, N., and Thomas, E. D., 1986, Risk factors for cytomegalovirus infection after human marrow transplantation, J.Infect.Dis., 153:478.

Miller, R. A., Oseroff, A. R., Strattle, P. T., and Levy, R., 1983, Monoclonal antibody therapeutic trials in seven patients with T-cell lymphoma, Blood, 62:988.

Mocarski, E. S., Pereira, L., and Michael, N., 1985, Precise localization of genes on large animal virus genomes: Use of λgt11 and monoclonal antibodies to map the gene for a cytomegalovirus protein family, Proc.Natl.Acad.Sci.USA., 82:1266.

Nose, M., and Wigzell, H., 1983, Biological significance of carbohydrate chains on monoclonal antibodies, Proc.Natl.Acad.Sci.USA., 80:6632.

Nowinski, R., Boglund, C., Lane, J., Lastrum, M., Bernstein, I., Young, W., Hakomori, S., Hall, L., and Cooney, M., 1980, Production of antibody to tetanus toxoid by continuous human lymphoblastoid cell lines, Science, 199:1439.

Olsson, L., and Kaplan, H. S., 1980, Human-human hybrids producing monoclonal antibodies of predefined antigenic specificity, Proc.Natl. Acad.Sci. USA., 77:5429.

Pande, H., Baak, S. W., Riggs, A. D., Clark, B. R., Shively, J. E., and Zaia, J. A., 1984, Cloning and physical mapping of a gene fragment coding for a 64-kilodalton major late antigen of human cytomegalovirus, Proc.Natl.Acad.Sci.USA., 81:4965.

Paryani, S. G., Arvin, A. M., Koropchak, C. M. Wittek, A. E., Amylon, M. D., Dobkin, M. B., and Budinger, M. D., 1984, Varicella zoster antibody titers after the administration of intravenous immune serum globulin or varicella zoster immune globulin, Am.J.Med., 76:124.

Plotkin, S. A., 1985, CMV vaccines in: "The Herpes viruses, vol. 4," B. Roizman and C. Lopez, eds., Plenum Press, New York.

Scheid, A., and Choppin, P. W., 1977, Two disulfide-linked polypeptide chains constitute the active F protein of paramyxovirus, Virology, 80:54.

Sly, W. S., Sekhon, G. S., Kennett, R., Bodmer, W. F., and Bodmer, J., 1976, Permanent lymphoid lines from genetically marked lymphocytes: Success with lymphocytes recovered from frozen storage, Tiss. Antigens., 7:165.

Steinitz, M., Klein, G., Koskimies, S., and Makela, O., 1977, EB virus-induced B lymphocyte cell lines producing specific antibodies, Nature, 269:420.

Takahashi, M., Asano, Y., Kamiya, H., and Baba, K., 1986, Experience with a live varicella vaccine in: "Human Herpes virus Infections: Pathogenesis, Diagnosis, and Treatment, C. Lopez and B. Roizman, eds., Raven Press, New York.

Tamura, G. S., Dailey, M. O., Gallatin, W. M., McGrath, M. S., Weissman, I. L., and Pillemer, E. A., 1984, Isolation of molecules recognized by

monoclonal antibodies and antisera: The solid phase immunoisolation technique, Anal.Biochem., 136:458.
Vildé, J. L., Bricaire, E., Leport, C., Renaudie, M., and Brun-Vezinet, F., 1986, comparative trial of acyclovir and vidarabine in disseminated varicella-zoster virus infections in immunocompromised patients, J.Med.Virol., 20:127.
Wade, J. C., Hintz, M., McGuffin, R. W. Springmeyer, S. C., Connor, J. D., and Meyers, J. D., 1982, Treatment of cytomegalovirus pneumonia with high-dose acyclovir, Am.J.Med., 73:249.
Zimmermann, U., 1986, Electrical breakdown, electropermeabilization and electrofusion, Rev.Physiol.Biochem.Pharmacol., 105:175.

MONOCLONAL ANTIBODIES TO GRAM-NEGATIVE BACTERIA

Colin M.M. Stern

Consultant Paediatrician
St. Thomas's Hospital
London SE1 7EH

INTRODUCTION

The need to identify and classify micro-organisms, either out of scientific curiosity, clinical need or industrial necessity, has been a major pre-occupation of bacteriologists and clinicians. Many methods have been used successfully for these purposes, although the utilization of the exquisite specificity of the immune response has often given valuable results. The immune response has evolved adaptively in the face of environmental pressure exerted, in part, by micro-organisms and one arm of this response rests upon the production of antibody. However, the polyclonality of the immunoglobulins generated during an immune reaction rendered the production of specific immune reagents a tedious process of cross-absorption.

The successful preparation of monoclonal antibodies by Kohler and Milstein promised a great leap forward in bacterial taxonomy, but, in the ten years of research that have followed, it has to be said that little has been achieved. Recently, more papers describing reagents with valuable patterns of reactivity are appearing, but few publications show new applications of monoclonal methodology in the field, or describe ways in which the preparation of a reagent has been usefully employed.

There are, I believe, two main reasons for this. Firstly, there is, too often, a professional dislocation between those scientists with the training and expertise to generate a reagent and those with the practical involvement in the field to employ it to best advantage. Secondly, new methods of RNA and DNA recombination have been developed and are being introduced in a competitive way, sometimes without adequate thought to the most suitable method.

I have been asked to discuss the present situation with respect to those monoclonal antibodies (MCABs) that have been described as reacting with antigens expressed by Gram-negative bacteria. I propose, first, to review the technical aspects of this field and then list those organisms for which reagents have been made and their applications. A representative review of the current literature is included.

METHODOLOGY

The presently-employed methods of MCAB production and the detection systems used have been discussed previously in this symposium. There are some peculiarities which apply to the generation of appropriate clones against bacteria.

Insufficient attention has been paid to the question of antigen preparation in this context. Usually, it has been assumed that only MCABs directed against cell surface epitopes are likely to be of interest (Mutharia et al, 1985). Obviously, that will depend upon the purpose for which it was intended. Furthermore, the screening out of MCABs which react with seemingly-irrelevant antigens requires meticulous selection of control antigens. Although the MCAB searcher wishes to detect only clones of interest, it is easy to construct a selection system so rigorous that one can throw the baby out with the bath-water.

When reactivity against a clearly-defined and pure bacterial antigen is required, such as flagellin or an exotoxin, the problems are fewer. When an MCAB is needed to detect an epitope common to a range of serotypes, it is often best to employ whole live or killed bacteria as immunogens and select for MCABs reacting with the desired group.

When working with pathogenic bacteria, a further subtle problem arises. Those bacterial epitopes most immunogenic in man may excite little or no response in the species immunized, often a mouse. It is possible to miscontrue experimental data about immunogenicity. It is probably more appropriate to employ other methods, such as the Epstein-Barr virus transformation of human peripheral blood lymphocytes, taken from an infected patient at an appropriate stage of their illness (Crawford).

Dr. Ekins has reviewed the peculiarities of the antigen binding shown by MCABs. In spite of these problems, it seems to me that utility is the guiding principle and it is sometimes irrelevant to become too concerned about the low affinity MCABs may possess. If an optimum detection system is chosen and the net spread as widely as is practicable, it is wisest to use a detection system which most nearly approximates the physical conditions under which the reagent will be required to bind.

For some purposes, such as taxonomy, a range of detection systems may be employed. In clinical medicine, however, MCABs may need to bind in plasma, urine, cerebro-spinal fluid or on tissue sections. MCABs may be needed to target a drug to an organism and even those monoclonals with low affinity may function satisfactorily, since, together with the drug, they may be internalized by the bacteria.

MCABs RAISED AGAINST BACTERIA

Table 1 described those bacteria against which monoclonals have been raised. Although it is not extensive, there has been a significant increase in the number of published papers over the last two years. I have included the references for these reagents and, for the sake of clarity, will not reiterate the authors here.

MCABs have been raised against Gram-negative organisms, but there is a dearth of applied research using them. Notable exceptions include the studies by Sugasawara (1985) on Neisseria meningitidis and that by Heelels and Viji (1985) on Neisseria gonorrhoea pili. However, Table 2 lists the common pathogenic Gram-negative organisms in man and it can be seen that there are only five common species for which there are not yet any

Table 1. Organisms against which Monoclonal Antibodies have been raised

Neisseria meningitidis
Neisseria gonorrhoea
Clostridium difficile
Bordatella pertussis
Streptococcus, Group A
Streptococcus, Group B
Haemophilus influenzae
Shigella flexneri
Pneumocystis carinii
Treponema pallidum
Brucella abortus
Pseudomonas aeruginosa
Pseudomonas cepacia
Escherichia coli
Mycobacterium tuberculosis
Bacteroides intermedium
Legionella pneumophila
Vibrio cholera
Yersinia pestis

described reagents: Proteus; Salmonella; Campylobacter, Petronella and Francisella.

A comparison of the methods used by each research team does not reveal any major differences between them and a protocol which embodies the technical philosophy underlying these studies is shown in Fig. 1. As yet, there is insufficient field work to comment on the usefulness of these MCABs or, in most cases, to compare reagents.

MCABs AGAINST PSEUDOMONAS

Dr M. B. Slade spent three years working with me on the production of MCABs against Pseudomonas species and has presented his findings at several meetings (1981, 1982). Recently, Lana and his colleagues (1985) have also raised anti-Pseudomonal MCABs and have carried out analogous studies on the spectrum of binding of their MCABs on different Pseudomonads, classified into RNA homology groups and biotypes. A comparison of the reactivities of these different reagents is shown in Table 3.

I have attempted to compare their results with ours in Table 3. The binding of their MCABs seems to adhere strictly to separations established by conventional taxonomic methods, whereas our results revealed the presence of an atypically reacting species, Pseudomonas putida B, which failed to bind MCAB HH.002, a reagent which was specific for a surface antigen found only on other pseudomonads. Lana's group have not described a MCAB with binding restricted to Pseudomonas species and it does not seem that either team has picked clones which react with a common epitope.

Both Lana's group and ourselves have discriminated some specific reactivities from binding to cell wall antigens. They have, in common with others, detected a clone, SBF8A1, which binds to an endotoxin antigen which is not species specific, as well as other anti-LPS MCABs which appear to bind to serotypic determinants of varying species distribution. We also detected some anti-LPS MCABs, but also described four clones which react with alginate, the co-polymer of mannuronic and guluronic acid which is

Table 2. Gram-Negative Organisms and the Infections they cause. The asterisks show those against which Monoclonals have been made

ORGANISM	DISEASE
*Neisseria meningitidis	Septicaemia; meningitis
*Neisseria gonorrhoea	Conjunctivitis; gonorrhoea
*Escherichia coli	Urinary tract infections; enteritis
*Pseudomonas	Opportunistic infections
*Klebsiella	
Proteus	
*Bacteroides	
Salmonella	Enteritis; typhoid
*Shigella	
Campylobacter	
*Vibrio cholera	Cholera
*Yersinia pestis	Plague
*Haemophilus influenzae	Septicaemia; meningitis
*Bordatella pertussis	Whooping cough
*Brucella	Brucellosis
Petronella multocida	Infected dog bites
Francisella tularensis	Tularaemia
*Legionella pneumophila	Legionnaire's disease

present in small amounts upon the surface of all species of Pseudomonas aeruginosa, but which is greatly increased in amount when such organisms become 'mucoid' strains. I have summarized these results in Table 4.

Table 5 shows a typical series of restricted bindings of MCABs, taken from our own data. How should we interpret these varying data taxonomically? It seems to me that we are simply describing the phenotypic expression of antigens which, in some instances, may indicate genetic divergence at some point, but, in others, may reflect only non-transcription. Here we can see the relative values of MCABs and DNA transcription techniques in bacterial taxonomy and identification. Comparisons of the presence of genes coding for cell wall antigens with the expression of their products is necessary to resolve this point. Clearly, bacterial taxonomy will continue to depend upon the comparison of various methods of identification.

Although unequivocal DNA fingerprints will be the basis of classification, phenotypic changes occur which will require the use of MCABs to distinguish them. Examples of such situations include the mucoid strains of Pseudomonas aeruginosa (already mentioned) and the differences between virulent and attenuated bacteria, such as the wild and in vitro forms of beta-haemolytic streptococcus, Group B (Easton, personal communication).

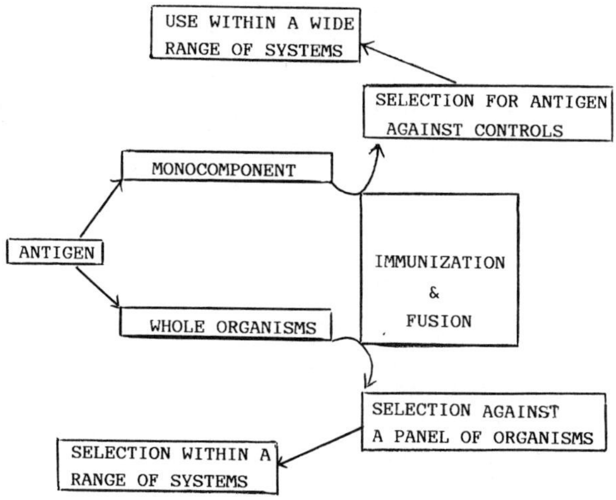

Fig. 1. A comparison of immunization patterns in the raising of monoclonal antibodies against bacteria.

Table 3. Pooled Reactivities of Two Monoclonal Antibodies, Reacting With Pseudomonas Aeruginosa, From Lana et al, (1985) and Slade et al, (1981), ? - Signifies Reactivity Unknown. Fractional Values Show Proportion of Positive Isolates.

RNA HOMOLOGY GROUP	TYPE	NUMBER REACTING H2	HH.026
Fluorescens	aeruginosa	+	26/26
	fluorescens	+	13/13
	syringa	+	3/3
	putida A	+	2/2
	putida B	?	0/5
	alkaligenese	+	1/1
	pseudo-alkaligenes	?	1/1
	stutzeri	+	3/3
Acidivorans	testeroni	?	0/3
Pseudomallei	cepacia	?	0/2
	marginata	?	0/1
Diminuta	Diminuta	?	0/1
Others	Maltophila	-	0/2
	Xanthoma campestris	-	0/1
	Escherichia coli	-	0/1
	Klebsiella	-	0/2
	Proteus	-	0/2

Table 4. Patterns of Reactivity of Anti-Pseudomonas Monoclonals

Antigens	Lana et al	Slade et al
Cell walls	+	+
'Serotypic'	+	+
LPS - Carbohydrate	+	+
- Lipid	+	+
Alginate	-	+

Table 5. The Distribution of Reactivity of Monoclonals Against Pseudomonas Isolates

	Pseudomonas aeruginosa types						
Monoclonal	217M	010	016	02a	011	PA103	RPMS
64	+	+	+	-	-	-	-
219	-	-	-	+	-	-	-
507	-	-	-	-	+	+	+
529	-	-	-	-	+	+	-

CLINICAL APPLICATIONS

Our MCABs were produced with the aim of detecting infection by Pseudomonas in immuno-compromised children, such as low birthweight newborn babies, patients with cystic fibrosis, or those receiving immuno-suppressive drugs. However, pari passu with these studies, we have developed the chromogenic substrate assay for endotoxin, based upon the use of amoebocyte lysate from Limulus polyphemus, to a level of sensitivity which has proved to be of considerable use in the detection of such infections. This advance has helped us to select a population of babies whose likelihood of infection is greater than 95%.

Table 6 describes the reactivities of sera taken from a small group of babies with raised LPS levels and from whom a clinically significant growth of Pseudomonas was obtained. A combination of MCABs 219, 02a and 204 was used, as this melange showed reactivity with all the isolates of Pseudomonas aeruginosa that we had tested - some 56 in all. Although in the cases of sera number 112, 236, 731 and 1088, binding seems to indicate the presence of antigen, the sensitivity of the test is rather low. Sample 478 showed no evidence of antigen.

It is possible that refinements of our immuno-assay may lead to a greater proportion of specific identifications. However, at present I believe that the detection of LPS, when combined with the platelet count, band count and in the presence of glycosuria, offers the most practical means of identifying infection in this group.

Some authors, such as Edberg (1983) have suggested that the value of MCABs in clinical diagnosis may be limited and that nucleic acid hybridization techniques are likely to replace them. A sober comparison of their relative strengths and weaknesses reveals that both methods will be needed.

Table 6. The Proportion of Infected Newborn Infants in whom Pseudomonas Antigen was Detected. Infected-X Indicates Infants not Infected with Pseudomonas. Infected-Ps shows Babies Infected with Pseudomonas.

Samples	N	Percentage Positive Tests 219	02a	204	Mixed
Controls	26	13.6	11.2	9.4	17.3
Non-infected	41	21.1	16.5	19.7	28.5
Infected-X	31	27.4	30.4	31.2	37.8
Infected-Ps	5	60.0	60.0	60.0	80.0
LPS -ve	3	66.7	66.7	66.7	100.0

Provided the DNA methods reach an acceptable level of sensitivity, they are likely to prove the most reliable means of bacterial identification when native DNA, usually in whole organisms, is present. However, bacteria which have been attacked and significantly damaged may only be detectable by MCABs which have been chosen to react with bacterial components more resistant to degradation. This constraint may apply more rigorously in specific biological fluids, such as cerebro-spinal fluid, or urine. The desire to detect a bacterial product, such as an exotoxin, may also dictate the use of MCAB-based methods.

Clearly, the complementary use of both of DNA hybridization and MCABs will be the most effective approach.

REFERENCES

Carlin, N. I. A., and Lindberg, A. A., 1985, Monoclonal antibodies specific for the O-antigen of Shigella flexneri and Shigella sonnei: immunochemistry, characterization and clinical usefulness," in: "Monoclonal Antibodies against Bacteria, A.J. Macario and E.C. de Macario, pp. 146-54, Academic Press.

de Macario, E. C., and Macario, A. J., 1986, "Monoclonal antibodies for bacterial identification and taxonomy," ASM News, 49:1.

Edberg, S. C., 1983, Principles of nucleic acid hybridization and comparison with monoclonal antibody technology for the diagnosis of infectious diseases, Yale J.Biol.Med, 58:425.

Frank, D. W., and Parker, C. D., 1984, Interactions of monoclonal antibodies with pertussis toxin and its sub-units," Inf and Immunity 46:195.

Gabay, J., Schenckman, S., Desaymard, C., and Schwarz, M., 1985, Monoclonal antibodies and the structure of bacterial membrane proteins, in: "Monoclonal Antibodies against Bacteria," A.J.Macario, and E.C. de Macario. pp.249-82, Academic Press.

Gigliotti, F., Stokes, D. C., Cheatham, A. B., Davis, D. S., and Hughes, W. T., 1986, Development of murine monoclonal antibodies to pneumocystis carinii," J.Infect.Dis., 154:315.

Gmur R., and Wyss, C., 1985, Monoclonal antibodies to characterize the antigenic heterogeneity of bacteroides intermedium, in: "Monoclonal Antibodies Against Bacteria," A.J.Macario and E.C. de Macario, eds., pp. 210-16, Academic Press.

Heelels, J. E., and Viji, M., 1985, "Monoclonal antibody against gonococcal pili: use in the analysis of gonococcal immunogenicity and virulence," in: "Monoclonal Antibodies Against Bacteria," A.J.Macario and E.C. de Macario, eds., pp. 91-103, Academic Press.

Holman, P. J., Schwing, G., and Douglas, J. T., 1985, "Development of monoclonal antibodies to Brucella cell surface antigens, in: "Monoclonal Antibodies against Bacteria," A.J.Macario and E.C. de Macario, pp. 82-110, Academic Press.

Holme, T., and Gusterson, B., 1985, Monoclonal antibodies against group and type specific antigens of Vibrio Choleri O:I, in: "Monoclonal Antibodies against Bacteria, A.J. Macario and E.C. de Macario, ed., pp. 186-191, Academic Press.

Hyypia, T., Jalava, A., Larsen, S., Terho, P., and Hukkanen, V., 1985, "Detection of Chlamydia trachomatis in clinical specimens by nucleic acid spot hybridization, J.Gen.Microbiol, 131:975.

Joly, J. R., McKinney, R. M., and Watkins, I. D., 1985, "Monoclonal antibodies to Legionella pneumophila: possible applications in clinical diagnostic tests and epidemiological investigations, in: "Monoclonal antibodies against bacteria," A.J. Macario and E.C. de Macario, eds, pp. 112-130, Academic Press.

Kamata, Y., Kozaki, S., Nagai, T. and Sakaguchi, G., 1985, Production of monoclonal antibodies against clostridium botulinum type E derivative toxin, FEMS Microbiology Letts., 26:305.

Lana, J. S., Mutharia, L. M., and Hancock, R. E. W., 1985, Application of monoclonal antibodies to the study of the surface antigens of pseudomonas aeruginosa," in: "Monoclonal Antibodies against Bacteria," A.J. Macario and E.C. de Macario, pp. 143-157, Academic Press.

Mitchison, D. A., and Coates, A. R. M., 1985, Monoclonal antibodies in bacteriology, in: "Monoclonal Antibodies in Clinical Medicine," A.J. McMichael and J.W. Fabre, eds., pp. 305-312, Academic Press.

Mutharia, L. M., Lana, J. S., and Hancock, R. E. W., 1985, Use of monoclonal antibodies in the study of various antigens of Gram-negative bacteria, in: "Monoclonal Antibodies against Bacteria," A.J. Macario and E.C. de Macario, eds., pp. 131-142, Academic Press.

Norgard, M. V., Marchitto, K. S., and Cox, D. L., 1986, Phenotypic expression of the major 47 kiloDalton surface immunogen of Treponema pallidum in virulent tissue-cultured treponemes, J.Gen.Microbiol., 132: 1775.

Oguma, K., Syuto, B., Kubo, S., and Hiroo, T., 1985, Analysis of antigenic structure of clostridium botulinum type Cl and D toxins by monoclonal antibodies, in: Monoclonal Antibodies Against Bacteria," A.J. Marcario, and E.C. de Macario, eds. pp. 159-184, Academic Press.

Okada, K., Mitsuyama, M., Miyake, S., and Amato, K., 1987, Monoclonal antibodies against the haemolysin of Vibrio vulnificus, J.Gen.Microbiol., 133:2279.

Sethi, K. K., 1985, Monoclonal antibodies against legionella pneumophila serogroup I antigens: characterization and their potential applications, in: "Monoclonal Antibodies against Bacteria," A.J.Macario and E.C. de Macario, eds., pp. 217-227, Academic Press.

Sheppard, A. J., Church, D., and Hughes, M., 1984, Production and characterization of monoclonal antibodies to tetanus toxin, Inf and Immunity, 43:710.

Soderstrom, T., 1985, Escherichia coli capsule and pili: serological, functional, protective and immunoregulatory studies with monoclonal antibodies, in: "Monoclonal Antibodies Against Bacteria," A.J. Macario and E.C. de Macario, eds., pp. 185-212, Academic Press.

Sugasawara, R. J., 1985, The use of monoclonal antibodies for detecting and serotyping Neisseria meningitidis, in: "Monoclonal Antibodies Against Bacteria, A.J. Macario and E.C. de Macario, eds. pp. 61-91, Academic Press.

Zumein, D. H., Munder, F. I. L., and O'Neill, S. P., 1985, Industrial application of monoclonal antibodies against bacteria, in: "Monoclonal Antibodies Against Bacteria," A.J.Macario and E.C., de Macario, eds., pp. 283-320, Academic Press.

CLINICAL APPLICATIONS OF MONOCLONAL ANTIBODIES AGAINST MYCOBACTERIA

A.R.M. Coates, P. Baird, H. Nicholai and Y. Nitzan*

Department of Medical Microbiology, The London Hospital Medical College, Turner Street, London E1 2AD, UK
*On leave from: Dept. of Life Sciences, BAR-ILAN University Ramat Gau, Israel

INTRODUCTION

The most important mycobacterial diseases are tuberculosis which is caused by *Mycobacterium tuberculosis* and leprosy, due to *M. leprae*. There are about 40 million cases of tuberculosis and leprosy in the world. Tuberculosis still kills more people than any other infectious disease, leading to 2-3 million deaths each year.

Monoclonal antibodies (MABs) against mycobacteria have led to a significant advance in our knowledge of mycobacterial immunology. The most dramatic effect of these MABs has been to improve knowledge about mycobacterial antigens, and this may result in the development of new vaccines. Improvements in diagnosis may also arise from the use of MABs.

The first set of MABs against mycobacteria were generated by Coates and his colleagues (1981) against *M.tuberculosis*. These MABs are known as the 'TB' series and are now well-characterized. This paper will primarily describe the clinical application of the TB series and other MABs against *M.tuberculosis*. The principles of these applications are also true for *M.leprae*.

Clinically useful MAB-based diagnostic tests for tuberculosis have not yet emerged. Attempts to detect mycobacterial antigen in clinical specimens with a MAB have not yielded a test with satisfactory specificity and sensitivity. An antibody-antibody competition assay (Hewitt et al, 1982) which detects elevated antibody levels in active disease might be of value in patients with smear negative disease who cannot be diagnosed by microscopy, but may be of less use in smear positive disease because the assay is less sensitive and specific than microscopy. The advantage of such a competition test over other assays is that it gives greater specificity than tests which are based on unpurified antigens, and does not require the laborious purification of antigens.

Vaccine development is the area which may benefit most from MABs against *M.tuberculosis*. *M. bovis* BCG is used as the vaccine against childhood tuberculosis worldwide, but has the following problems:

1. The efficacy of this vaccine seems to range from 0% (Indian Council of Medical Research, 1980) in the Chingelput area near Madras in India, to 70% (Medical Research Council, 1956; Medical Research Council, 1963) in British schoolchildren.

2. Epidemiological evidence (Styblo and Meijer, 1976) suggests that the vaccine does not effectively control the spread of the disease because it protects children who do not have the infectious pulmonary type of disease. It is the adults with 'open' pulmonary tuberculosis who spread the disease.

3. The BCG vaccine has a number of severe side-effects (Lotte et al, 1978) which may increase in areas such as Central Africa where the incidence of childhood infection with the human immunodeficiency virus is rapidly rising.

A worldwide effort is now directed at developing a new vaccine for tuberculosis using MABs and recombinant DNA technology. Whilst it is possible that this recombinant vaccine will be safer than BCG, it is not likely that it will be more effective or more immunogenic than BCG.

DIAGNOSIS

Attempts to develop effective diagnostic tests using MABs are aimed at antibody and antigen detection.

Antibody Detection

Diagnosis of pulmonary tuberculosis is easily achieved by direct microscopy of stained organisms in sputum of 35% of patients in London, U.K. and 82% of patients in Madras, India, who are culture positive (Mitchison, 1968). The reason for the low proportion of smear-positive cases in U.K. is probably due to patients presenting with less severe disease than individuals in India. In the smear-negative culture-positive cases, diagnosis takes 4-6 weeks and in a further large group who had active disease both smear and culture are negative (Hong Kong Chest Service/Tuberculosis Research Centre, Madras/British Medical Research Council, 1984). It is in these smear-negative tuberculosis patients whose serology has been little studied, that an effective serological test would be of most value. Even a test with a low sensitivity rate of 30-40% would be useful in this group so long as it was highly specific.

In order to study the incidence of antibodies against different mycobacterial antigens in smear-negative patients we measured the antibody response in sera against SDS-PAGE separated nitrocellulose-bound *M.tuberculosis* protein antigens. Sera from 34 smear-negative tuberculosis patients and 36 healthy controls bound to many bands in the extract (Fig. 1). Sera from both patients and controls reacted with 10, 14, 30, 38, 45 and 65 kdal antigens. In the same assay MAB TB68 bound to the 14 kdal, TB77 to 30 kdal, TB71 to 38 kdal and TB 78 to the 65 kdal antigen. Discrimination between patients and controls on the basis of binding to individual antigens was not possible but a combination of results of antigens was more productive. For instance, if a result was judged to be positive when a serum reacted with all 3 antigens of the 65, 45 and 38 kdal antigen, or binding to the 10 kdal antigen alone, 13 (38%) out of 34 smear-positives were positive but all the controls were negative. These experiments were repeated three times with the same results for this set of sera, but cautious interpretation is required until more testing of other sera has been completed. Nevertheless, on the basis of this data smear-negative patients appear to produce antibodies to a wider range of MAB-defined antigens than control patients.

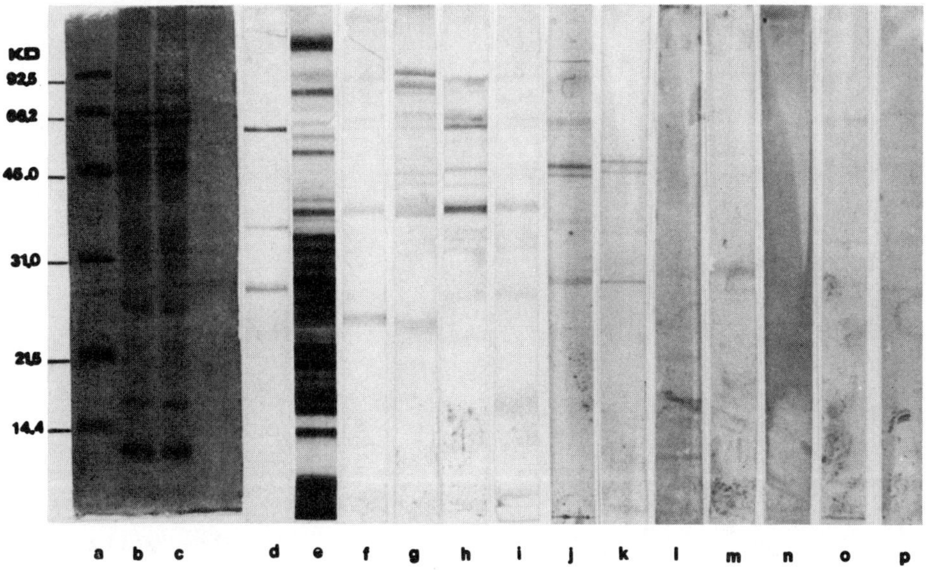

Fig. 1. Immunoblots of SDS-PAGE separated M.tuberculosis H37Rv antigens with human sera. Track a, SDS-PAGE separated molecular weight markers, stained with Coomassie Brilliant Blue. b & c, M.tuberculosis antigens in SDS-PAGE stained with Coomassie Brilliant Blue; d-p are strips of nitrocellulose filter papers containing SDS-PAGE separated antigens which have been incubated with; d, MABs TB78 detects the 65 kdal, TB71 the 38 kdal, TB77 the 30 kdal and TB68 the 14 kdal (faintly) molecules; e, polyclonal rabbit anti-BCG serum; f-i, sera from patients with smear-positive tuberculosis; j-m, sera from smear-negative tuberculosis patients; n-p, sera from healthy controls.

SDS-PAGE immunoblotting is not suitable as a routine laboratory screening test. An alternative system, the MAB-based antibody-antibody competition assay (Hewitt et al, 1982) is being developed. This assay can detect anti-M.tuberculosis antibody in the sera of smear-positive tuberculous patients and it is possible that such a test could also be suitable for smear-negative patients. The method is more specific (Ivanyi et al, 1983) than tests which use crude, unpurified antigen alone, although no more sensitive. The sensitivity of the test varies from MAB to MAB, the best results being obtained with MAB TB72 (Hewitt et al, 1982) combined with other MABs. In smear-positive patients it is unlikely that this serological test will replace direct staining and microscopy of M.tuberculosis in clinical specimens because microscopy is better established, more sensitive, more specific, probably cheaper and easier to perform than the current antibody-antibody competition assays.

Further evidence on the use of this assay in bacteriologically well-characterized smear-negative culture positive individuals is required.

Antigen Detection

MABs have also been used to develop an antigen detection test for tuberculosis (Chandramuki et al, 1985). In this assay the antigen reacts with a MAB ML34 which is bound to sheep red blood cells. Using reverse passive haemagglutination the test was used to detect antigen in the cerebrospinal fluid of 89 Indian patients with tuberculous meningitis and

127 controls. Of 25 culture positive cases 22 (88%) were positive for the test and out of 64 culture negative cases with a presumptive diagnosis of tuberculous meningitis 47 (73%) were positive. Unfortunately the test was also positive in 4 (21%) out of 19 Indian patients with pyogenic meningitis and 2 (2%) of 108 patients with other neurological conditions. These results indicate that this MAB based test is quite sensitive but that the specificity is too low for it to be useful in a clinical setting. It is possible that if another MAB were used in this test, the specificity might improve.

Shared Antigens

In tuberculosis anti-DNA antibodies have been shown to occur (Lindquist et al, 1970) in 28 (47%) out of 60 patients with pulmonary disease. Whilst the functional significance of these cross-reacting antibodies is unknown there are practical implications for diagnosis. For instance, if a test for tubercle antigen in clinical specimens used a MAB which binds to human tissue in addition to *M.tuberculosis*, a low degree of specificity might result.

Studies of MABs against *M.tuberculosis* have shown that some bind to DNA and this was particularly high for TB71, 73 and 77 (Shoenfeld et al, 1986). The binding could be inhibited by polynucleotides. Furthermore anti-DNA human MABs derived from patients with systemic lupus erythematosis bound to purified *M.tuberculosis* glycolipids A, B and C and preincubation of MABs with ssDNA inhibited binding. These experiments suggest that the anti-DNA antibodies which are known to occur in tuberculosis cross-react with *M.tuberculosis* antigens.

VACCINE DEVELOPMENT

Protective immunity against tuberculosis with the vaccine *M.bovis* BCG is cell-mediated. Available evidence suggests that *M.tuberculosis*-specific T lymphocytes mediate protection (Lefford et al, 1973; Collins et al, 1975). The activated lymphocytes secrete macrophage activating factors such as interferon-γ which stimulate macrophages to inhibit the replication of or to kill the bacteria (Khor et al, 1986).

There are at least 60 antigens in *M.tuberculosis* which are recognized by polyclonal antigen using two-dimensional cross-immunoelectrophoresis (Roberts et al, 1971). Although there has been some success in purifying these molecules it is still unknown which antigen(s) are involved in stimulating protective immunity. The reason for the slow progress is that it is difficult to obtain a very pure antigen in large quantities using conventional techniques.

The application of MAB technology to the identification and purification of *M.tuberculosis* antigens has dramatically increased progress in this field. The limitless supply of MABs has enabled purification of reasonable quantities of antigen by immuno-affinity chromatography, protein sequencing of such antigens and the cloning and DNA sequencing of the antigen genes. Although there is still no way to directly purify antigens which stimulate T cells, protein molecules which induce antibody or MABs usually contain epitopes which will also stimulate T lymphocytes. In practice this is true for the MAB-defined *M.tuberculosis* antigens many of which are now known to stimulate T cells.

Identification of the Antigens of M.Tuberculosis with Monoclonal Antibodies

The first set of MABs against mycobacteria were made against M.tuberculosis in 1981 by Coates and his colleagues. These are known as the 'TB' series of MABs the details of which are set out in Table 1. These MABs bind to SDS-polyacrylamide gel-separated antigens of molecular mass 65 (TB78), 38 (TB71, TB72), 32 (TB77), 19 (TB23) and 14 (TB68) kdal. Other laboratories have now made more MABs (Table 1), some of which bind to the same antigens and others bind to 71, 45, 40-50, 25-60, 23, 20-80, 17, 15 and 10 kdal antigens (Shou et al, 1985; Kadival and Chaparas, 1987; Ivanyi et al, 1983; Daniel et al, 1984; Morris et al, 1985; Young et al, 1984; Minden et al, 1984; Kolk et al, 1984; Gillis and Buchanan, 1982).

Protein Purification with Monoclonal Antibodies

After the identification of M.tuberculosis antigens with MABs, the next question was whether these antigens stimulated T cell mediated immunity. This needed the purification of quite small amounts of antigen by immunoaffinity chromatography. The first of these antigens which was identified as possessing a T lymphocyte determined epitope was the 14 kdal antigen which was affinity purified with MAB TB68. This molecule stimulated the proliferation of a human T helper cell clone (Matthews et al, 1985). Furthermore, this T cell clone secreted a macrophage activating factor which stimulated the human macrophage-like cell line U937 to release the mycobacteriocidal substance hydrogen peroxide (Andrew et al, 1984). Hydrogen peroxide release is closely associated with antimycobacteriocidal activity and killing can be abolished by the addition of catalase (Walker and Lowrie, 1981). This suggested that the TB68 antigen not only stimulated T cells, but also induced the correct type of T cell response which results in an appropriate macrophage mediated mycobacteriocidal response. Other antigens such as the 10 kdal (Minden et al, 1986), and the 65 kdal (TB78) (Emmrich et al, 1986) molecules have also proven to bear T lymphocyte epitopes.

Purification of the TB23 antigen was carried out by immuno-affinity chromatography (Nitzan and Coates, in preparation). The specific monoclonal antibody was coupled to CNBr activated sepharose to form the structure Seph-NH-MAB TB23. The TB23 antigen was eluted from the affinity column by 0.1 M citric acid and 1% sodium lauryl sulfate at pH = 2.5. When

Table 1. Molecular Mass (kdal) of M.tuberculosis Antigens

65 (Multiband)	38	32	19	14	71	45-60	45	40-50	25-60	17	15	10	
TB78	TB72 TB71	TB77	TB23	TB68									(Coates et al,1981)
IIH9 IIC8													(Gillis & Buchanan, 1982)
ML30								ML02 ML34					(Ivanyi et al, 1983)
									TB-C1				(Daniel et al, 1984)
			F29-47	F23-49 F24-2			F29-29						(Kolk et al, 1984)
												SA12	(Minden et al, 1984)
											B9D		(Young et al,1984)
	HYT28	HYT27	HYT6		HAT1								(Schou et al, 1985)
										MB17			(Morris et al, 1985)
	HGT3a					HGT1 HGT2	HGT6 (2 bands)		HGT4				(Kadival & Chaparas,1987)

transferred onto a Gel Filtration HPLC Column (7.5 x 300 mm, Beckman TSK GEL 3000) TB23 antigen has been shown to consist of two molecular weight species (Fig. 2), a high molecular weight protein of about 120 kdal (elution time 16 min) and a low molecular weight protein of about 20 kdal (elution time 37.5 min). In the presence of 2-ME and heat the 120 kdal protein was converted to the 19 kdal protein indicating that the 120 kdal protein is a multi-subunit complex consisting of six low molecular weight subunits. Both protein species bind MAB TB23 in ELISA. The 19 kdal TB23 antigen was further purified by a reverse-phase HPLC column (2 x 250 mm Ultrasphere). Elution was done by increasing the concentration of acetonitrile from 0 to 100%. One main peak can be shown (Fig. 3) when the concentration of MeCN reaches 71% indicating a high hydrophobic nature of the protein. The hydrophobicity of the protein was confirmed by the amino-acid analysis of TB23 antigen which showed that out of the 150 estimated amino-acids found in the protein 40% are of hydrophobic nature. Preliminary experiments indicated that the TB23 antigen stimulates T cell mediated skin hypersensitivity in guinea-pigs.

This protein may be suitable for attempted N-terminal gas-phase micro-sequencing as has been achieved for the 10 kdal BCG-a antigen (Minden et al, 1986). This illustrates an important potential use of MABs, because a nucleotide probe can be made which is complementary to the amino-acid sequence (see Fig. 4). This probe, when radiolabeled, could be utilized to select the antigen from a library of *M.tuberculosis* genomic DNA. This approach is likely to complement the cloning of genes by other methods such as direct expression, because the latter may not work for all antigens and the protein sequence should confirm the DNA sequence, the reading frame and the site of the N-terminus. Furthermore the known sequence of a protein can be used to make a synthetic peptide (see Fig. 4). Such a peptide was made on the basis of the N-terminal sequence of BCG-a and stimulates T cell mediated delayed-type hypersensitivity (Minden et al, 1986).

Cloning of M.tuberculosis Antigen Genes

An effective approach to make large quantities of a subunit vaccine is by using recombinant DNA technology. Young and his colleagues (1985) cloned the *M.Tuberculosis* 65 kdal antigen with MAB TB78 using direct expression, and also claim to have cloned the 14 (TB68) and 19 kdal TB23 antigens. They fractionated genomic *M.Tuberculosis* DNA (see Fig. 4) into small pieces which were then inserted into the vector bacteriophage gt11 under the control of bacterial gene signals to ensure expression. This recombinant vector was grown overnight on antibiotic-containing agar, and the surviving organisms expressed the fragments of DNA as a fusion protein connected to *E.coli* β-galactosidase. The entire repertoire of colonies were transferred by replica-plating onto nitrocellulose filter paper, the bacteria were lysed and were incubated with MABs against *M.tuberculosis* antigens. The colonies which reacted with the antibody were picked out of the original plate, and purified so that single colonies were obtained. These single, purified colonies and their direct progeny are called clones and contain one species of *M.tuberculosis* DNA. The cloning of the 65 kdal antigen has also been achieved by others (Emmrich et al, 1986) using polyclonal antisera. Cloning of other *M.tuberculosis* genes is also being undertaken.

Shinnick (1987) has determined the entire DNA sequence of the 65 kdal antigen and has deduced the amino-acid sequence of this molecule. This protein is predicted to contain 540 amino-acids, and a second long open reading frame encoding a protein of 517 amino-acids is thought to occur downstream to the 65 kdal protein. The protein sequence shows homology

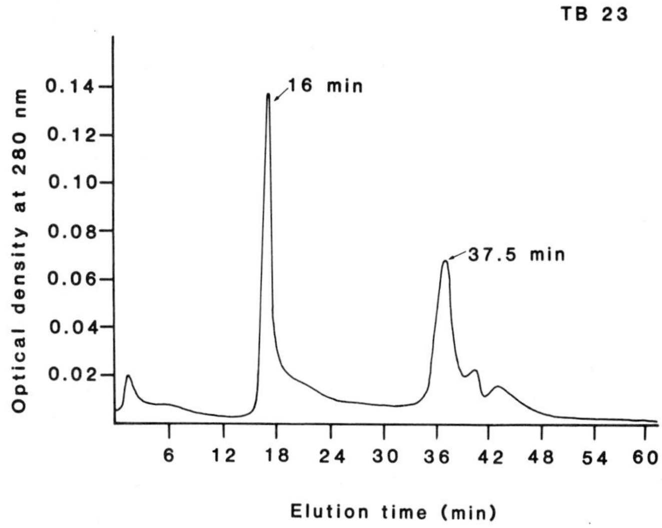

Fig. 2. Optical density of MBT323 immunoaffinity purified antigen fractions eluted from a gel filtration HPLC column.

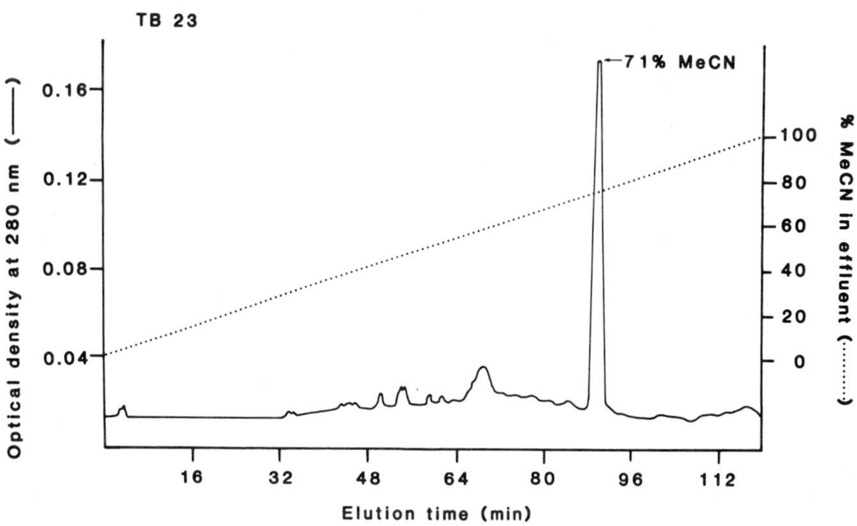

Fig. 3. Purification of 19 kdal antigen using reverse-phase HPLC. —— = optical density of fractions; --- = %MeCN in effluent.

with the mRNA stability gene (ams+) (Chanda et al, 1985) of E.coli and so may have a similar function in mycobacteria.

Analysis of T Lymphocyte Determined Epitopes of M.Tuberculosis Determined Antigens

Several strategies (see Fig. 4) can be tried in order to identify the parts of the antigen which contain T cell determined epitopes. For instance, Mehra and colleagues (1986) made a mini-library of the 65 kdal M.leprae gene and expressed fragments of the DNA in the gtll/E.coli system.

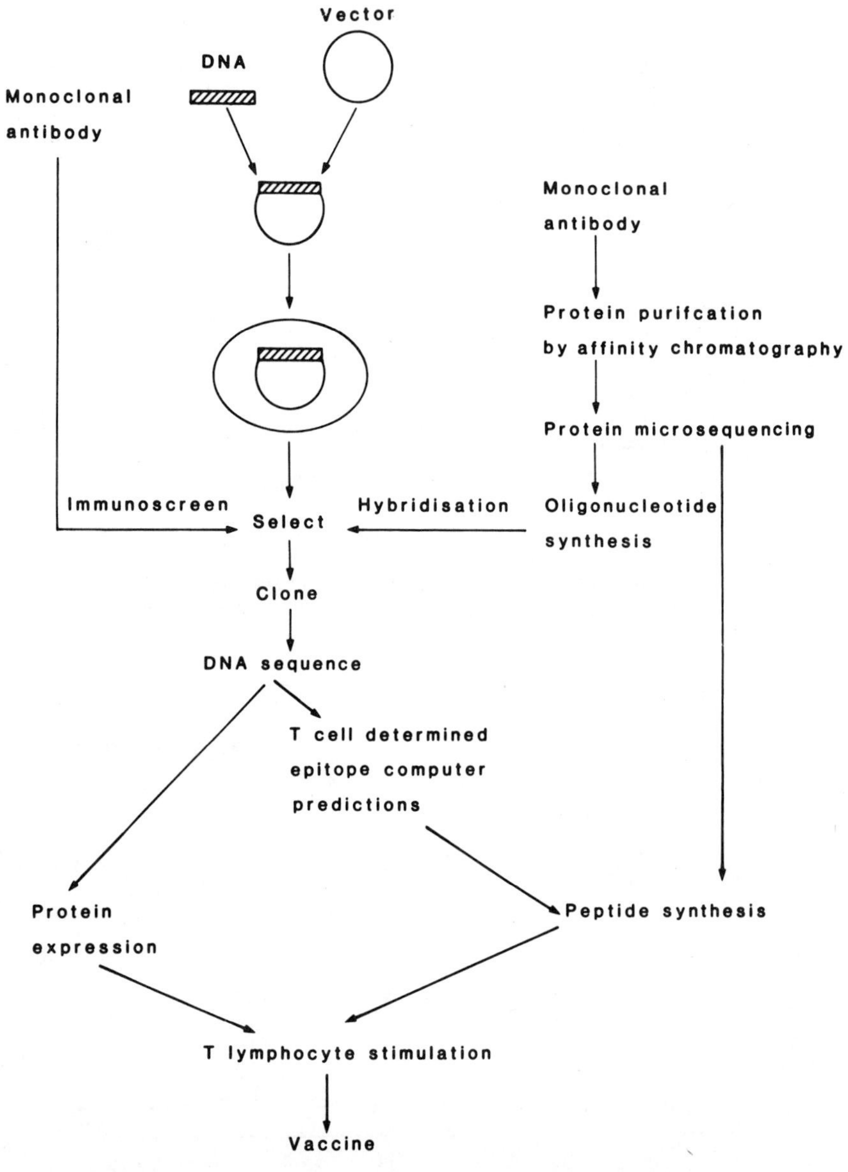

Fig. 4. Approaches to the development of a new recombinant vaccine for tuberculosis using monoclonal antibodies, recombinant DNA technology and synthetic peptides.

T cell defined epitopes were localized to areas of the protein (Lamb et al, 1987). Using a different complementary route, Lamb and his colleagues (1987) have used a computer programme (Rothbard et al, 1986) to predict which 65 kdal M.leprae peptides might stimulate T cells. These predictions are based on a template which consists of a charged or glycine residue, followed by two hydrophobic amino-acids and then a polar or glycine residue or another hydrophobic amino-acid followed by a polar or glycine group. They showed that two regions 112-132 and 437-459 which are conserved between M.leprae and M.tuberculosis are human T cell epitopes. In Fig. 5 we show our predictions of T cell determined epitopes for the 65 kdal M.tuberculosis molecule. We also used the template approach developed by

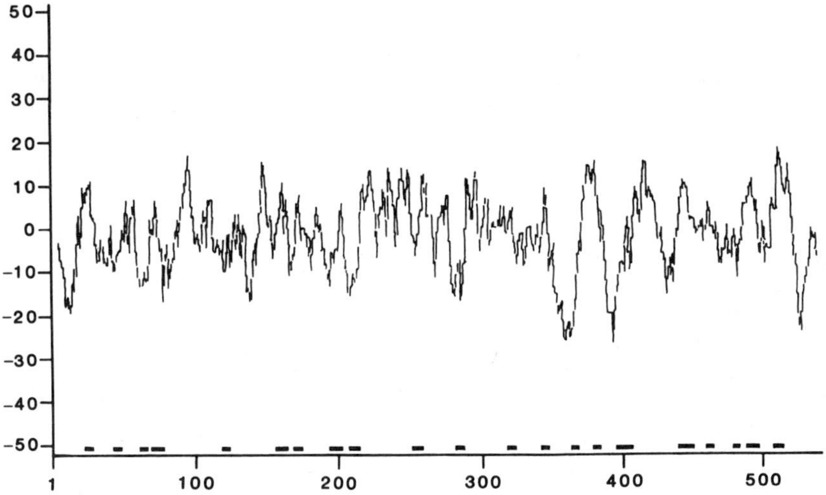

Fig. 5. Predicted T lymphocyte determined epitopes, and hydropathic index of M.tuberculosis 65 kdal antigen from amino-acid 1-540. —— = hydropathic index; ▬▬ = T cell defined sites.

Rothbard and colleagues (1986) and found that there were 21 distinct areas which may contain T cell epitopes. Most of the predicted areas were partly or predominantly hydrophobic as measured by the hydropathic index (Kyte and Doolittle, 1982).

It is now possible to test these cloned gene products in animal models of tuberculosis to assess their protective efficacy as a vaccine. Optimal stimulation of T cell-mediated immunity is required, and the introduction of the genes into recombinant vectors such as vaccinia virus (Mackett et al, 1982) or Salmonella typhimurium (Hoiseth and Stocker, 1981) may be tried, because both of these systems efficiently present foreign antigens to T lymphocytes.

CONCLUSION

Monoclonal antibodies against M.tuberculosis have led to greater knowledge of the antigens of M.tuberculosis. In combination with recombinant DNA technology, these MABs are being used to attempt to develop a new vaccine for tuberculosis. Several antigen genes have been cloned, and DNA and amino-acid sequences for some of them have been determined. This will result in the production of large quantities of pure recombinant M.tuberculosis protein, recombinant vectors containing single mycobacterial genes, and synthetic peptides which will greatly enhance the prospects for vaccine development.

MABs against mycobacteria may also result in improved diagnositc tests for active disease. Although there are no satisfactory serological tests for tuberculosis as yet, an antibody-antibody competition assay for the detection of raised antibody levels in patients could be useful if it were developed for smear-negative disease. Antigen detection with monoclonal antibodies has not proven successful as yet.

Acknowledgements

We thank Dr. Lucinda Hall and Dr. Corrine Ong for helpful discussions. We are also grateful to Mrs. Paula Edwards for typing the manuscript.

REFERENCES

Andrew, P. W., Rees, A. D. M., Scoging, A., Dobson, N., Matthews, R., Whittal, J. T., Coates, A. R. M., and Lowrie, D.B., 1984, Secretion of macrophage-activating factor distinct from interferon-γ by human T cell clones, Eur.J.Immunol., 14:962.

Chanda, P. K., Ono, M., Kuwano,M., and Kung, H-F., 1985, Cloning, sequence analysis and expression of alteration of the mRNA stability gene (ams+) of Escherichia coli, J.Bact., 161:446.

Chandramuki, A., Allen, P. R. J., Keen, M., and Ivanyi, J., 1985, Detection of mycobacterial antigen and antibodies in the cerebrospinal fluid of patients with tuberculous meningitis, J.Med.Microbiol., 20:239.

Coates, A. R. M., Hewitt, J., Allen, B. W., Ivanyi, J., and Mitchison, D. A., 1981, Antigenic diversity of Mycobacterium tuberculosis and Mycobacterium bovis detected by means of monoclonal antibodies, Lancet, ii:167.

Collins, F. M., Congdon, C.C., and Morrison, N. E., 1975, Growth of Mycobacterium bovis (BCG) in T lymphocyte-depleted mice, Infect.Immun., 11:57.

Daniel, T. M., Gonchio, N. J., Katzman, J. A., and Olds, G. R., 1984, Specificity of Mycobacterium tuberculosis antigen 5 determined with mouse monoclonal antibodies, Infect.Immun., 45:52.

Emmrich, F., Thole, J., van Embden, J., and Kaufmann, S. H. E., 1986, A recombinant 64 Kilodalton protein of Mycobacterium bovis Bacillus Calmette-Guerin specifically stimulates human T4 clones reactive to mycobacterial antigens, J.Exp.Med., 163:1024.

Gillis, T. P., and Buchanan, T. M., 1982, Production and partial characterization of monoclonal antibodies to Mycobacterium leprae, Infect. Immun., 37:172.

Hewitt, J., Coates, A. R. M., Mitchison, D. A., and Ivanyi J., 1982, The use of murine monoclonal antibodies without purification of antigen in the serodiagnosis of tuberculosis, J.Immunol.Methods, 55:205.

Hoiseth, S. K., and Stocker, B. A. D., 1981, Aromatic dependent Salmonella typhimurium are non-virulent and effective live vaccines, Nature, 291:238.

Hong Kong Chest Service/Tuberculosis Research Centre, Madras/British Medical Research Council, 1984, A controlled trial of 2-month, 3-month and 12-month regimens of chemotherapy for sputum-smear-negative pulmonary tuberculosis, Am.Rev.Respir.Dis., 130:23.

Indian Council of Medical Research, 1980, Tuberculosis prevention trial, Madras, Indian J.Med.Res., 72 (suppl) July:1.

Ivanyi J., Krambovitis, E., and Keen, M., 1983, Evaluation of a monoclonal antibody (TB72) based serological test for tuberculosis, Clin.Exp. Immunol., 54:593.

Ivanyi, J., Sinha, S., Aston, R., Cussell, D., Keen, M., and Sengupta, U., 1983, Definition of species specific and cross-reaction antigenic determinants of Mycobacterium leprae using monoclonal antibodies, Clin.Exp.Immunol., 52:528.

Kadival, G. V., and Chaparas, S. D., 1987, Production, characterization and species specificity of five monoclonal antibodies to Mycobacterium tuberculosis, J.Clin.Microbiol., 25:76.

Khor, M., Lowrie, D. B., Coates, A. R. M., and Mitchison, D. A., 1986, Recombinant interferon-gamma and chemotherapy with isoniazid and rifampicin in experimental murine tuberculosis, Brit.J.Exp.Pathol., 67:587.

Kolk, A. H. J., Ho, M. L., Klatser, P. R., Eggelte, T. A., Kuijper, S., De Jonge, S., and Van Leenwen, J., 1984, Production and characterization of monoclonal antibodies to Mycobacterium tuberculosis, M. bovis (BCG) and M.leprae, Clin.Exp.Immunol., 58:511.

Kyte, J., and Doolittle R. F., 1982, A simple method for displaying the hydropathic character of a protein, J.Mol.Biol., 157:105.

Lamb, J. R., Ivanyi J., Rees, A. D. M., Rothbard, J. B., Howland, K., Young R. A., and Young, D. B., 1987, Mapping of T cell epitopes using recombinant antigens and synthetic peptides, EMBO J., 6:1245.

Lefford, M. J., McGregor, D. D. and Mackaness, G. B., 1973, Immune response to Mycobacterium tuberculosis in rats, Infect.Immun., 8:182.

Lindquist K. J., Coleman, R. E., and Osterland, C. K., 1970, Autoantibodies in chronic pulmonary tuberculosis, J.Chronic Dis., 22:717.

Lotte, A., Wasz-Hockert O., Poisson N., and Dumitrescu, D., 1978, Complications induced by BCG vaccination: a retrospective study, Bull.Int. Union.Tuberc., 53:114.

Mackett, M., Smith, G. L., and Moss, B., 1982, Vaccinia virus: a selectable eukaryotic cloning and expression vector, Proc.Natl.Acad.Sci.USA., 79:7415.

Matthews, R., Scoging A., and Rees, A. D., 1985, Mycobacterial antigen-specific human T-cell clones secreting macrophage activating factors, Immunology, 54(1):17.

Medical Research Council, 1956, BCG and vole bacillus vaccines in the prevention of tuberculosis in adolescents, Br.Med.J., 1:413.

Medical Research Council, 1963, BCG and vole bacillus vaccines in the prevention of tuberculosis in adolescence and early adult life, Br.Med.J., 1:973.

Mehra, V., Sweetser D., and Young, R. A., 1986, Efficient mapping of protein antigenic determinants, Proc.Natl.Acad.Sci.USA., 83:7013.

Minden, P., Kelleher, P. J., Freed, J. H., Neilsen, L. D., Brennan, P. J., McPherson, L., and McClatchy, J. K., 1984, Immunological evaluation of a component isolated from Mycobacterium bovis BCG with a monoclonal antibody to M.bovis BCG, Infect.Immun., 46:519.

Mitchison, D. A., 1968, Examination of sputum by smear and culture in case-finding, Bull.Int.Un.Tuberc., 41:139.

Morris, J. A., Thorns, C. J., and Woolley, J., 1985, The identification of antigenic determinants on Mycobacterium bovis using monoclonal antibodies, J.Gen.Micro., 131:1825.

Roberts, D. B., Wright, G. L., Affronti, L. F., and Reich, M., 1972, Characterization and comparison of mycobacterial antigens by two-dimensional immunoelectrophoresis, Infect.Immun., 666:564.

Rothbard, J. B., Townsend, A., Edwards, M., and Taylor, W., 1986, In: "Modern Trends in Human Leukaemia, VII," Elsevier.

Schou, C., Yuan, Z. L., Andersen, M. B., and Bennedsen, J., 1985, Production and partial characterization of monoclonal hybridoma antibodies to Mycobacterium tuberculosis, Acta.Path.microbiol.Immunol. Scand.Section C., 93:265.

Shinnick, T. M., 1987, The 65-kilodalton antigen of Mycobacterium tuberculosis, J.Bact., 169:1080.

Shoenfeld, Y., Vilner, Y., Coates, A. R. M., Rauch, J., Lavie, G., Shaul, D., and Pinkhas, J., 1986, Monoclonal anti-tuberculosis antibodies react with DNA, and monoclonal anti-DNA autoantibodies react with Mycobacterium tuberculosis, Clin.Exp.Immunol., 66:255.

Styblo, K., and Meijer, J., 1976, Impact of BCG vaccination programmes in children and young adults on the tuberculosis problem, Tubercle., 57:17.

Walker, L., and Lowrie, D. B., 1981, Killing of Mycobacterium microti by immunologically activated macrophages, Nature, 293:69.

Young, D. B., Kharnolkar, S. R., Barg, L. L., and Buchanan, T. M., 1984, Generation and characterization of monoclonal antibodies to the phenolic glycolipid of Mycobacterium leprae, Infect.Immun., 43:183.

Young, R. A., Bloom, B. R., Grosskinsky, C. M., Ivanyi, J., Thomas D., and Davis, R. W., 1985, Dissection of Mycobacterium tuberculosis antigens using recombinant DNA, Proc.Natl.Acad.Sci.USA., 82:2583.

USE OF MONOCLONAL ANTIBODIES IN MALARIA

Judith A. Deans

Department of Immunology
UMDS, Guy's Hospital Medical School
London SE1 9RT

INTRODUCTION

Malaria constitutes a major health problem in many tropical and sub-tropical countries; it is estimated that in addition to the large reservoir of chronic malaria infections, approximately 100 million cases of acute clinical malaria occur annually and substantial mortality from falciparum malaria occurs especially in young, non-immune persons in tropical Africa. Malaria control measures have been impaired by the spread of drug-resistant falciparum malaria (now confirmed in more than 40 of the 100 countries in which malaria occurs) and insecticide resistant vectors.

Infection is initiated by the bite of an infected female anopheline mosquito which inoculates motile sporozoites into the bloodstream. The sporozoites are localized in hepatic parenchymal cells, where they undergo exoerythrocytic schizogony. Merozoites released from exoerythrocytic schizonts invade red blood cells, thereby initiating the cyclically proliferative stage of the parasite life cycle with which the clinical symptoms of malaria are associated. After a period of purely asexual replication of erythrocytic parasites, a proportion of newly invaded merozoites differentiate into male and female gametocytes; sexual reproduction follows the ingestion of gametocytes into the gut of the mosquito. The complicated malaria life cycle and the corresponding antigenic complexity of the parasites have hindered the development of satisfactory immunodiagnostic tests and the isolation of protective antigens which might comprise useful vaccines. The application of hybridoma technology to malaria research has provided invaluable tools for the analysis and isolation of malaria antigens. In this paper I will attempt to illustrate the use of monoclonal antibodies for identifying and isolating putative protective malaria antigens, for defining serological diversity and for developing novel immunodiagnostic tests.

PROTECTIVE ANTIGENS

Protective Sporozoite Antigens

Vaccination with irradiated sporozoites confers protection against viable sporozoite challenge; the immunity so engendered is both stage and species specific. Immune serum from protected individuals neutralized sporozoite infectivity and reacted with the entire surface of the sporozoite. Immunoprecipitations of surface radiolabeled sporozoites with immune sera revealed that the membrane contains a single major immunogenic protein. Monoclonal antibodies raised against sporozoites recognize the same circumsporozoite (CS) protein and neutralize sporozoite infectivity both in vitro and in vivo; Fab fragments of monoclonal antibodies to CS proteins also neutralize parasite infectivity and prevent attachment of the parasite to hepatocytes in vitro. CS proteins of different plasmodial species have similar biosynthetic, structural and immunological properties. CS proteins are stage specific and are synthesized in large amounts in mature salivary gland sporozoites. There are two precursor forms of the CS protein, of higher molecular weight and isoelectric point; the membrane form of the CS protein is generated by sequential removal of two small basic peptides from the precursor. The molecular weight of the CS proteins of different plasmodium species vary between 40 and 60 kDa and isoelectric point 4 and 5. Studies employing panels of monoclonal antibodies to CS proteins of P.falciparum, P.vivax and P.knowlesi revealed that these antigens contain a single immunodominant region with two or more identical epitopes. The CS genes from P.knowlesi, P.falciparum, P.vivax and P.cynomolgi have been cloned and sequenced; all contain a large domain consisting of tandemly repeated sequences of identical amino acids, in P.knowlesi this domain consists of twelve amino acids repeated twelve times (Godson et al, 1983). All monoclonal antibodies which neutralize sporozoite infectivity are directed against the repeats of the various CS proteins; most or all anti-sporozoite antibodies in polyclonal antiserum also recognize the repetitive epitope. The repeat subunit of P.falciparum CS protein is simply asparagine-alanine-asparagine-proline (NANP) (Enea et al, 1984, Dame et al, 1984). Monoclonal antibodies were used to establish the minimum number of NANP tandem repeats that can form an epitope. Peptides containing between two and five tandem repeats of NANP were tested for their ability to inhibit the reactions of various monoclonal antibodies with the native CS protein. Although $(NANP)_2$ was a poor inhibitor, $(NANP)_3$, $(NANP)_4$ or $(NANP)_5$ were equally effective, suggesting that the epitope recognized by the monoclonal antibodies was contained within $(NANP)_3$ (Zavala et al, 1985). The anti-sporozoite response of humans living in endemic areas has been shown to be mainly or exclusively directed against $(NANP)_3$. Two vaccines containing the P.falciparum NANP repeats have been prepared, and clinical trials of their safety and immunogenicity in adult volunteers are currently underway in the USA. One vaccine consists of a chemically synthesized twelve mer $(NANP)_3$ linked to a tetanus toxoid carrier molecule. The other vaccine is based on the repetitive epitope expressed by recombinant DNA in Escherichia coli. The adjuvant used in both vaccines is aluminium hydroxide. The results of these Phase I clinical trials are awaited with interest. For recent reviews of protective sporozoite antigens, see Nussenzweig and Nussenzweig (1985 and 1986).

Protective Erythrocytic Stage Antigens

The identification of protective blood-stage malaria antigens has been much more difficult due to the antigenic complexity of the erythrocytic merozoite; the merozoite surface comprises at least six major surface antigens (identified by surface radio-iodination) and an unknown number of minor components. Antigens recognized by monoclonal antibodies which confer protection on passive transfer, or inhibit multiplication of the parasite in vitro, however, constitute obvious vaccine candidates.

A class of high molecular weight proteins has been identified that is common to all Plasmodium species studied to date; immunofluorescence and immunoelectronmicroscopy using monoclonal antibodies have localized them on the outer surface of schizonts and merozoites. The first of these to be identified was the 230 kDa P.yoelii antigen: monoclonal antibodies directed against it conferred protection on passive transfer (Majarian et al, 1984), and immunization with purified protein also conferred protection (Holder and Freeman 1981) (interestingly the monoclonal antibody used to affinity purify the antigen was non-protective on passive transfer). Two monoclonal antibodies directed against the 250 kDa P.chabaudi protein inhibited parasitaemia in vivo (Boyle et al, 1982), and certain of the monoclonal antibodies directed against the 250 kDa antigen of P.knowlesi (Epstein et al, 1981) and the 195 kDa antigen of P.falciparum (Pirson and Perkins 1985) block invasion in vitro. A polyvalent antiserum to the 230 kDa P.yoelii protein cross reacted with the P.chabaudi 250 kDa and P.falciparum 195 kDa proteins, suggesting that these malaria species contain homologous proteins (Holder et al, 1983). These high molecular weight proteins are synthesized by mature trophozoites and schizont, they are processed at the time of merozoite formation to peptides of lower molecular weight that appear on the surface of extracellular merozoites (Freeman and Holder 1983). Saimiri monkeys immunized with monoclonal antibody-affinity purified P.falciparum 195 kDa antigen exhibited partial protection (Perrin et al, 1984, Hall et al, 1984). The complete structure of the gene coding for the P.falciparum 195 kDa protein has now been deduced (Holder et al, 1985b) and synthetic peptides corresponding to parts of it have been made. Monkeys immunized with one of the synthetic peptides were partially protected from a challenge infection that induced a fulminating infection in control animals (Cheung et al, 1986).

Another group of blood stage antigen which have been shown to be protective are rhoptry antigens. A monoclonal antibody directed against a 235 kDa P.yoelii rhoptry protein conferred protection on passive transfer, and immunization with the monoclonal antibody-affinity purified protein also protected mice against an otherwise lethal challenge (Holder and Freeman 1981). The antigen was precisely located within the rhoptries by immunoelectronmicroscopy. Monoclonal antibodies directed against a P.falciparum 41 kDa rhoptry protein inhibited the multiplication of the parasite in vitro; immunization with this purified protein induced a partial immunity in Saimiri monkeys (Perrin et al, 1985). Several monoclonal antibodies that react with the apical region of P.falciparum merozoites with a dot-like immunofluorescence pattern have been produced, these monoclonal antibodies immunoprecipitate either a 140 kDa protein or peptides of 80, 66 and 42 kDa molecular weight (Holder et al, 1985a, Schofield et al, 1986): some inhibit the multiplication of P.falciparum in vitro; whether these antigens are protective in vivo remains to be established.

Monoclonal antibodies against a 140 kDa major P.knowlesi merozoite surface protein (precursor 143 kDa MW) also blocks merozoite invasion. Rhesus monkeys immunized with purified 140 kDa protein were partially protected against a potentially lethal challenge; however the partially protected monkeys developed chronic erythrocytic infections. The 140 kDa protein was not present in its original form in these relapsing parasites but was represented in multiple, cross-reacting, alternate antigenic forms (David et al, 1985).

Monoclonal antibodies directed against a very minor P.knowlesi merozoite surface antigen (44/42 kDa MW), which is formed from a 66 kDa precursor protein, were also inhibitory. Fab and $F(ab')_2$ fragments of these monoclonal antibodies were also active in inhibition. These monoclonal antibodies react with an important but labile conformational epitope on the 66 kDa antigen, a polyclonal antiserum raised against a preparation of the antigen in which the conformational epitope had been destroyed was completely non-inhibitory (Deans and Jean 1987). Rhesus monkeys immunized with monoclonal antibody-affinity purified 66 kDa protein (in which it was estimated that less than 10% of the molecules were conformationally intact), showed variable protection; two of the six vaccinated monkeys demonstrated a sterile immunity, while the other four rhesus required drug cure. Although the protection afforded in this vaccination trial was only partial, it was achieved with saponin as adjuvant. Vaccinations with other purified proteins from blood-stage simian and human malarias that protected monkeys from fatal infections were administered in FCA, other adjuvants were ineffective.

Monoclonal antibodies that react with a wide variety of other P.falciparum blood-stage antigens have been prepared; how many of these antigens are protective remains to be established.

Protective sexual stage Antigens

Transmission-blocking antibodies are effective against two stages of parasite development; anti-gamete antibodies can prevent fertilization while antibodies directed against zygotes and ookinetes can prevent subsequent development in the mosquito. Monoclonal antibodies have defined three P.falciparum gamete surface antigens (MW 230, 48 and 45 kDA) which are shed from the zygote surface following fertilization (Carter et al, 1984, Vermeulen et al, 1985). Monoclonal antibodies which react with the 48 and 45 kDa antigens block fertilization with a variable efficiency which relates to the epitope with which they react. Blocking antibodies do not react with the proteins under reducing conditions, indicating that the conformational epitopes are maintained by disulfide bonds. All the antibodies that block fertilization react with both the 48 and 45 kDa antigens, the relationship between the two proteins is however unknown. P.vivax transmission-blocking monoclonal antibodies react with gamete surface antigens of 36 and 42 kDA MW (Mendis et al, 1987). Monoclonal antibodies that passively block transmission of P.yoelii have been used to affinity purify antigens from microgametes; the predominant 42 kDa antigen and minor components of 67, 59, 57 and 35 kDa were isolated; active immunization with these proteins (without adjuvant) produced a marked reduction in transmission to the vector (Harte et al, 1985). Following gametogenesis in P.falciparum a 25 kDa protein is synthesised which replaces the 230, 48 and 45 kDa proteins on the zygote and ookinete surface membranes. Monoclonal antibodies to this antigen do not prevent fertilization and ookinete formation, but they do stop subsequent development in the mosquito. Monoclonal antibodies directed against the 23 kDa antigen on P.berghei zygotes and ookinetes block the interaction between the ookinete and the mosquito gut wall. It has recently been demonstrated that anti-gamete monoclonal antibodies and antibodies in immune human sera can have two opposite effects on transmission, blocking and enhancing. The idea of inducing transmission blocking immunity by vaccination requires re-evaluation in view of the infectivity-enhancing effect of anti-gamete antibodies at sub-neutralizing concentrations.

ANTIGENIC VARIATION AND ANTIGENIC DIVERSITY

P.knowlesi parasites display antigenic variation of an erythrocyte surface antigen as defined by antibody-mediated cell agglutination (the SICA test). This variant antigen is a malaria protein that changes both in size (MW) 185 - 225 kDa) and antigenicity. Natural isolates of P.falciparum express a major glycoprotein of MW approximately 200 kDa which exhibits size diversity and expresses antigenically conserved as well as diverse epitopes, as defined by a panel of monoclonal antibodies (Howard et al, 1986). Similar antigenic and structural diversity has been demonstrated (by the use of monoclonal antibodies) for a P.chabaudi antigen of approximately 250 kDa (Newbold et al, 1984). A 140 kDa P.knowlesi merozoite surface protein was chosen for vaccination studies because antibodies to it blocked invasion in vitro. This protein was invariant in parasites from chronically infected monkeys. However, following challenge of immunized monkeys with cloned parasites, variants expressing proteins of markedly different MW (165-76 kDa) and of different reactivity with monoclonal antibodies to the 140 kDa protein had replaced the original cloned population (Klotz et al, 1987).

Antigenic diversity has also been demonstrated for gamete and sporozoite components. Monoclonal antibodies directed against the surface of P.falciparum gametes detected antigenic differences between different isolates of the parasite (Graves et al, 1985). Repeat sequence epitopes on the circumsporozoite protein of P.falciparum appear to be invariant in natural isolates, however, in P.knowlesi and P.cynomolgi different isolates of each species fail to cross-react with monoclonal antibodies against sporozoite repeats (Cochrane et al, 1985).

IMMUNODIAGNOSTIC TESTS

Anti-CSP monoclonal antibodies have been used in two-site immunoradiometric (IRMA) and enzyme-linked immunosorbent (ELISA) assays to detect sporozoites in anopheline mosquitoes. Unlike current dissection techniques, these tests can detect sporozoites in dried mosquitoes long after collection and identify the species of malaria sporozoite. The tests can quantify sporozoites in individual or pools of mosquitoes, and can detect sporozoite densities of less than 100 (Zavala et al,1982). These simple and rapid tests have been used to determine the P.falciparum sporozoite rate in female anopheline mosquitoes in The Gambia (Collins et al, 1984), and to determine the Anophelese species infected with and therefore potentially capable of transmitting P.vivax and P.falciparum malaria in Para, Brazil (Arruda et al, 1986).

Immunodiagnostic tests which reliably detect anti-malarial antibodies could also be used for epidemiological studies, the lack of standardized reagents for such tests has hitherto been their major limitation. Recently, two micro-ELISA methods employing a monoclonal-antibody purified P.falciparum rhoptry polypeptide of 41 kDa, and a 31-1 fusion polypeptide corresponding to a P.falciparum merozoite surface antigen permitted the precise determination of antimalarial antibodies in both adults and children from the Gambia. Results obtained using a crude parasitized red blood cell extract as antigen were, by comparison, unreliable, especially in young children (Gabra et al, 1986).

REFERENCES

Arruda, M., Carvalho, M. B., Nussenzweig, R. S., Maracic, M., Ferreira, A. W., and Cochrane, A. H., 1986, Potential vectors of malaria and their different susceptibility to Plasmodium falciparum and Plasmodium vivax in northern Brazil identified by immunoassay, Am.J. Trop.Med.Hyg., 35:873.

Boyle, D. B., Newbold, C. I., Smith, C. C., and Brown, K. N. 1982, Monoclonal antibodies that protect in vivo against Plasmodium chabaudi recognize a 250,000 - Dalton parasite polypeptide, Infect.Immun., 38:94.

Carter, R., Miller, L. H., Rener, J., Kaushal, D. C., Kumar, N., Graves, P. M., Grotendorst, C. A., and Gwadz, R. W., 1984, Target antigens of malaria transmission-blocking immunity, Philos.Trans.R.Soc.London B., 307:99.

Cheung, A., Leban, J., Shaw, A. R., Merkli, B., Stocker, J., Chizzolini, C., Sander, C., and Perrin, L. H., 1986, Immunization with synthetic peptides of a Plasmodium falciparum surface antigen induces anti-merozoite antibodies, Proc.Natl.Acad.Sci.USA, 83:8328.
Crochrane, A. H., Gwadz, R. W., Ojo-Amaize, E.,Hii, J., Nussenzweig, V., and Nussenzweig, R. S., 1985, Antigenic diversity of the circumsporozoite proteins in the Plasmodium cynomolgi complex, Molec. Biochem.Parasit., 14:111.
Collins, F. H., Zavala, F., Graves, P. M., Cochrane, A. H., Gwadz, R. W., Akoh, J., and Nussenzweig, R. S., 1984, First field trial of an immunoradiometric assay for the detection of malaria sporozoites in mosquitoes, Am.J.Trop.Med.Hyg., 33:538.
Dame, J. B. WIlliams, J. L., McCutchan, T. F., Weber, J. L., Wirtz, R. A., Hockmeyer, W. T., Maloy, W. L., Haynes, J. D., Schneider, I., Roberts, D., Sanders, G. S., Reddy, E. P., Diggs, C. L., and Miller, L. H., 1984, Structure of the gene encoding the immunodominant surface antigen on the sporozoite of the human malaria parasite Plasmodium falciparum, Science, 225:593.
David, P. H., Hudson, D. E., Hadley, T. J., Klotz, F. W., and Miller, L. H., 1985, Immunization of monkeys with a 140 kilodalton merozoite surface protein of Plasmodium knowlesi malaria: appearance of alternate forms of this protein, J.Immunol., 134:4146.
Deans, J. A., and Jean, W. C., 1987, Structural studies on a putative protective Plasmodium knowlesi merozoite antigen, Molec.Biochem. Parasit., In press.
Enea, V., Ellis, J., Zavala, F., Arnot, D. E., Asavanich, A., Masuda, A., Quakyi, I., and Nussenzweig, R. S., 1984, DNA cloning of Plasmodium falciparum circumsporozoite gene: amino acid sequence of repetitive epitope, Science, 225:628.
Epstein, N., Miller, L.H., Kaushal, D. C., Udeinya, I. J., Rener, J., Howard, R. J., Asofsky, R., Aikawa, M., and Hess, R. L.,1981, Monoclonal antibodies against a specific surface determinant on malaria (Plasmodium knowlesi) merozoites block erythrocyte invasion, J. Immunol., 127:212.
Freeman, R. R., and Holder, A. A., 1983, Surface antigens of malaria merozoites: a high molecular weight precursor is processed to an 83,000 mol wt form expressed on the surface of Plasmodium falciparum merozoites, J.Exp.Med. 158:1647.
Gabra, M. S., Grossiord, D., Perrin, L. H., Shaw, A., Cheung, A., and McGregor, I. A., 1986, Defined Plasmodium falciparum antigens in malaria serology, Bull.WHO 64:889.
Godson, G. N., Ellis, J., Svec, P., Schlesinger, D. H., and Nussenzweig, V., 1983, Identification and chemical synthesis of a tandemly repeated immunogenic region of Plasmodium knowlesi circumsporozoite protein, Nature 305:29.
Graves, P. M., Carter, R., Burkot, T. R., Rener, J., Kaushal, D. C., and Williams, J. L., 1985, Effects of transmission-blocking monoclonal antibodies on different isolates of Plasmodium falciparum, Infect. Immun., 48:611.
Hall, R., Hyde, J. E., Goman, M., Simmons, D. L., Hope, I. A., Mackay, M., Scaife, J., Merkli, B., Richle, R., and Stocker, J., 1984, Major surface antigen gene of a human malaria parasite cloned and expressed bacteria, Nature, 311:379.
Harte, P. G., Rogers, N., and Targett, G. A. T., 1985, Vaccination with purified microgamete antigens prevents transmission of rodent malaria, Nature 316:258.
Holder, A. A., and Freeman, R. R., 1981, Immunization against blood-stage rodent malaria using purified parasite antigens, Nature 294:361.
Holder, A. A., Freeman, R. R., and Newbold, C. I., 1983, Serological cross-reaction between high molecular weight proteins synthesized in blood

schizonts of Plasmodium yoelii, Plasmodium chabaudi and Plasmodium falciparum, Molec.Biochem.Pasasit, 9:191.

Holder, A. A., Freeman, R. R., Uni, S., and Aikawa, M., 1985a, Isolation of a Plasmodium falciparum rhoptry protein, Molec.Biochem.Parasit, 14:293.

Holder, A. A., Lockyer, M. J., Odink, K. G., Sandhu, J. S., Riveros-Moreno, V., Nicholls, S. C., Hillman, Y., Davey, L. S., Tizard, M. L. V., Schwarz, R. T., and Freeman, R. R., 1985b, Primary structure of the precursor to the three major surface antigens of Plasmodium falciparum merozoites, Nature, 317:270.

Howard, R. J., McBride, J. S., Aley, S. B., and Marsh, K., 1986, Antigenic diversity and size diversity of P.falciparum antigens in isolates from Gambian patients. II. The schizont surface glycoprotein of molecular weight 200,000, Parasit,Immunol., 8:57.

Klotz, F. W., Hudson, D. E., Coon, H. G., and Miller, L. H., 1987, Vaccination-induced variation in the 140 kD merozoite surface antigen of Plasmodium knowlesi malaria, J.Exp.Med., 165:359.

Majarian, W. R., Daly, T. M. Weidanz, W. P., and Long, C. A., 1984, Passive immunization against murine malaria with an IgG3 monoclonal antibody, J.Immunol., 132:3131.

Mendis, K. N., Peiris, J. S. M., Premawansa, S., Udagama, P. V., Munesinghe, Y., Ranawaka, M., Carter, R., and David, P. H., 1987, Immune modulation of parasite transmission in Plasmodium vivax malaria: anti-gamete antibodies can both block and enhance transmission, UCLA Symposium 42:417.

Newbold, C. I., Schryer, M., Boyle, D. B., McBride, J. S., McLean, A., Wilson, R. J. M., and Brown, K. N., 1984, A possible molecular basis for strain specific immunity to malaria, Molec.Biochem.Parasit., 11:337.

Nussenzweig, V., and Nussenzweig, R. S., 1985, Circumsporozoite proteins of malaria parasites, Cell, 42:401.

Nussenzweig, V., and Nussenzweig, R. S., 1986, Development of a sporozoite malaria vaccine, Am.J.Trop.Med.Hyg., 35:678.

Perrin, L. H., Merkli, B., Gabra, M. S., Stocker, J. W., Chizzolini, C., and Richle, R., 1985, Immunization with a Plasmodium falciparum merozoite surface antigen induces a partial immunity in monkeys, J.Clin.Invest., 75:1718.

Perrin, L. J., Merkli, B., Loche, M., Chizzolini, C., Smart, J., and Richle, R., 1984, Antimalarial immunity in Saimiri monkeys: immunization with surface components of asexual blood stages, J.Exp.Med., 160:441.

Pirson, P. J., and Perkins, M. E., 1985, Characterization with monoclonal antibodies of a surface antigen of Plasmodium falciparum merozoites, J.Immunol., 134:1946.

Schofield, L., Bushell, G. R., Cooper, J. A., Saul, A. J., Upcroft, J. A., and Kidson, C., 1986, A rhoptry antigen of Plasmodium falciparum contains conserved and variable epitopes recognized by inhibitory monoclonal antibodies, Molec.Biochem.Parasit., 18:183.

Vermeulen, A. N., Ponnudurai, T., Beckers, P. J. A., Verhave, J. P., Smits, M. A., and Neuwissen, J. H. E. Th., 1985, Sequential expression of antigens on sexual stages of Plasmodium falciparum accessible to transmission-blocking antibodies in the mosquito, J.Exp.Med., 162:1460.

Zavala, F., Gwadz, R. W., Collins, F. H., Nussenzweig, R. S., and Nussenzweig, V., 1982, Monoclonal antibodies to circumsporozoite proteins identify the species of malaria parasite in infected mosquitoes, Nature 299:737.

Zavala, F., Tam, J. P., Hollingdale, M. R., Cochrane, A. H., Quakyi, I., Nussenzweig, R. S., and Nussenzweig, V., 1985, Rationale for development of a synthetic vaccine against Plasmodium falciparum malaria, Science, 228:1436.

PRODUCTION AND USE OF BOVINE AND OVINE MONOCLONAL ANTIBODIES

D.J. Groves and B.A. Morris

AFRC Antibody Development Group
Department of Biochemistry, University of Surrey
Guildford, Surrey GU2 5XH

INTRODUCTION

As part of a research initiative by the Agricultural and Food Research Council to investigate the potential for immunomodulation of the reproductive performance of farm animals, this research group has been examining approaches to improving the fecundity of sheep and cattle by passive immunization against the gonadal steroid hormones.

Fecundity is defined as the number of young born per female allocated to be mated (Land et al, 1982). It covers all stages of the reproductive process from maturation of the Graafian follicle, through ovulation fertilization, conception and pregnancy, to parturition. In polyzygous animals, such as sheep and pigs, most of the variation in fecundity is due to differences in the number of eggs shed, i.e., ovulation rate. There have been many attempts to increase this parameter using nutritional, genetic and pharmacological means with varying degrees of success.

More recently, immunological modulation of the negative feed-back control of the hypothalamic-pituitary gonadal axis has been undertaken (Scaramuzzi and Hoskinson, 1984); the response depending on the method of immunization and the hormone used for the immunogen.

Ovulation rate is in part determined by the number of follicles which are brought to maturation in a particular oestrous cycle and which subsequently rupture to release their ova. Whilst the precise mechanisms determining ovulation rate still remain to be elucidated, it is known that these two processes are in part under the control of the gonadotrophins, follicle stimulating hormone (FSH) and luteinising hormone (LH) respectively, which are both secreted by the pituitary (Fig. 1). The release of the two hormones is itself regulated by gonadotrophin releasing hormone (GnRH) which is produced by the hypothalamus. The developing follicles produce a number of gonadal steroid hormones as well as inhibin, a polypeptide hormone. These collectively exert a negative feedback effect on the hypothalamic-pituitary axis to limit the amount of gonadotrophin produced, and consequently, the ovulation rate.

Variation in prolificacy, the number of young born per pregnancy, between different breeds of the same species has been associated with variations in the sensitivity of the hypothalamus to the negative feedback

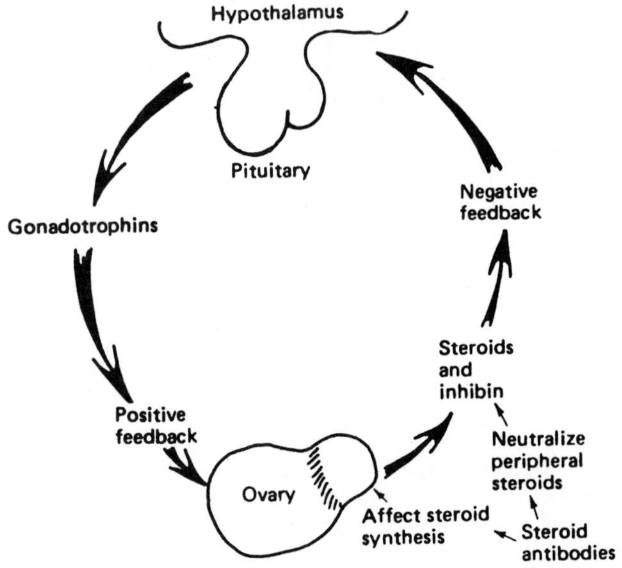

Fig. 1. Interaction of hypothalamo-hypophysio-gonadal axis in the control of ovulation. (From Webb et al, 1984).

effects of the gonadal hormones. Land (1976) found that a prolific breed of sheep was less sensitive to the negative feed-back effects of oestrogen than one of lower prolificacy. This finding suggested that if the negative feed-back in less prolific breeds could be blocked or inactivated, such action might result in a higher ovulation rate.

One way in which this has been achieved has been by the immuno-modulation of the gonadal steroid hormones using specific antibodies (Scaramuzzi, Davidson and Van Look 1977; Land et al, 1982). The antibodies react with the hormone in the peripheral circulation to form a biologically inactive antigen-antibody complex. Their effectiveness depends on their affinity for the hormone in question, which should be greater than 10^{-9}M, in order to ensure that as much of the circulating hormone as possible is complexed and that once formed, the complex does not dissociate to release the active hormone (Webb, Land, Pathijara and Morris, 1984).

Immunization against a particular hormone can be achieved in one of two ways; either by (i) active immunization, in which the animal under treatment produces the necessary antibodies following an injection of the appropriate immunogen, or (ii) passive immunization, in which antibodies are first produced either in vivo, as an antiserum or as ascites, or in vitro from cell cultures and then, after isolation and purification, injected into the recipient animals.

The advantages of active immunization are that the cost of the immunogen is usually relatively inexpensive and that the immunization itself is a simple procedure. However, for the treatment to be effective, the animal has first to be primed and then, a few weeks later, boosted in order to produce a state of autoimmunity, the magnitude of which is unpredictable.

Passive immunization, on the other hand, is effective immediately following the first injection of antibodies. It allows a calculated dose to be given at a particular stage of the oestrous cycle and so eliminates the widely variable immune and biological responses seen in actively immunized animals (Scaramuzzi et al, 1977). The effects of passive immunization are of short duration and are governed by the half-lives of the antibodies involved (approximately 15-16 days for allogeneic antibodies in sheep and cattle (Clayton, Morris and Rhind, unpublished observations)). In addition, one can give mixtures of different antibodies in precise proportions and, as can be seen later, one can be selective in the treatment being given, according to the individual animal's condition. Finally, in passive immunization, there are no exogenous veterinary residues from the immunogen or adjuvant, since the therapeutic agent is an endogenous molecule.

That the use of passive immunization against the gonadal steroids to increase fecundity is feasible in practice has been shown by land et al, (1982) with Welsh Mountain ewes, whose normal lambing rate is approximately 1.0. These workers treated separate groups of ewes with ovine antisera to androstenedione, oestradiol, oestrone and testosterone, as well as with a mixture of all four antisera. All five treatments were successful in increasing ovulation and, subsequently, lambing rates (Fig. 2). Anti-oestradiol antiserum was the most effective in increasing ovulation rate, but it unfortunately reduced conception rate. The mixture of antisera produced the largest increase in lambing rate, an additional 0.5 lambs per ewe compared to that of the control group.

In a later experiment, Rhind et al, (1985) showed that the magnitude of the response of Scottish Blackface ewes to passive immunization with anti-testosterone antiserum was very dependent on the body condition of the ewe during the two weeks prior to mating. A significant increase in ovulation, conception and lambing rates occurred in those ewes which were passively immunized in the thin and moderate body condition groups compared with their respective controls. However, there was no significant increase in these parameters in the ewes in the fat body condition suggesting that there was a maximum level to which fecundity could be increased by passive immunization and that this level had already been reached by the non-immunized fat ewes.

These results also demonstrated that passive immunization with anti-testosterone antiserum could also be used as a feed sparing treatment in moderate to thin ewes in the period prior to tupping when they would normally be 'flushed' by being fed a higher energy diet.

Recently the Group's efforts have been directed primarily to applying this strategy for increasing cattle fecundity in order to produce twin calves for beef production. Despite a popular impression of a beef surplus in the UK, demand is likely to exceed supply in the near future. Between 65% and 70% of UK home-produced beef originates from dairy herds. A variety of economic and social factors have resulted in a decrease of 2-5% per annum in the national dairy herd over recent years. This has been brought about by an increase in the use of higher yielding breeds, such as Holstein Friesians, or pharmacological stimulation of milk production using somatotrophin, resulting in greater efficiency, the introduction of milk quotas by the EEC, the reduction in milk consumption in line with recommendations from the UK Committee on Medical Aspects of Food Policy (COMA) and the import of cheaper UHT milk from the continent. Increasing twinning in cattle would go some way towards off-setting the effects of the declining number of dairy cows as well as improving the efficiency of beef herds.

The immunization of cattle against steroids has been attempted in a number of preliminary experiments. Active immunization of cows against androstenedione resulted in an increase in ovulation rate from 1.0 to 1.36 (Wise and Schandbacher, 1983), with three of the four double ovulations

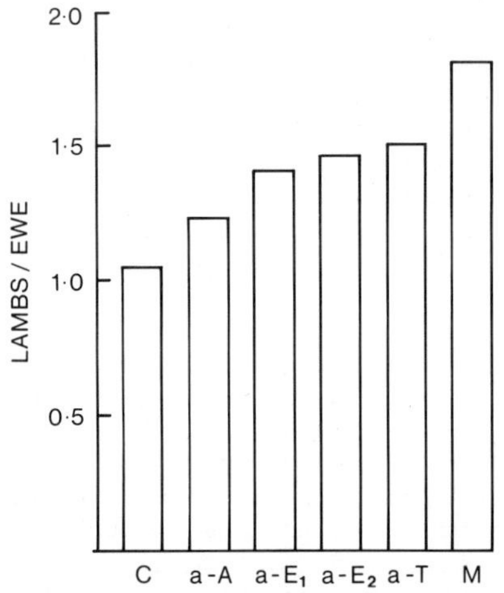

Fig. 2. Mean number of lambs born alive per Welsh Mountain ewe treated with antisera to gonadal steroids. (C: Control; a-A: Anti-androstenedione; a-E1: Anti-oestrone; a-E2: Anti-oestradiol; a-T: Anti-testosterone; M: Mixture).

resulting in twin pregnancies. On the other hand, the same workers found active immunization against oestradiol to be ineffective in inducing double ovulations. However, the cows concerned continued to cycle and become pregnant. In contrast, Price, Morris and Webb (1987), studying the effects of active immunization in heifers, found that the animals became acyclic after boosting and remained so for many months whilst the antibody levels remained elevated. When three of the immunized animals resumed cyclicity, double ovulations were observed in two individuals, one for two cycles and the other for a single cycle. Some multiple ovulations have been induced in cattle using ovine anti-testosterone serum (Webb et al, 1984). Again, the response was very variable and there was a serious potential for anaphylactic response if repeated dosing was used. Because of the weak endogenous immune response in cattle homogeneic passive immunization requires the production of large volumes of antisera to achieve the desired effect. Murine monoclonal antibodies would provide a very poor alternative as immunomodulators for use in cattle due to their low affinities which make the resulting antigen-antibody complexes susceptible to dissociation. The potential for anaphylactic reaction also exists. For these reasons, it was found necessary to explore the potential for producing bovine and ovine antisteroid antibodies using hybridoma technology.

The basis of monoclonal antibody production is the fusion of two cell types - the myeloma, which is potentially immortal, but is also deficient in an enzyme necessary for nucleic acid synthesis in the presence of aminopterin, and the sensitized B-lymphocyte, which has a finite life span but contains the gene coding for the enzyme, hypoxanthine (guanine) phosphoribosyl transferase (HGPRT), along with genes coding for antibody production and packaging.

Cell fusion studies were first developed in the 1960s as a method for the genetic analysis of animal cells. Interspecific fusions were first carried out by Weiss and Green (1967) who observed differential chromosome loss from the parenteral lines. Since Kohler and Milstein's (1975) original work on the immortalization of antibody-secreting cell lines interspecific cell hybridization has increasingly been used in attempts to derive monoclonal antibodies from species for which suitable myeloma lines are not readily available. Initial cross-species fusions involved mouse myeloma cells and spleen cells from rats immunized against murine antigens (Trowbridge, 1978). However, the paucity of suitable established human myeloma lines, combined with the obvious demand for human monoclonal antibodies led to early attempts at fusing lymphocytes from immunized individuals with murine myeloma cells (Schwaber and Rosen, 1978). Initial success, at least in the short term, has prompted the extension of this approach to the production of monoclonal antibodies from domestic species, especially cattle and sheep.

The first report of the immortalization of sheep antibody production was by Tucker et al, 1981, using a direct fusion between ovine spleen or lymph node lymphocytes and murine NS1 cells. Although the resulting hybrids grew well in culture, they soon lost the capacity to produce immunoglobulin, presumably due to chromosome loss. Tucker and her co-workers (Tucker et al, 1984) subsequently used a technique pioneered in human x murine fusions by Ostberg and Pursch (1983) and by Teng et al, (1983) to improve the stability of their hybrids. Interspecific hybridomas, obtained from previous fusions, were sensitized to aminopterin and then fused with more lymphocytes from the immunized animal to produce second generation heterohybridomas. This approach has overcome some of the problems associated with interspecific fusions, especially the low fusion efficiency and the instability of the resulting hybrids. The use of heteromyeloma fusion partners has been exploited to generate human (Ichimori et al, 1985); Hirata and Sugawara, 1987), chimpanzee (Van Meurs and Jonker, 1986), cattle (Tucker et al, 1984; Groves, et al, 1987; Anderson et al, 1986, 1987) and sheep (Anderson et al, 1986) antibodies and the number of generations of refusion are being investigated to see if any of the benefits continue to accrue on repeated fusion.

Previous reports of the isolation of bovine monoclonal antibodies, using first generation fusions, have been primarily concerned with the provision of isotype preparations for sequencing and serology studies (Guidry et al, 1986). Lymphocytes have generally been obtained from the spleen or lymph nodes of these animals.

In this study first and second generation interspecific fusions have been used to generate heterohybridomas secreting ovine and bovine monoclonal antibodies.

MATERIALS AND METHODS

Generation of Heteromyelomas

The murine myeloma line used in the first generation fusions was NS1-1/AG4-1. The second generation fusions necessitated the production of bovine and ovine heteromyeloma fusion partners.

These heteromyeloma lines were isolated from both bovine x murine and ovine x murine fusions. In both cases NS1 mouse myeloma cells were fused to peripheral blood lymphocytes from either cattle or sheep, immunized against testosterone-3-carboxymethyloxime-ovalbumin. Cell lines secreting antibodies to testosterone were isolated but several of these lines subsequently ceased secreting. A number of these were cloned by limiting dilution into media containing 8-azaguanine (Sigma). This purine selects for cells which are deficient in the enzyme HGPRT, and therefore sensitive to aminopterin. These cells were tested frequently for aminopterin sensitivity and maintained in azaguanine medium for 60 days. The resulting cell lines can be substituted for the murine myeloma in fusions since they are both aminopterin sensitive and potentially immortal.

Cell fusion

Peripheral blood lymphocytes were collected from the jugular vein of animals previously immunized against steroid-protein conjugates.

The lymphocytes were separated using ficoll density gradient centrifugation (Lymphopaque, Nyegaard, UK) and washed with Hank's balanced salts solution. Lymphocytes were then fused to either mouse myeloma or heteromyeloma cells at a ratio of between 4:1 and 8:1. The cells were mixed, pelleted and drained before being resuspended in 0.8 ml of PEG 1500 (BDH) over 2 minutes. This was immediately followed by gradual dilution with 10 ml of Leibovitz L15 (Gibco) medium over 6 minutes. The fused cells were washed twice with L15 and resuspended in Dulbecco's modified Eagle's medium (DMEM) supplemented with 10% Myoclone foetal calf serum (FCS) (Gibco) and hypoxanthine, thymidine and aminopterin (HAT medium). The cells were plated out over a monolayer of mouse thymocyte 'feeder' cells in 24-well tissue culture plates (Nunc). Plates were then incubated in a CO_2 gassed incubator at 37°C and media was replaced, when exhausted, with further HAT media. After 7 days media was replaced with HT supplemented medium only.

Screening

All cultures were screened for specific antibody production from 7 days after fusion. Most secreting colonies were initially detected in the first three weeks, but screening continued for about 2 months. The screening assays developed were rapid and simple ELISAs.

The PVC flexible plates (Dynatech) were rinsed with an alkaline carbonate/bicarbonate coating buffer (pH9.6) and coated with the relevant steroid-protein conjugate. These conjugates were produced using the mixed anhydride reaction, except for the oestradiol conjugate which was purchased from Sigma. The carrier proteins (chicken lgG or BSA) used in the preparation of the coating conjugates were tested to ensure that there was no cross-reactivity with either antibodies against the carrier protein used for the immunogen (ovalbumin) or with the enzyme-labelled anti-immunoglobulin antibody. The wash buffer at all stages was phosphate buffered saline containing 0.5% gelatin and 0.5% Tween 20 to reduce non-specific binding effects.

Supernatant samples were transferred directly from the tissue culture plates to assay plates and were assayed undiluted. On each plate a dilution series of a polyclonal, non-specific antisera against the relevant steroid was included as a quality control. After incubation and washing, the plates were incubated with horseradish peroxidase (HRP) labelled anti-species antibodies. In the bovine assays, this was a rabbit anti-bovine IgG (Dakopatts), cleaned up prior to use on a Sephadex G200 column. For the ovine assays, an affinity purified donkey anti-sheep IgG, (Guildhay Antisera), was used. Following incubation and washing, the enzyme was visualized using orthophenylenediamine.

Colonies detected as secreting specific antibodies in two consecutive assays were rapidly cloned by limiting dilution over mouse thymocyte feeder cells. Antibody secretion was monitored regularly throughout production. Careful scrutiny of secretion rates, followed by prompt cloning if a significant fall was detected, resulted in no secreting lines being lost during culture after the initial cloning. After 3 or 4 clonings, antibody secretion rates remained relatively constant for any given conditions and medium.

Characterization

After clones were established as stable secretors, further characterization of the cells and their secreted product was carried out (Table 1). Of the five lines established to date, the fusion efficiency (clones per lymphocyte fused) has been between 1 x 10^{-7} and 2 x 10^{-6}. This compares with typical values for a murine x murine system of between 1 x 10^{-7} and 1 x 10^{-5}. The specific fusion efficiency was 5 -10% of the overall fusion efficiency.

The specificity of the antibodies was established in a liquid phase radioimmunoassay using a tritiated tracer, in order to reflect the in vivo situation as closely as possible. Displacement studies using potentially cross-reacting steroids demonstrated that both ovine and bovine antibodies raised against testosterone were primarily specific for that steroid, although they exhibited a high degree of cross-reactivity with DHT. The bovine antibody raised against oestradiol was surprisingly more specific for oestrone than for oestradiol.

The class of the ovine monoclonal antibodies was established by the use of an anti-IgG antiserum in the screening assays and also by checking the cryogenic stability of the antibodies. Bovine antibody sub-class was established using competitive ELISA (Groves et al, 1987) with rabbit anti-bovine sub-class IgG_1, IgG_2 and IgM antibodies (Jackson) and bovine sub-class standards from Miles. The assay was completed with HRP-labelled donkey anti-rabbit IgG (Guildhay Antisera).

The affinities of the antibodies were calculated using the same RIA system as that for the cross-reactivity determinations, from equilibrium binding studies performed at 4°C according to the methods of Scatchard (1949) and Rosenthal (1967). Affinities are equal to, or better than the best polyclonal antisera.

Secretion rates which have been determined were comparable to those reported in murine x murine systems.

Karyotyping studies have revealed that most of the first generation clones appear to have only a few more chromosomes than the myeloma line. Both bovine x (bovine x murine) heterohybridoma lines, however, have an average chromosome complement of about 120, compared to the bovine complement of 60 and the murine one of 40; of these 8-10% are definitely of bovine origin.

Both bovine cell lines have been successfully adapted to a wide range of different media in order to facilitate purification and/or reduce costs. Both cell growth and antibody secretion vary markedly with different basal media and supplements (Fig. 3). DMEM or RPMI 1640 supplemented with 5% newborn calf serum (Gibco) gave a useful, and reproducible, balance between cellular replication, antibody production and medium costs.

Cell line sterility was routinely checked using nutrient agar at 20 and 37°C and screened for mycoplasma contamination using a Gibco 'Mycotect' kit. Production stocks were cultured in the absence of antibiotics.

Interspecific hybridomas will not produce ascites in normal BALB/c mice. Batches of the heterohybridoma cells were therefore injected IP into pristane-primed Onu, genetically athymic mice. Ascitic tumors became established between 30 and 50 days after the initial injection. Subsequent passages of ascitic cells exhibited a shorter and more uniform delay between injection and harvesting, but resulted in decreased antibody

Table 1

Line	O/MT	O/MP.1A.9	B/MT.4A.17	B/ME.8.10
Species	Sheep	Sheep	Cattle	Cattle
Immunogen	Testosterone-3CMO-Ovalbumen	Progesterone-11-Hemisuccinate-BSA	Testosterone-3CMO-Ovalbumen	Oestradiol-6CMO-Ovalbumen
Stability	6 Months	2 Months (ongoing)	23 Months	20 Months
Specificity	Testosterone	Progesterone *	Testosterone	Oestrone/ Oestradiol
Mab Class	Ovine IgG	Ovine IgG	Bovine IgG1	Bovine IgG1
Dissociation Constant (K_D)	7.63×10^{-12}M	NT	2.5×10^{-11}M	5.33×10^{-10}M
Secretion Rate	2.5 ng/ml	NT	3.4 μg/10^6cells /24 h	20 μg/10^6cells /24 h
Chromosome Number	NT	NT	111	130
Fusion Partner	NS1	NS1	B/MF-2	B/MF-10

* Tested in solid phase only.

levels. In the initial passage, yields were variable in volume (0.5 to 5 ml/animal) and antibody concentration (5 to 200 μg/ml).

DISCUSSION

Many of the current applications of polyclonal antisera in the clinical and veterinary fields could be more readily served by the development of non-specific monoclonal antibodies (see James and Bell, 1987; Sherman and Markham, 1986). Current technological limitations in the generation of non-specific myeloma lines, together with the generally poor secretion rates achieved with the EBV transformation system, dictate that most non-rodent monoclonals must be generated using interspecific fusions. The fusion efficiencies of murine myeloma cells and non-murine lymphocytes can be significantly lower than homogeneic fusions, and the secretion rates achieved are generally lower. However, it has been demonstrated that interspecific fusions are a viable technique for the generation of non-murine Mabs, provided sufficient attention is paid to early cloning of secreting lines. Refusion techniques to increase the non-murine chromosome complement have been shown to improve the fusion efficiency compared to that of murine myelomas.

The demonstrable success of passive immunization to improve fecundity using polyclonal antisera has made the production of monoclonal antibodies an obvious step. Initial problems in the generation of stable cell lines secreting bovine or ovine antibodies have been overcome. Large scale production in a variety of static and suspension systems and cell reactors has been successfully carried out, as has ascitic production in athymic mice.

In conclusion, monoclonal antibodies derived from sources other than mouse and rat have a great, but relatively unexploited, potential. Current technology effectively limits the production of non-murine Mabs to interspecific fusions. Many of the reported disadvantages of interspecific fusions such as the low fusion efficiency and the poor stability of the resulting clones, can be overcome by the use of refusion techniques.

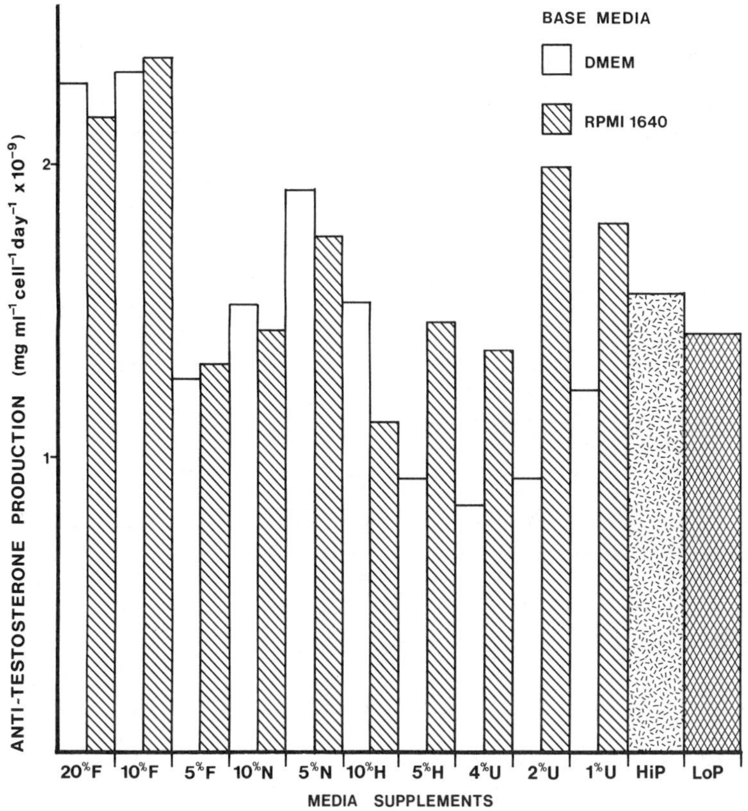

Fig. 3. Antibody production by B/MT cells cultured in different basal media and supplements. (DMEM: Dulbecco's modified Eagle's Medium (Gibco); RPMI 1640 (Gibco); HiP: Serum-free hybridoma medium, high protein (Gibco); LoP: Serum-free hybridoma medium, low protein (Gibco); F: Myoclone foetal calf serum (Gibco); N: Newborn calf serum (Gibco); H:Donor horse serum (Imperial Laboratories); U: Ultroser G, serum substitute (Gibco).

This research group have used, and is continuing to use, a combination of direct interspecific fusion and refusion methods to generate a panel of ovine and bovine antibodies for use in improving the fecundity of domestic animals. These Mabs should present a viable alternative to the use of exogenous synthetic hormones in the control of animal production and allow the exploitation of routes of immunological control not currently available with polyclonal or murine antibody preparations.

Acknowledgements

We would like to express our gratitude to Professors V. Marks and R. E. Spier, in whose departments this work was carried out. The work was supported by grant AG 90/9 from the A.F.R.C. to B.A.M.

REFERENCES

Anderson, D. V., Clarke, S. W., Stein, J. M., and Tucker, E. M., 1986, Bovine and ovine monoclonal antibodies to erythrocyte determinants, produced by interspecies hetero-myelomas, Biochem.Soc.Trans., 14:72.

Anderson, D. V., Tucker, E. M., Powell, J. R., and Porter, P., 1987, Bovine monoclonal antibodies to the F5 (K99) pilus antigen of E. coli, produced by murine/bovine hybridomas, Vet.Immunol.Immunopath., 15:223.

Groves, D. J., Morris, B. A., and Clayton, J., 1987, Preparation of a bovine monoclonal antibody to testosterone by interspecies fusion, Res.Vet.Sci., 43:253.

Guidry, A. J., Srikumaran, S., and Goldsby, R. A., 1986, Production and characterization of bovine x murine hybridomas, in: "Methods in Enzymology," Vol. 121, J.J. Langonne and H.V. Vunakis, eds., Academic Press, New York, pp. 244-265.

Hirata, Y., and Sugawara, I., 1987, Characterization of mouse-human hybridoma as a useful fusion partner for the establishment of mouse-human-human hybridoma secreting anti-tetanus toxoid human monoclonal antibody of IgM or IgG class, Microbiol.Immunol., 31:231.

Ichimori, Y., Sasano, K., Itoh, H., Hitotsumachi, S., Kimura, Y., Kaneko, K., Kida, M., and Tsukamoto, K., 1985, Establishment of hybridomas secreting human monoclonal antibodies against tetanus toxin and hepatitis B virus surface antigen, Biochem.Biophys.Res.Comm., 129:26.

James, K., and Bell, G. T. 1987, Human monoclonal antibody production: current status and future prospects, J.Immunol.Methods., 100:5.

Kohler, G., and Milstein, C., 1975, Continuous culture of fused cells secreting antibody of predefined specificity, Nature, 256:595:.

Land, R. B., 1976, The sensitivity of the ovulation rate of Finnish Landrace and Blackface ewes to exogenous oestrogen, J.Reprod.Fert., 48:217.

Land, R. B., Morris, B. A., Baxter, G., Fordyce, M., and Forster, J., 1982, Improvement of sheep fecundity by treatment with antisera to gonadal steroids, J.Reprod.Fert., 66:625.

Ostberg, L., and Pursch, E., 1983, Human x (mouse x human) hybridomas stably producing human antibodies, Hybridoma, 2:361.

Price, C. A., Morris, B. A., and Webb, R., 1987, Reproductive and endocrine effects of active immunization against a testosterone conjugate in the heifer, J.Reprod.Fert., 81:149.

Rhind, S. M., Gunn, R. G., Morris, B. A., Clayton, J., Leslie, I. D., and Gittus, G., 1985, Effect of passive immunization against testosterone on the reproductive performance of Scottish Blackface ewes in different levels of body condition at mating, Anim.Prod., 41:97.

Rosenthal, H. E., 1967, A graphic method for the determination and presentation of binding parameters in a complex system, Analyt.Biochem., 20:525.

Scaramuzzi, R. J., and Hoskinson, R. M., 1984, Active immunization against steroid hormones for increasing fecundity, in: "Immunological Aspects of Reproduction in Mammals," D.B.Crighton, ed., Butterworths, London, pp. 445-474.

Scaramuzzi, R. J., Davidson, W. G., and Van Look, P. F.A., 1977, Increasing ovulation rate in sheep by active immunization against an ovarian steroid androstenedione, Nature, 269:817.

Scatchard, G., 1949, The attractions of proteins for small molecules, Ann.N.Y. Acad.Sci., 51:660.

Schwaber, J. F., and Rosen, F. S., 1978, Induction of human immunoglobulin synthesis and secretion in somatic cell hybrids of mouse myeloma and human B lymphocytes from patients with agamma globulinemia, J.Exp.Med., 148:974.

Sherman, D. M., and Markham, R. T. F., 1986, Current and future applications of monoclonal antibodies against bacteria in veterinary medicine, in: "Monoclonal Antibodies Against Bacteria," Vol. III.,

A.J.L. Macario and E.C. de Macario, eds, Academic Press, Orlando, pp.295-340.

Teng, N. N. H., Lam, K. S., Riera, F. C., and Kaplan, H. S., 1983, Construction and testing of mouse-human heteromyelomas for human monoclonal antibody production, Proc.Nat.Acad.Sci.USA., 83:7308.

Trowbridge, I. S., 1978, Interspecies spleen-myeloma hybrid producing monoclonal antibodies against mouse lymphocyte surface glycoprotein T200, J.Exp.Med., 148:313.

Tucker, E. M., Dain, A. R., Wright, L. J., and Clarke, S. W., 1981, Culture of sheep x mouse hybridoma cells in vitro, Hybridoma, 1:77.

Tucker, E. M., Dain, A. R., Clarke, S. W., and Donker, R. A., 1984, Specific bovine monoclonal antibody produced by a re-fused mouse/calf hybridoma, Hybridoma, 3:171.

Van Meurs, G. J. E., and Jonker, M., 1986, Production of primate monoclonal antibodies, J.Immunol.Methods, 95:123.

Webb, R., Land, R. B., Pathiraja, N., and Morris, B. A., 1984, Passive immunization against steroid hormones in the female, in: "Immunological Aspects of Reproduction in Mammals," D.B. Crighton, ed., Butterworths, London, pp.475-499.

Weiss, M. C., and Green, H., 1967, Human-mouse hybrid cell lines containing partial complements of human chromosomes and functioning human genes, Proc.Nat.Acad.Sci., USA, 58:1104.

Wise, T. H., and Schanbacher, B. D., 1983, Reproductive effects of immunizing heifers against androstenedione and oestradiol-17β, J.Reprod.Fert., 69:605.

THE PRODUCTION AND USES OF MONOCLONAL ANTIBODY VACCINES

R. E. Spier and D. Langley

Department of Microbiology
University of Surrey, Guildford
Surrey GU2 5XH, United Kingdom

INTRODUCTION

Monoclonal antibodies (MAbs) offer a number of ways of arriving at one of the primary goals of those who hold that the way ahead for vaccine development is the administration of a product which is completely defined in molecular terms. It must be said that most, if not all, of the vaccines which are currently licensed for use in man or his animals contain many thousands of different molecular species. It may be thought that the recently licensed Hepatitis B vaccine made from a genetically engineered yeast cell, which produces a version of the glycosylated surface antigen of the infectious virus, is a vaccine which is as pure as is possible under the modern conditions of concentration and purification. Yet even if the purity of this material were to be quoted at 99% or thereabouts, there is much room for the existence of many different molecular species in such a highly purified material.

Do, then, the totally chemisynthetic vaccines offer the ultimate in defined molecular species for vaccine manufacture? Again the answer has to be qualified because the unknown impurities inherent in any operation which requires multiple reactions and which, in spite of the high yields of the individual reactions, must necessarily yield by-products of known and unknown natures.

In spite of the difficulties which are inherent in the production of a quantity of a defined molecular species it is well to consider the possible advantages such a preparation would possess. In the first place the quality control or assurance of the product material would be much simplified and the material could be admitted to the market with a higher degree of confidence of its efficacy and safety. Secondly, it is more likely that a relationship between the nature of the vaccine product material and the way it expresses its prophylactic effects can be determined. This is advantageous as it then becomes possible "rationally" to change the nature of the prophylactic and expect with surety beneficial consequences. Also it becomes possible to predict circumstances when such a vaccine should be contraindicated, thus the deleterious effects of some attempts at prophylaxis can be obviated. Thirdly the manufacturing process which leads to the immunogen can be so carefully defined and controlled that each and every production run can be planned with the certainty of generating a calculatable amount of useable vaccine.

In addition to the purity argument there is also the fear that the more conventional vaccines which are either based on a live attenuated pathogen or can, in a certain percentage of the vaccinees revert to, the more pathogenic organisms from which they were derived. The change of a single amino acid of the Polio type III capsid protein can change the vaccine virus to one which is pathogenic.

The use of killed vaccines is not without its problems. Inadequate inactivation can result in an outbreak of disease as was demonstrated in the Foot-and-Mouth Disease outbreak in the Isle of White in 1978. Such fears are echoed in many other virus and bacterial vaccines and in particular there is great concern over the nature of materials which are potential immunogens for those diseases such as AIDS which have a very high mortality. For these reasons vaccines which are independent of any trace or hint of pathogenic organisms have become the objective of both society and the vaccine manufacturers.

The production of some immunogens such as those involved in Hepatitis B infections and those of particular diseases of parasites such as the trypanosomes or the nematodes which cause schistosomiasis preclude the conventional approach to the development of vaccines. For such cases the manufacture of adequate quantities of immunogenic materials requires the exploitation of the new techniques which have become available for use in the last ten years; those of genetic engineering and hybridoma production.

How do MAbs meet the requirements which have been set out above and thereby lead to an advance in the currently operative techniques of vaccine manufacture? The outstanding characteristic of the MAb is that it consists of a single molecular species. However, it is generated by an animal cell, (although yeasts and other microorganisms can and have been engineered to produce such molecules), and therefore one can expect that the product MAb will be contaminated by cellular proteins, (from cells which have disintegrated or which have sloughed off vesicles or other subcellular components), and by the materials which are present in the culture medium. As will be recounted below, the culture media which can be used to maintain the hybridoma cells which produce the MAb's is often both highly defined, low in total protein content as well as relatively poor in terms of the number of components and their exoticity, so that it is expected that it is practicable to arrive at a product which is of the order of 99% pure with relative ease. Given the prospect of readily generating materials with such high levels of purity, can these compounds be used for prophylactic purposes? Subsequent sections will consider how such immunogens can be construed and, in the event of their practical utility, how they can be made in quantity.

MONOCLONAL ANTIBODY VACCINES

Vaccines consisting of almost pure monoclonal antibody offer a number of advantages over conventional vaccines;

(a) an almost completely defined nature,
(b) absence of any connection with a live pathogen,
(c) Availability of the techniques of production on a large scale,
(d) high degree of versatility,
(e) safe processes for manufacture,
(f) inherent compatibility with the immune system,
(g) can be made without having to overcome the technical problems of cultivating an organism which has not been adapted to growth in vitro.

As in the case of conventional vaccines there are two ways in which MAb's can be used for vaccines. They may be presented to the vaccinee as preformed antibody and hence act in a passive way as, for many years, the standard vaccine for Hepatitis has been delivered. Alternatively MAb's may be used as "active" vaccines. To achieve such an effect the vaccine is designed to elicit a response from the immune system which then translates the stimulus and amplifies the response which leads to the generation of pathogen neutralizing levels of antibody, as for a typical organism based immunogen, and then there follows the process which results in the priming of the memory cells of the immune system.

The Passive Vaccines

Passive vaccines are thought to be required for a number of human diseases. These include those caused by Cytomegalovirus, Human Immunodeficiency virus (HIV) and Herpes Simplex viruses. The two routes which are under investigation for such vaccines are those based on the generation of either antibodies made in the cells of non-human species, typically mouse or rat, and those that are ostensibly human. Cross species antibodies have been shown to be of value in the control of the immune response when organ transplantation operations are undertaken. However, much of the emphasis of present work for passive vaccines for humans centers on the use of human MAb's.

Although hybridoma cells secreting large amounts of antibody per millilitre of culture medium are readily produced, (productivities of 100 to 500 micrograms of antibody per ml. of medium per day are often quoted), hybridomas which feature human cells as one of the partners are generally more than one order of magnitude less productive, (less than 1 to 5 micrograms per ml. are reported figures). Such productivities seem to be independent of the way in which the hybridoma cells are produced. The most common methods involve the fusion of a human myeloma cell line with a circulating blood lymphocyte, (removal of the spleen of a hyperimmunized individual has not been practiced generally), while a second method requires the transformation of a B-lymphocyte by an agent such as Epstein-Barr virus or most recently by engineering into the lymphocytes the oncogenic genes myc and/or ras.

More recently, the productivity of the mouse hybridomas has been exploited in two ways. In the first way heterohybridomas have been formed between mouse hybridomas and human lymphocytes in an attempt to promote the higher productivity of the heterohybrid. Secondly mouse cells have been engineered with chimeric genes comprising of DNA made up from the gene coding for the constant region of a human antibody with the gene which codes for the variable region derived from a mouse cell line. In this way it has been possible to use the hyperimmunisation and spleen excision techniques which are generally not possible in human subjects. The latest developments along these lines (although the work was not initially directed at the production of a prophylactic, further developments of the idea could be used to that end), are the production of a MAb whose composition is bifunctional, in that each of its two combining sites is designed to react with a completely different antigen.

Monoclonal Antibodies as Active Vaccines

The infectious agents which challenge the immune system may be considered as collections of reactive groups with which antibody molecules react. The active groups on the antigen are called epitopes while the corresponding groups on the reacting antibody molecule are called paratopes. The collection of paratopes on an antibody molecule combined with those other reactive groups in the hypervariable area of the antibody molecule which can be identified with antisera or antigens are collectively called idiotopes and the antibody molecule as a whole is characterized by such a collection of groups as an idiotype. Other parts of the antibody molecule react with antisera and when injected into responsive animals provoke an immune response. Thus there are the determinants of the isotype of the antibody which are characteristics of the class of immunoglobulin involved (IgA or IgG or IgG1 or IgG2A or IgM etc.). The allotype of determinant is one which is recognizable between animals of the same species while the xenotype is a reflection of the differences between molecules which is caused by their origin in an animal of a different species. Theoretically and, to a large measure in practice also, antibodies which are autologous, that is molecules which are made by an individual animal and which are then injected into another animal of the same genotype, (syngeneic), generate a response in the recipient animal which is due to the only "different" area of the innoculated molecule which is the area defined by the idiotype. Alternatively, were an antibody molecule derived from one animal and injected into an animal of a different species then the recipient animal would mount an antibody response to all the xenotypic determinants which would include those of the idiotype as well as allotype and possibly isotype. Having introduced the required terms it is possible to describe how an antibody molecule which is normally one of the main products of the vaccination process, can itself become an active vaccine.

Following the stimulation of an animal by the injection of an antigen the recipient mounts an immune response, one component of which is the production of a variety of antibody molecules most of which are in some way capable of reacting with the active groups (epitopes) or the presented antigen. Were such antibody molecules to be removed from the so stimulated animal and injected into a second animal of the same species then that second animal would make antibodies to the injected antibodies which would recognize the major differences which are those at the site of the idiotype. This latter site is the one which is responsible for the combination with the antigen. The second animal would therefore generate antibodies which would be the mirror image of that site on the injected antibody which combined with the original antigen. In which case the configuration of the combining site of the anti-idiotypic antibody has a strong chance of resembling the antibody stimulating groups of the original antigen. Were this antibody molecule to be removed from the second animal and injected into a third animal it would be expected that the latter animal would mount an antibody response which would be similar to, if not identical with, the response of the first animal to the original antigen. In this way an antibody molecule has been made to mimic a particular antigen.

The description of the way in which the anti-idiotype antibody is made and the way in which it can be used to stimulate an immune response has been in terms of the normal polyclonal response which is the product of whole animal experimentation. However, were the antibody molecules described above to have been made from hybridomas then the way opens up to the production of such material in large-scale manufacturing operations. Indeed for each of the two antibody separations required, it is practicable to use a MAb. It is however necessary to sift through all the possible MAb's normally formed to find the ones which are the most reactive in the specific system under investigation.

It must also be understood that the account given above is an inevitable simplification. There is little doubt that T-cells are involved in the response to injected antibodies as it is through such agencies that the antibody response is controlled in the first place. It is also the case that the most efficacious responses are those which can be generated from the coinjection of a number of defined antibody molecules so that, in a deliberate manner, both arms, humoral and cellular, of the immune response can be invoked.

It is also important to realize that the response which can be obtained to the injection to an anti-idiotype antibody is dependent on a number of factors:

(a) the amount of the material injected; this controls whether there is to be a response at all and in the event that there is a response whether that response is stimulatory or repressive to the development of the immune state,
(b) the age of the animal at the time of stimulation,
(c) the site of injection of the anti-idiotype antibody and the immunostimulation formulation in which it is administered,
(d) the degree of foreigness of the material and
(e) the extent to which the injected material induces cross reactions.

Given all these difficulties and reservations there is little doubt that mice can be protected against infection by trypanosomes by the method of vaccination with an anti-idiotype antibody molecule preparation. Some protective efficacy has also been demonstrated by the stimulation of rabies virus neutralizing antibodies in rabbits by the injection of the appropriate anti-idiotype antibody. The production of antibodies to the hepatitis B virus surface glycoprotein has been enhanced by the prior injection of an anti-idiotype antibody to that virus. Indeed such priming of the immune system to a number of immunogens by anti-idiotype antibodies seems to be one way in which such material may move into common useage in the near future. In addition to the priming activity it is also possible that the coadministration of anti-idiotype antibody with original antigen can yield responses, in a model system involving phosphorylcholine tagged pneumococcal antigens, which are twice as large as the sum of the responses to the individual stimulants. Furthermore, the attachment of the anti-idiotype antibody to a non-specific stimulant of the immune system such as keyhole limpet haemocyanin can also evoke a strong response to the antibody moiety of the conjugate.

Work is also in hand with attempts to produce;

(a) cross reactive responses to influenza viruses,
(b) reactions to Reovirus haemagglutinin,
(c) Polio virus capsid proteins,
(d) Herpes Simplex glycoprotein,
(e) Rhino virus proteins,
(f) Feline Leukemia virus glycoprotein and
(g) Human Immunodeficiency virus envelope glycoprotein.

In view of the exciting possibilities attending on the use of this new concept of prophylaxis it is worthwhile to consider how such materials can be made in the quantities required by the vaccine manufacturers using either existing techniques or the methods which are issuing in profusion from the biotechnology companies who are seeking to make their contribution to the advancement of mankind.

MONOCLONAL ANTIBODY PRODUCTION TECHNIQUES

Overall Efficiency

There are a number of techniques whereby MAb's may be produced. Each has its advantages and disadvantages. There is as yet no consensus as to which of the in vitro techniques is to be preferred and indeed the literature is replete with productivity data which have little relevance to the "bottom line" which recounts the actual cost of making a defined amount of material at a stated degree of purity with a quoted measure of biological activity per unit mass of material. At present we have a situation wherein the following parameters are quoted;

(a) the yield of product per unit animal cell, (or per unit or measure of biomass), tells the reader about the physiology of the cells and the degree to which those cells are performing in relation to some "animal cell biomass ultimate capability"; it does not necessarily relate to economic or bioprocess performance,

(b) the yield of product per unit of medium used expresses one of a number of important factors which may be relevant to the overall economics of the process but which may also relate to the physiology of the cell, or when a recycle or make up system is used, it may reflect a more materially efficient process; clearly the two factors of overall raw material cost and the simplicity of the composition of the medium stream are some indication of the likely cost of the downstream processing operation,

(c) the yield of product per unit volume of bioreactor per unit of time is a measure of the efficiency of one of the key elements in the production system; however, this factor can be misleading because in some bioreactors where the cells are held at high cell densities, (in excess of 10^8 cells/ml), either a large or small volume of medium may be made to flow through the cell containing part of the system; under such circumstances it is more relevant to the determination of the space requirements of the bioreactor system to consider the volume of all the components which have been used to hold and feed the mass of animal cells in the system,

(d) the cost of a unit weight of the final product at a defined state of purity and activity is the most valuable determinant of the efficiency of a process, yet it is a parameter which is seldom determined and often difficult to acquire; it is critically dependent on the performance of the cell and on medium utilization efficiencies as well as on the nature of and performance of the downstream concentration and purification operation.

It is clearly necessary to arrive at a common way of evaluation of the alternative bioprocesses which are available and in a manner whereby those who make the decisions on the nature of the bioprocesses they chose to adopt are provided with figures they can rely upon.

In Vivo Methods

While much of the MAb material used in the diagnostic tests has been made by the in vivo route, (mouse ascites tumors), the problems of the quality of the raw material derived from this source in terms of its high degree of contamination with rodent proteins and antibodies, as well as its putative carcinogenic potential, the high cost of labour needed for this manual operation and the unethical nature of the procedures militate strongly against this method becoming a preferred technique for the production of MAb's for human injectible prophylactics.

In Vitro Methods

There are at present two main contenders for the preferred production system. One is based on the principle of providing each of the cells in the bioreactor with an environment which is completely homogeneous whereas the alternative technique allows the development of concentration gradients within the bioreactor such that some of the cells experience different conditions than their bioreactor cohabitants.

Homogeneous Systems

Two bioreactor types vie for dominance in this area. The one is based on the mixing of the bioreactor contents using a stirring mechanism whereas the second system relies on the use of air bubbles to achieve the same effect and to aerate the medium at the same time. Both of these approaches tend to be characterized by short batch times, (figures of up to 10 days are given), even when some attempt is made to continually perfuse the system with fresh medium and the product streams are relatively dilute. There is also a greater tendency for such systems to be more demanding of medium containing added growth factors, be they whole serum or defined materials such as insulin, transferrin or interleukin 2. Were whole serum to be used in addition to the costs involved the protein load on the downstream processing operation leads to increases in cost per unit of material generated. In the event that a low protein plus specific growth factor route is chosen then the cost of the specialized materials becomes a significant part of the final bill for the product.

Inhomogeneous Systems

There has presently become available a bewildering array of alternative systems for the cultivation and exploitation of hybridoma, (and other), types of animal cells in culture. In general the cells are immobilized either in containers or in the matrix of solid or semi solid materials. As they are not moving about the bulk of the bioreactor freely, the cells experience gradients in the supply of raw materials as well as in the concentration of waste materials and products. The kinds of alternatives available can be appreciated from the offered listing;

(a) Immured systems in which the cells are held within a container;

i hollow microspheres made of polyornithene and polylysine,
ii immuration on the "shell side" of a hollow fibre bundle contained in an outer case,
iii immuration between sheets of parallel planar membranes,

(b) Entrapped systems wherein the cells are held within the interstices of a material or matrix of a gel,

i entrapment within alginate, agarose or collagen particles which move freely through the body of the bioreactor,
ii entrapment within the crevasses of ceramic materials,
iii entrapment within an open sponge like matrix of polyester.

In each of the above cases run lengths of several months have been quoted and it is the experience of the investigators that there is little need for the complex or costly growth factors which compound the purification and economic problems. However, each of the systems mentioned above has disadvantages as well as advantages. Some of them are difficult to scale-up while others incur problems during the setting up phase. While the immured systems are generally freely available from the equipment manufacturers the entrapped alternatives are often covered by patents which renders their use less free.

It is clear that from this welter of opportunities there are at present many options for the manufacture, on commercial scale, of MAb's.

Thus once it has become clear as to which of the MAbs is required for a prophylactic purpose the production techniques can be applied to satisfy the demand.

CONCLUSIONS

1. MAbs can be used for prophylactic purposes and can offer advantages over the use of conventional vaccine materials.

2. Much remains to be discovered as to which MAbs are the most appropriate and the methods by which they can be most effectively applied.

3. The techniques for the large-scale production of MAbs are available; it remains to be determined which one(s) are the most efficient for factory floor use.

4. There is yet much room for investigative and synthetic experimentation to arrive at optimal solutions to these problems.

REFERENCES

"Animal Cell Biotechnology," 1985, R.E. Spier and J.B. Griffiths, eds., Vols, 1 and 2 (Vol. 3, 1987) Academic Press, London.

"Developments in Biological Standardisation," v. 42, 46, 50, 55, 60, 66, W. Hennessen, R.E. Spier and J.B. Griffiths, eds., Kager, Basle.

Hay, F. C., Thanavala, Y., and Roitt, I. M., 1984, Idiotype Vaccines, in: "Immune Intervention," v. 1. p. 117-138, I. M. Roitt, ed., Academic Press, London.

Klausner, 1987, Quadromas yield bispecific antibodies, Bio.Technol. v. 5. p. 195.

Lambert, P. H., 1986, Anti-idiotypic vaccines, in: "Progress Towards Better Vaccines," p. 88-95, R. Bell and G. Torrigiani, eds., Oxford University Press, Oxford.

"Large Scale Mammalian Cell Culture," 1985, J.Feder and W. Tolbert, eds., Academic Press, London.

McCullough, K. C., and Langley, D., 1985, Anti-idiotope vaccines: can they exist? Vaccine, v.3:59.

van Brunt, J., 1987, More hope for anti-idiotypic vaccines, Bio.Technol. v. 5. p. 421.

ANTIBODY-TARGETED PLASMINOGEN ACTIVATORS

Edgar Haber, Marschall S. Runge, Christoph Bode*, Elizabeth E. Branscomb, Janet M. Schnee**, Thomas Quertermous and Gary R. Matsueda

Massachusetts General Hospital and Harvard Medical School
Boston, Massachusetts 02114

INTRODUCTION

Survival after myocardial infarction is dramatically improved if a plasminogen activator can be administered within the first hours following chest pain. In a multicenter study of over 11,000 patients, mortality was nearly halved in those who received streptokinase (GISSI, 1986). Unfortunately, the administration of streptokinase, as well as that of the more recently studied tissue-type plasminogen activator (tPA), is often complicated by severe bleeding (Collen et al, 1984; TIMI Study Group, 1985) and in approximately 1% of patients, by cerebral hemorrhage - a commonly fatal event (Sheehan et al, 1987). This lytic state is thought to be caused by the destruction of plasma fibrinogen and clotting factors V and VIII (Laffel and Braunwald, 1984).

In an effort to improve the risk/benefit ratio of thrombolytic therapy, we decided to construct plasminogen activators that were characterized by greater selectivity for fibrin. The vehicle for effecting this enhancement of affinity and specificity was a high-affinity antibody specific for fibrin that does not cross-react with its precursor, fibrinogen.

Fibrin-specific antibodies can be raised by immunizing with peptides representing epitopes that are not accessible on fibrinogen (Pacella et al, 1983; Hui et al, 1983; Hui et al, 1986). For example, the amino termini of the α and β chains of fibrin are two new epitopes that are generated after thrombin releases fibrinopeptides A and B from fibrinogen. With synthetic amino-terminal peptides linked to protein carriers, our laboratory has produced high-affinity antibodies that bind uniquely to fibrin in vitro, as well as to thrombi in vivo (Hue et al, 1985; Liau et al, 1987).

To test the feasibility of using these antibodies as vehicles for targeting plasminogen activators to fibrin, we have constructed several

Present address: * Medizinische Klinik III, Kardiologie, Universität, Heidelberg, Bergheimerstrasse 58, 6900 Heidelberg, Federal Republic of Germany
** Dartmouth Medical School, Box 449, Hanover, New Hampshire 03756

plasminogen activator-antifibrin antibody molecules. Whole antifibrin antibody or its Fab has been linked to plasminogen activator, by both chemical and biosynthetic conjugation strategies. Another approach we have taken was to use a bispecific antibody as a bridge between the plasminogen activator and fibrin.

CHEMICAL CONJUGATES OF FIBRIN-SPECIFIC ANTIBODIES AND PLASMINOGEN ACTIVATORS

Urokinase or tPA were linked covalently to a monoclonal antibody specific for the amino terminus of the β chain of human fibrin by means of the unidirectional cross-linking reagent N-succinimidyl 3-(2-pyridyldithio) propionate (SPDP). The fibrinolytic potency of the conjugates at equal amidolytic activities was compared with that of the native plasminogen activators in an assay measuring lysis of ^{125}I-labeled fibrin monomer covalently linked to Sepharose CL-4B. Urokinase was the least potent, and the potency of tPA was ten times greater than that of urokinase. However, the fibrinolytic potency of the urokinase- and tPA-antifibrin antibody conjugates (and of a urokinase-antifibrin Fab conjugate) was 250 times greater than that of urokinase and 25 times greater than that of tPA (Bode et al, 1985; Bode et al, 1987a; Bode et al, 1987b; Runge et al, 1986). Fig. 1 compares the relative potencies of urokinase and a chemical conjugate between it and an antibody specific for the amino terminus of the β chain of fibrin (antibody 59D8) (Bode et al, 1987a). The enhanced fibrinolysis was fully inhibited by the amino terminal peptide that had been used as immunogen (β peptide), indicating that the enhancement was dependent on antigen binding (Fig. 2).

In fresh frozen human plasma (Lijnen et al, 1984), differences are not so dramatic, perhaps because of accumulation of inhibiting peptide fragments.

Conjugates between tPA or urokinase and antifibrin antibody produced a 3.2- to 4.5-fold enhancement in clot lysis over that of the respective unconjugated plasminogen activator. Although the antibody conjugates of urokinase and tPA were equipotent in the simplified assay described above, in the plasma assay the tPA-antifibrin antibody conjugate was more potent than the urokinase-antifibrin antibody conjugate. Whether this difference is related to an increased susceptibility in the urokinase conjugate to plasminogen activator inhibitors or to some other plasma component remains to be determined (Runge et al, 1986).

Of considerable interest is the apparent relative increase in the specificity of the plasminogen activator-antibody conjugates. Antibody-conjugated tPA and urokinase consumed less fibrinogen, α_2-antiplasmin and plasminogen than did the unconjugated activators, at equipotent thrombolytic concentrations (Runge et al, 1986). A likely explanation is that plasmin is preferentially generated in the local environment of the thrombus and is subsequently inactivated after diffusing into the supernatant plasma.

In a quantitative in vivo thrombolysis mode, in which a thrombus is formed in the jugular vein of a rabbit (Collen et al, 1983), the activity of a tPA-antifibrin antibody 59D8 conjugate (tPA-59D8) was compared with that of tPA alone and with that of a conjugate between tPA and a digoxin-specific monoclonal antibody (tPA-antidigoxin). After correction for spontaneous thrombolysis (10.9 ± 2.5 percent), tPA-59D8 was shown to be 2.8 to 9.6 times more potent than tPA alone (Fig. 3). Unconjugated tPA and tPA-antidigoxin were equipotent. At equivalent thrombolytic concentrations, tPA-59D8 degraded less fibrinogen and consumed less α_2-anti-

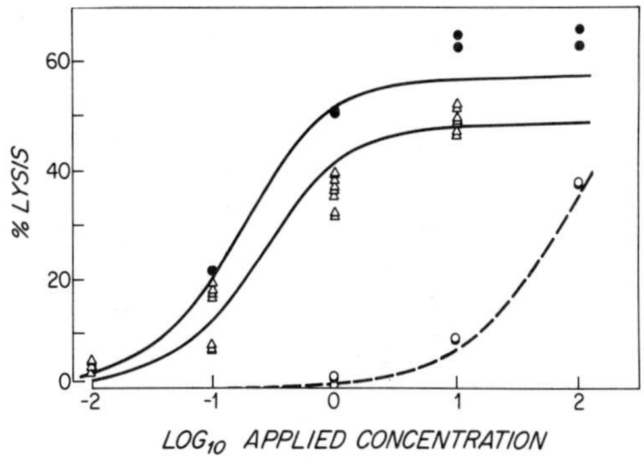

Fig. 1. Fractional lysis (expressed as percent lysis) versus applied concentration of activator for three sets of experiments where fractional lysis is represented after 15 h of the fibrinolysis reaction. Plasminogen activator: urokinase alone (open circle), antifibrin antibody-urokinase conjugate (open triangle), and antifibrin Fab'-urokinase conjugate (closed circle). Curves were fitted to the data points by a computer program. (Reprinted from Bode et al, 1987a).

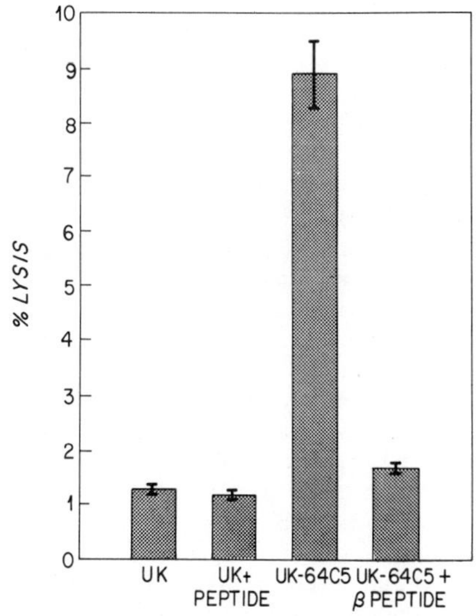

Fig. 2. Inhibition of Fibrinolysis by β peptide. Urokinase (UK) and antifibrin antibody-urokinase conjugate (at equal peptidolytic concentrations) were incubated with 5 mg/ml of β peptide for 60 min and then tested in the fibrinolysis assay in the presence of 5 mg/ml of β peptide. These were compared with urokinase and antifibrin antibody-urokinase conjugate (of equal peptidolytic potency) which had not been exposed to β peptide, in the same assay. The percent lysis was determined as described. Standard deviations are shown. (Reprinted from Bode et al, 1987a).

plasmin than did tPA alone (Runge et al, in press). These results suggest that tPA can be efficiently directed to the site of a thrombus by conjugation to an antifibrin monoclonal antibody, resulting in both more potent and more selective thrombolysis.

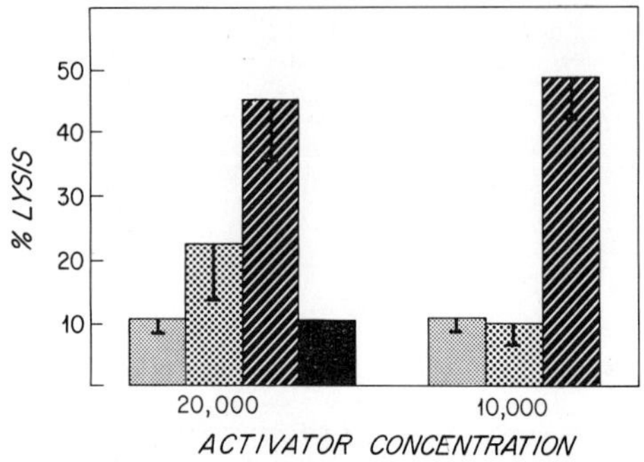

Fig. 3. Thrombolytic potency of tPA, tPA-59D8 and tPA-antidigoxin. tPA (coarse speckle), tPA-59D8 (cross-hatched), tPA-antidigoxin (solid) and control (infusion of saline, fine speckle) were compared in the rabbit jugular vein model at 20,000 units and 10,000 units per animal. Error bars represent the standard deviations. (Reprinted from Runge et al, in press.).

RECOMBINANT FUSION PROTEIN COMPRISING A FIBRIN-SPECIFIC ANTIBODY SITE AND THE B CHAIN OF tPA

The chemical coupling reaction between plasminogen activator and antibody has a relatively low yield (about 30%) and the product is not homogeneous because there are many potential coupling sites on both proteins. We were also concerned that irrelevant parts of the antibody molecule, such as the Fc, or of the plasminogen activator, such as the kringle, growth factor and finger regions, were present in the chemical conjugate and might, together with the cross-links, provide an unnecessary antigenic burden if these compounds were used in clinical therapy. For these reasons we decided to construct a recombinant protein that contained an antibody combining site, a hinge region for flexibility, and only the catalytic part (the B chain) of tPA.

The rearranged 59D8 heavy chain gene was cloned and combined in the expression vector pSV2gpt with sequence coding for a portion of the γ2b constant region and the catalytic B chain of tPA (Schnee et al, in press). This construct was transfected into heavy chain loss variant cells derived from the 59D8 hybridoma. The recombinant protein was purified by affinity chromatography and analyzed by electrophoresis, which revealed a 65-kD heavy chain-tPA fusion protein that is secreted in association with the 59D8 light chain in the form of a 170-kD disulfide-linked dimer. A chromogenic substrate assay showed the fusion protein to have 70 percent of the peptidolytic activity of native tPA and to activate plasminogen as efficiently as tPA (Fig. 4). In a competitive binding assay, reconstituted antibody was shown to have a binding profile similar to that of native 59D8. Thus by recombinant techniques we have produced a novel hybrid protein capable of high affinity fibrin binding and plasminogen activation.

CHEMICAL CONJUGATE OF A FIBRIN-SPECIFIC ANTIBODY AND A tPA-SPECIFIC ANTIBODY

tPA is present in normal plasma at a concentration of approximately 5 ng/ml. Although this concentration is far below that necessary for effective thrombolysis, a 10- to 50-fold concentration at the site of a thrombus might result in a therapeutically effective level. To accomplish this aim, we constructed a bifunctional antibody containing one combining

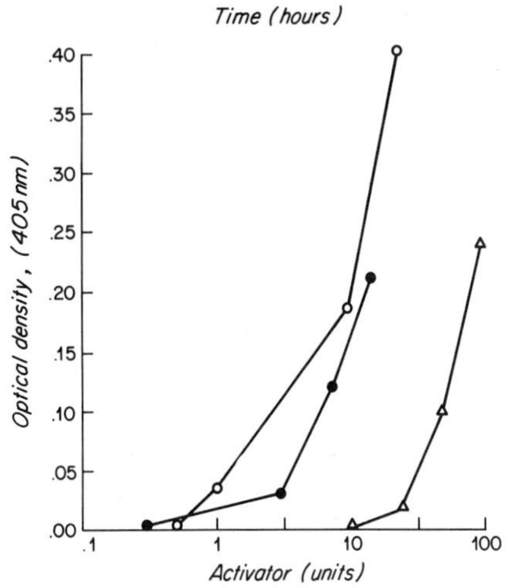

Fig. 4. Plasminogen activation was examined using the colorimetric substrate S-2251, which is not cleaved by enzymes other than plasminogen. The assay was conducted with varying activities of melanoma tPA standard (open circles), recombinant protein (filled circles), and bovine trypsin (open triangles) in the presence of plasminogen. The activity of each activator was standardized by measuring its ability to cleave the chromogenic substrate S-2288. (Reprinted from Schnee et al, in press.)

site specific for fibrin and another specific for tPA. It was important that the tPA-specific site bind but not inactivate the enzyme.

This type of bispecific antibody was first synthesized by conjugating and iminothiolane-modified anti-tPA monoclonal antibody (TCL8) to a pyridyldithio-modified antifibrin monoclonal antibody (59D8) (Runge et al, 1987). Sequential addition of the bispecific antibody and tPA to a plasma clot assay solution resulted in enhanced clot lysis. Two hundred units of tPA in the presence of bispecific antibody produced 23.1 ± 0.1% lysis, 400 units 29.5 ± 0.1% lysis. When tPA was added to the solution by itself, 200 units produced 1.8% lysis, 400 units 19% lysis. Despite the increased potency associated with the inclusion of bispecific antibody, fibrinogen and α_2-antiplasmin levels at equal tPA concentrations did not differ. Thus, it is possible to use and antibody conjugate with specificities for both tPA and fibrin to concentrate tPA (added separately) to the site of a thrombus in plasma.

BIOSYNTHETICALLY PRODUCED BISPECIFIC ANTIBODIES THAT ARE BOTH FIBRIN AND tPA-SPECIFIC

Chemically cross-linked antibody conjugates suffer the same disadvantages as do chemically cross-linked plasminogen activator-antibody conjugates (discussed above). A biosynthetically produced bispecific antibody would eliminate many of these drawbacks.

Bispecific antibodies were prepared in two ways (Branscomb et al, 1987). First, polyethylene glycol-mediated fusions were performed with two hybridoma cell lines: 59D8, which produces a monoclonal antibody specific for the β chain of human fibrin, and TCL8, which produces a monoclonal antibody specific for the catalytically active B chain of tPA. TCL8 cells were treated with 6-thioguanine to select for hypoxanthine phosphoribosyl transferase-deficient (HPRT-minus) variants and then fused with thymidine kinase-deficient 59D8 cells. One cell line, F36.23, possessed both anti-human fibrin and anti-human tPA immunoreactivities. A second approach yielded another bispecific antibody, F32.1. This cell line was selected after fusing TCL8 HPRT-minus cells with spleen cells from a mouse immunized with a synthetic peptide representing the amino terminus of the fibrin α chain. To remove unproductive chain recombinants and each of the parental antibodies, bispecific antibodies were purified from culture supernatants by serial affinity chromatography selective for antibodies able to bind both fibrin and tPA. Affinity-purified F36.23 and F32.1 retained antifibrin and anti-tPA activities and enhanced the fibrinolytic potency of tPA by a factor of 10.

Of the strategies described here, the recombinant fusion and biosynthetic approaches show the most promise for further refinement and, ultimately, clinical testing. The biologically produced bifunctional antibody with both plasminogen activator and fibrin specificities is of special interest, because it may effect thrombolysis by concentrating endogenous plasminogen activators (tPA or prourokinase) at the site of a fibrin deposit.

REFERENCES

Bode, C., Matsueda, G. R., Hui, K. Y., and Haber, E., 1985, Antibody-directed urokinase: a specific fibrinolytic agent, Science 229:765.

Bode, C., Runge, M. S., Newell, J. B., Matsueda, G. R., and Haber, E., 1987a, Characterization of an antibody-urokinase conjugate: a plasminogen activator targeted to fibrin, J.Biol.Chem., 262:10819.

Bode, C., Runge, M. S., Newell, J. B., Matsueda, G. R., and Haber, E., 1987b, Thrombolysis by a fibrin-specific antibody Fab'-urokinase conjugate, J.Mol.Cell.Cardiol., 19:335.

Branscomb, E. E., Matsueda, G. R., and Haber, E., 1987, Bispecific antibodies that bind both tissue plasminogen activator (t-PA) and fibrin (abstract), Clin.Res., 35:264A.

Collen, D., Stassen, J. M., and Verstraete, M., 1983, Thrombolysis with extrinsic (tissue-type) plasminogen activator in rabbits with experimental jugular vein thrombosis, J.Clin.Invest., 71:368.

Collen, D., Topol, E. J., Tiefenbrunn, A. J., Gold, H. K., Weisfeldt, M. L., Sobel, B. E., Leinbach, R. C., Brinker, J. A., Ludbrook, P. A., Yasuda, I., Bulkley, B. H. Robison, A. K., Hutter, A. M., Jr., Bell, W. R., Spadaro, J. J., Jr., Khaw, B. A., and Grossbard, E. B., 1984, Coronary thrombolysis with recombinant human tissue-type plasminogen activator: a prospective, randomized, placebo-controlled trial, Circulation, 70:1012.

Gruppo Italiano per lo studio della Streptochinasi nell'Infarto miocardico (GISSI), 1986, Effectiveness of intravenous thrombolytic treatment in acute myocardial infarction, Lancet, I:397.

Hui, K. Y., Haber, E., and Matsueda, G. R., 1983, Monoclonal antibodies to a synthetic fibrin-like peptide bind to human fibrin but not fibrinogen, Science, 222:1129.

Hui, K. Y., Haber, E., and Matsueda, G. R., 1985, Immunodetection of human fibrin using monoclonal antibody 64C5 in an extracorporeal chicken model, Thromb.Haemostas., 54:524.

Hui, K. Y., Haber, E., and Matsueda, G. R., 1986, Monoclonal antibodies of predetermined specificity for fibrin: a rational approach to monoclonal antibody production, Hybridoma, 5:215.

Laffel, G. L., and Braunwald, E., 1984, Thrombolytic therapy: a new strategy for the treatment of acute myocardial infarction, N.Engl.J.Med., 311:710.

Liau, C. S., Haber, E., and Matsueda, G. R., 1987, Evaluation of monoclonal antifibrin antibodies by their binding to human blood clots, Thromb. Haemost., 57:49.
Lijnen, H. R., Marafino, B. J., and Collen, D., 1984, In vitro fibrinolytic activity of recombinant tissue-type plasminogen activator in the plasma of various primate species, Thromb.Haemostas., 52:308.
Pacella, B. L., Jr., Hui, K. Y., Haber, E., and Matsueda, G. R., 1983, Induction of fibrin-specific antibodies by immunization with synthetic peptides that correspond to amino termini of thrombin cleavage sites, Mol.Immunol., 20:521.
Runge, M. S., Bode, C., Matsueda, G. R., and Haber, E., 1986, Tissue-type plasminogen activator-antifibrin monoclonal antibody conjugate enhances clot lysis in plasma (abstract), Circulation, 74(Suppl II): II-246.
Runge, M. S., Bode, C., Matsueda, G. R., and Haber, E., 1987, Antibody enhanced thrombolysis (1) capture of endogenous tissue plasminogen activator (tPA) by a heteroantibody duplex and (2) direct targeting by an antifibrin-tPA conjugate in vivo (abstract), Clin.Res., 35:643A.
Runge, M. S., Bode, C., Matsueda, G. R., and Haber, E., In press, Antibody-enhanced thrombolysis: targeting of tissue plasminogen activator in vivo, Proc.Natl.Acad.Sci., USA.
Schnee, J. M., Runge, M. S., Matsueda, G. R., Hudson, N. W. Seidman, J. G., Haber, E., and Quertermous, T., In press, Construction and expression of a recombinant antibody-targeted plasminogen activator, Proc.Natl.Acad.Sci., USA.
Sheehan, F. H., Braunwald, E., Canner, P., Dodge, H. T., Gore, J., Van Natta, P., Passamani, E. R., Williams, D. O., and Zaret, B., 1987, The effect of intravenous thrombolytic therapy on left ventricular function: a report on tissue-type plasminogen activator and streptokinase from the Thrombolysis in Myocardial Infarction (TIMI Phase I) trial. Circulation, 75:817.
The TIMI Study Group, 1985, The thrombolysis in myocardial infarction (TIMI) trial, N.Engl.J.Med., 312:932.

SECTION D
ANTIBODIES IN TUMOUR DIAGNOSIS AND THERAPY

MONOCLONAL ANTIBODIES IN THE DIAGNOSIS AND TREATMENT OF CANCER

Karol Sikora

Department of Clinical Oncology
Hammersmith Hospital
Du Cane Road, London W12 OHS

INTRODUCTION

There has been remarkable little progress in the management of patients with common solid tumors over the last 20 years. Surgery and radiotherapy have reached a peak of therapeutic effectiveness whilst the benefits of systemic treatment using drugs or hormones have reached a therapeutic plateau in tumors such as those of the breast, lung and colon. The reason for this is the lack of selectivity of currently available systemic agents. The closeness of the biochemical properties of tumor cells to their normal counterparts has made the successful development of anti-cancer agents an elusive quest.

The advent of monoclonal antibodies (MCAs) in the mid 70's led to considerable impetus for cancer research. Not only was there a new set of molecular flags to identify potential tumor specific molecules but also a revitalization of the old concept of the magic bullet. The essence of this is to use the antibody as a specific carrier for a powerful warhead. Over the last decade considerable effort has been put into devising suitable monoclonal antibodies for clinical use with varied success.

A major problem which has contributed to slow progress in terms of therapeutic gain has been the inability to identify tumor specific antigens. The search for these molecules was the holy grail of tumor immunologists in the pre-monoclonal era. It is now apparent that many molecules that can be recognized as tumor antigens are just quantitative differences in glycosylation patterns and not major changes in protein structure. For this reason the discriminatory power of most antibodies is not absolute.

ANTIBODIES TO CANCER CELLS

The molecular composition of the tumor cell surface differs from that of its normal counterpart. These differences can be detected by the effector arms of the immune response, both humoral and cellular. Although there is considerable evidence that the immune system can recognize the tumor cell surface as abnormal, there is little evidence that once a tumor has developed effector mechanisms are able successfully to inhibit neoplastic growth. Over the last century many investigators have tried to augment the specific response of the immune system to provide selective tumor destruction in both animal systems and patients with cancer (Curry, 1972). A major problem has been complexity of the immune system and our inability to break it down into its individual components.

Antibodies raised in a variety of animal species have been used in attempts to treat patients with cancer with little success. These sera have been prepared by immunizing animals with human tumor material and then absorbing the sera with human normal tissue to remove anti-human activity and, it was hoped, to leave behind specific anti-tumor activity. In most cases, however, such absorption removes most of the activity and therefore such sera are unlikely to be effective. The MCA technology yields unlimited quantities of pure reagents which avoid cross-reaction from contaminating antibodies and offers the best hope of providing reagents that will selectively recognize tumor cell surfaces. There have been many reports of the production of MCAs to human solid tumors (Ross et al, 1983). Whether a particular MCA will be clinically useful depends on several factors that can be measured in vitro; the degree of cross-reaction with normal tissue; the Ig class of the antibody; its affinity to target antigen; and the amount of free antigen in the serum potentially blocking its localization. Since predicting the possible effects in vivo of all these factors is difficult, it is important to assess monoclonal antibodies in carefully chosen clinical situations.

Such antibodies have other potential advantages over conventional antisera as well as their defined and reproducible specificity. They can be prepared as pure proteins and thus the total load of exogenous protein given to a patient is low. A comparison can be made of antibodies of similar specificity but varying in antibody class or affinity. Large batches of antibody can be made, allowing the comparison of each preparation by different investigators in different clinics. There are, of course, problems with MCAs. The workload involved in their production is enormous, with certain molecules of the tumor cell surface dominating on immunization, thus resulting in many MCAs to a single molecule. A further problem is activity against epitopes that are shared by several different disparate molecules, such as the sugar side chains of glycoproteins and glycolipids. Furthermore, by definition they cannot possess the range of properties found in antibodies of a polyclonal mixture with regard to complement fixation, antibody-dependent cytotoxic cell activation and a range of affinities. To select MCAs with particular functions or affinities requires even further effort.

DO HUMAN TUMOR ANTIGENS EXIST?

There is considerable evidence that the immune system is able to recognize tumor cells but almost nothing is known about the molecular nature and genetic specificity of the components on the cancer cell surface involved in this recognition. A variety of immunological assays have been used to examine components present on the cell surface. Several groups of antigens can be detected on tumor cells, some of which are shared with normal cells (Table 1). Normal transplantation antigens can be detected serologically and by cell-mediated lympholysis in vitro. These histocompatibility antigens are important in studying immune response to cancer for two reasons. The first is that unless syngeneic systems are used there will be confusion in the detection of antigens specific for transformed cells and not on normal counterparts. Secondly some tumor-specific transplantation antigens may in fact be modified forms of histocompatibility antigens. In a variety of animal systems tumor-specific transplantation antigens (TSTA) have been detected (Sikora and Lennox, 1982). These are defined by a rejection assay which involves the immunization of groups of syngeneic animals, either by excision of the growing tumor or by repeated injections of tumor cells made inactive by irradiation or by other means. After a suitable interval the animals are challenged with viable tumor cells and the growth of the resultant tumors is compared with that of a group of unimmunized animals. Several observations have been made from rejection assays. There appears to be a considerable range of immunogenicities amongst different experimental tumors. Chemically induced tumors often possess specific rejection antigens, i.e. immunization with one tumor confers immunity only to itself and not to others, even if induced by the same carcinogen in another location in the same animal. Clearly such assays are not possible with human tumors and so we do not know if human TSTAs exist. In virally induced experimental tumors the surface antigen characteristic of the transforming virus can, under certain circumstances, act as a transplantation antigen. Embryonic antigens are defined by anti-sera and cytotoxic lymphocytes prepared in syngeneic animals by injection of preparations of embryonic tissue. Such sera detect antigens on a wide variety of tumor cells which may well represent products expressed during the rapid growth of cells, both in embryogenesis and tumorigenesis. Normal tissue antigens which appear during differentiation can also be expressed in neoplastic cells (Bradstock et al, 1980). Under certain circumstances such antigens may appear tumor specific. An example is the idiotype of the immunoglobulin expressed on neoplastic B lymphocytes in patients with certain B-cell lymphomas.

STRATEGIES FOR THE PRODUCTION OF CLINICALLY USEFUL ANTIBODIES

The Perfect Antibody

Before describing laboratory strategies for producing and selecting suitable MCAs it is useful to consider the attributes of a perfect MCA for use in clinical oncology. Human MCAs have the attraction of not eliciting an immune response when administered to patients, with resulting abrogation of their effects and the problems of unpleasant immune complex phenomena. A perfect MCA should clearly be totally tumor specific with a high binding affinity to an antigen present in abundance on all tumor cells and no normal tissue. An IgG antibody has advantages compared with an IgM, as its relatively low molecular weight will allow free diffusion within the extracellular space. The MCA should be secreted in high concentration by a hybridoma line which should be adaptable for growth in serum-free medium to enable complete purification.

Immunization Schedules

Immunization schedules for preparing MCAs to human tumor surface antigens have varied widely. Very little detailed work has been performed on optimizing the schedules. Different groups have used different sources

Table 1. Antigens on Human Tumor Cells

Histocompatibility antigens
Tumour-specific transplantation antigens
Embryonic antigens
Differentiation antigens
Viral antigens

of tumor material, whole cells, membranes or solubilized products, either derived from cell lines or from clinical biopsy samples. Mice and rats have been immunized, using different timing schedules and different routes of immunization which will almost certainly produce different spectra of antibodies. In preparing human hybridomas no immunization is possible but there is a choice in the source of lymphocytes used for fusion.

Two techniques have been developed to enhance the frequency of antigen-specific hybridomas by enriching for antibody precursor or producing lymphocytes prior to cell fusion (Box et al, 1982). Spleen cells from animals immunized with the haptens dinitrophenyl and benzylpenicilloyl were cultured in vitro for four days in the presence of the immunizing antigen or transferred into x-irradiated syngeneic recipients, followed by an in-vivo antigen boost four days prior to their use in cell fusion. Both procedures decrease the total number of cells available for hybridization but increase the percentage of antigen-specific antibody secreting clones following cell fusion. Such techniques are now being investigated for the production of anti-tumor MCAs.

Fusion Methods

There are several hybridoma systems available from which anti-tumor MCAs can be prepared. The mouse system has predominated but rat hybridomas have recently become more popular. Several investigators have attempted to fuse human lymphocytes directly from patients with different types of malignant disease (Sikora et al, 1982). Such fusions have been performed with rat or mouse myelomas, so obtaining mixed species hybrids producing human monoclonal antibodies. Unfortunately, the stability of such hybrids is low, with preferential shedding of human chromosomes resulting in hybrids that lose immunoglobulin production. There are, however, several human myeloma lines now available for fusion. Human hybridomas established from such lines show no apparent preferential loss of chromosomes and thus immunoglobulin secretion is stable. The amount of immunoglobulin secreted is often low, in the order of 1 μg/ml (Sikora and Wright 1981). Human MCAs to tumors may well provide greater specificity than animal systems, although current experimental evidence is inconclusive. Furthermore, such antibodies allow the evaluation of the way in which the human immune system is able to see autologous tumor in vivo.

Screening Methods

Whichever immunization schedule and fusion system are used, the eventual spectrum of MCAs produced against any antigen will depend on the methods used for screening to find suitable immunoglobulins. There are three main screening methods: radioimmunoassay by binding to membranes or whole cell lines; immunohistology using either fluorescence or peroxidase techniques; and cytotoxicity either by the addition of complement or by antibody-dependent cell-mediated cytotoxicity. In searching for anti-human tumor MCAs, there are advantages in the use of immunohistology in a primary screen. As well as providing a rapid screening technique, this method also allows the acquisition of information about the exact location of an antigen and its distribution within a tumor. Binding radioimmunoassays provide the most rapid way of screening large numbers of supernatants for antibody activity but may not pick up all antibodies that have clinical use, such as those which activate cell mediated cytotoxicity.

Specificity Studies

Although it is relatively easy to construct sets of MCAs which have activity against a particular tumor, it is much more difficult to characterize specificity and produce antibodies reacting with defined target antigens. The most common method to characterize MCAs is to study binding to a panel of different cell lines. In a particularly elegant study, MCAs were raised by immunizing mice with a human melanoma cell line and tested for binding to a large panel of different tumor lines (Dippold et al, 1977). By increasing the sensitivity of these binding assays, apparent specificity often disappears. A good example of this is the work of Brown (Brown et al, 1981). The binding to cells and tissues of an apparently specific anti-melanoma antibody was investigated in a highly sensitive assay system using a second MCA (double determinant immunoassay). Its target antigen was shown to be present, although in small quantities on normal tissue as well as on tumor cells.

The search for MCAs with well-defined specificity is clearly essential if antibodies are to be of use for targeting drugs or toxic agents. For tumor localization, or in-vitro removal of tumor cells from bone marrow during autologous marrow transplantation, fine specificity may not be so essential for clinical usefulness.

DIAGNOSTIC USE

Immunohistology

The diagnosis of malignant disease is made by the clinical pathologist from a biopsy sample of tumor. Increasing complexity of the combinations of different therapies makes accurate pathological diagnosis essential in modern practice. In more than 90% of malignant biopsies, there is little doubt as to the nature of the tumor type and its malignant potential. unfortunately, the remaining 10% pose a problem. Anaplastic carcinoma, lymphoma, teratoma and undifferentiated sarcomas, may all present with very similar appearances on conventional microscopy, particularly when the size of biopsy is small. MCAs raised against tumors may be used to differentiate between different neoplasms. MCAs are now available for example, that can discriminate between different histological types of lung carcinoma (Cuttitta et al, 1981). Small cell carcinoma is known to metastasize rapidly in the brain and bones whilst other histological types remain localized within the chest for long periods of time. The availability of a MCA to identify small cell carcinoma may be of great use in a disease where often only limited amounts of material are available for pathological examination, such as bronchoscopic biopsies or clumps of cells in sputum samples.

Another clinical problem is the presence of neoplastic cells in a lymph node with no clinical evidence of primary disease. Such patients are subjected to an extensive laboratory and radiological work-up to search for the primary growth, often without result. Tissue type specific MCAs could be of use in determining the likely primary source of the malignancy. In non-Hodgkin's lymphoma, where at least 15 categories of disease have been defined many classification schemes exist based on morphology. There are now several panels of antibodies which are able to recognize different lymphocyte populations and provide information about the origin of the cells in a lymphoma sample (Nadler et al, 1983). This may help in the choice of the correct treatment for an individual patient.

ASSESSMENT OF TUMOR LOAD

Only a small number of cancer patients have disease which can be reliably assessed by clinical examination or diagnostic radiology. After removal of the primary growth there is often no evidence of tumor and therefore no way in which current techniques can be used to evaluate the re-growth of tumor until a mass of more than 1-2 cm arises (10^9). Certain relatively rare tumors shed products into the circulation, such as alphafetoprotein (hepatoma and teratoma), human chorionic gonadotrophin (choriocarcinoma and teratoma), CEA (Some colorectal and other gastro-intestinal carcinomas) and monoclonal immunoglobulin (multiple myeloma). Such shed tumor markers have been used to detect tumor masses of less than 10^8 cells, an order of magnitude less than can be detected by conventional investigation.

There are three clinical situations in which such knowledge is of use to the clinician. Firstly, the extent of treatment given to a primary lesion can be influenced by the knowledge that the lesion had already metastasized, as in carcinoma of the breast where more than half the patients have micrometastases at the time of first discovering a breast lump. Patients in which tumor markers remain elevated following excision of the primary could then go on to receive aggressive systemic therapy. Secondly, following treatment for a primary carcinoma, the ability to diagnose metastatic disease early may be advantageous, as systemic therapy can be given when tumor bulk is minimal and the patient's general health is still good. This is best illustrated in testicular teratoma where there is good evidence that recurrent disease of less than 2 cm in diameter responds better to chemotherapy than large tumor masses. Thirdly, the accurate measurement of tumor load provides objective criteria by which response to therapy can be monitored. This is of particular significance when considering duration of cytotoxic therapy or the development of resistance to particular forms of treatment.

There are now several reports of monoclonal antibodies being used to detect circulating markers. The detailed assessment of their role in patient management is currently under investigations (Darras, 1986).

TUMOR LOCALIZATION

Radiolabelled polyclonal antibodies against human neoplasms have been used to localize deposits of tumor by immunoscintigraphy. Several groups have used anti-CEA antibody to localize human colorectal carcinoma metastases in liver and lung. CEA however is produced by normal non-neoplastic tissue in small quantities and is not tumor specific. Furthermore, this antigen is often shed by a tumor into the peripheral blood and therefore any technique which images the distribution of this antigen will produce an image of the circulating blood pool. The development of monoclonal antibodies has several advantages. Their defined specificity may provide the contrast required for effective imaging and their production is reproducible on a vast scale. It may also be possible to use human monoclonal antibodies for such scans.

Several groups have not used labelled mouse and rat MCAs for tumor localization in man. The initial results are not spectacular, but much of the problem may reside in the physical limitations inherent in current scanning techniques. Mach and his colleagues (Mach et al, 1981) injected 14 patients with colorectal and pancreatic cancer with ^{131}I-labelled purified mouse monoclonal anti-CEA antibody. In 8 patients increased radioactivity was observed in the region of known deposits. To detect blood pool and secreted radioactivity a 99 m technecium scan was also performed. After subtraction of ^{99} Tc radioactivity from ^{131}I radioactivity a concentration of ^{131}I activity was found in the tumor site.

We have studied a family of monoclonal antibodies produced by the fusion of the rat myeloma line, Y3.123.AG. and lymphocytes obtained from the spleens of rats immunized with human colorectal carcinoma membrane

(Smedley et al, 1983). The supernatants from such hybrids were tested in a solid phase radioimmunoassay for activity against glutaraldehyde-fixed colorectal carcinoma cell lines. Positive hybrids were identified, cloned and bulk supernatants purified for subsequent investigation. From one immunization, 17 hybridoma clones were produced which showed activity against HT29 cells in a binding assay. Extensive immunoperoxidase and immunofluorescence studies on both normal colon and colorectal carcinoma were carried out and antibodies graded by the amount of staining produced on both normal and malignant tissue. In addition HT29 xenograft-bearing nude mice were injected intravenously with radio-iodinated antibody and scanned 48 hours afterwards. The animals were killed, organs were removed and uptake of ^{131}I antibody expressed in counts per milligram of tissue. The relative distribution of radioactivity between tumor and other organs could therefore be calculated (Table 2). From the results of binding assays, immunohistology and mouse xenograft immunoscintigraphy, it was possible to choose a MCA that was likely to be of clinical use for tumor localization.

We have now used this antibody, YPC.2/12.1, for tumor localization in patients with colorectal and other malignancies. The antibody was labelled with [^{131}I] iodine using a modification of the chloramine T method. 1.0 mg of purified monoclonal antibody labelled with 1.0 mCi of ^{131}I was injected intravenously; 48 hours later a second injection of ^{99m}Tc labelled human serum albumin and ^{99m}Tc-labelled pertechnetate was given. Patients were then scanned 30 minutes after the second injection and separate images of ^{131}I and ^{99m}Tc distribution recorded simultaneously. Computerized subtraction of the two images produced the best tumor image. There was good correlation between known clinical sites of disease and scanning results in 38 out of 52 patients with colorectal carcinoma. In some patients areas of disease were identified which had previously been undiagnosed; this could subsequently be confirmed by other radiological tests such as computerized tomography. We are currently investigating the use of radioimmunoscintigraphy with this and other monoclonal antibodies in the pre-operative assessment of patients undergoing radical surgery for colorectal carcinoma. The detection of unsuspected metastases would clearly influence the choice of operation and subsequent management of these patients.

Diagnosis can only be useful if it changes the management of an individual patient. So far, monoclonal imaging has not done this. One of the main reasons is the inability of systemic therapy for patients with metastases from one of the common solid tumors to improve survival. For those tumors where we do have effective strategies, radioimmunoscintigraphy cannot yet compete with the precise anatomical delineation given by computerized tomography or magnetic resonance imaging (Finan et al, 1986). Detecting necrotic cardiac muscle using anti-myosin monoclonals is of interest in correlating anatomical damage to function (Khaw et al, 1986), but again there is little evidence of impact on patient management. The quality of the information given by such scans is not adequate for choosing between infarct excision or conservative therapy. What are the problems? Most techniques involve giving a very small dose of antibody, usually one milligram, most of which continues to circulate in the blood pool. With most anti-tumor antibodies, less than 0.5% actually deposits on neoplastic cells, the rest is cleared by liver and the reticuloendothelial system. This has led to the use of secondary blood pool imaging agents allowing computerized subtraction of the circulating component. This further detracts from image clarity and can introduce considerable artifact (Sikora et al, 1983). A particularly ingenious attempt to get round the blood pool problem was to inject liposomes containing anti-murine antibody to mop up rapidly any free antibody in the blood (Barratt et al, 1983).

Table 2. MCA Killing Mechanisms

Complement cytotoxicity
Macrophage activation
Antibody-directed cell-mediated cytotoxicity
Natural killer cell activation
Immuno-regulation

Recently, there have been some remarkable advances in the technology of antibody design by a combination of gene shuffling and protein engineering. Molecules with high affinity for their target antigen can be designed (Roberts et al, 1987). Hybridoma cell lines producing an antibody of moderate activity can be specifically engineered to produce a much improved version. The molecule can be made smaller so it can penetrate easily into the extravascular space surrounding its target. Foreign species determinants can be replaced by human counterparts so humanising a mouse or rat product and reducing the risk of a secondary response. Furthermore, antibodies can be constructed to couple easily to a radioactive emitter for good imaging. This type of technology can considerably improve the amount of antibody depositing on a tumor and therefore the quality of image avoiding the need for subtraction. Whether an agent of widespread use will appear remains to be seen.

CANCER THERAPY

Arming Monoclonal Antibodies

Over the next few years it is likely that different monoclonal antibodies will find a routine place in the diagnosis and evaluation of cancer patients. Whether they will also have a role in delivering effective systemic therapy for metastatic disease remains to be seen. Serotherapy using conventional antibodies has been attempted for more than 50 years, so far with almost no success. Most of such sera have poor specificity and also a low titre against putative tumor antigens. Monoclonal antibodies may overcome both these problems and, in addition, the development of human monoclonal antibodies may prevent the immunological reactions against the administered antibodies with associated adverse side-effects. There are several mechanisms by which the administration of a suitable monoclonal antibody can result in tumor cell death (Table 3). In addition, drugs, toxins and radionuclides can be coupled to MCAs in such a way that their immunological activity remains unaltered (Table 4). The antibody therefore provides a targeting mechanism whilst the coupled agent acts as a war head. Indeed the level of complexity can be increased further by the coupling of drugs such as interferon which may trigger activity in natural-killer cells locally around the tumor (Embleton et al, 1982).

At present there is much published work on the use of MCAs either on their own or coupled with agents to destroy tumor cells in vitro. There is also evidence that such systems can be used to eradicate in vivo tumors in rodents (Thorpe et al, 1982). There are fewer reports of the clinical use of MCAs as therapeutic tools although many studies are now in progress. A fundamental problem in the therapeutic use of MCAs is the poor vasculature of many solid tumors. It may well be that the new immunotherapy will find a role only in the very early stages of malignant disease. Profound ethical problems surround the use of MCA therapy in these patients.

Table 3 Arming Monoclonal Antibodies

Drugs	Chlorambucil
	Adriamycin
	Methotrexate
	Neocarzinostatin
Toxins	Ricin
	Abrin
	Diphtheria Toxin
	Gelonin
Radionuclides	^{131}Iodine
	^{125}Iodine
	^{10}Boron
	^{211}Astatine
	^{90}Yttrium

In Vitro Destruction of Malignant Cells in Bone Marrow

Bone marrow provides a repository for tumor cells of a variety of types. Bone marrow can be relatively easily collected and stored from a patient whilst aggressive radiotherapy and chemotherapy can be administered at a dose which would normally be lethal to the patient because of marrow toxicity (Thomas et al, 1977). Following such therapy, the patient is given back his own marrow. The problem at present with this technique which has now been attempted in a variety of leukaemias and also some solid tumors, is that the re-infused marrow may contain tumor cells. In vitro treatment of a marrow suspension with a MCA may remove these cells. There are two problems with this approach; the first is that even the most aggressive radiotherapy and chemotherapy may not effectively destroy all tumor cells present in the body. The total dose of drugs and radiation is still limited by other rapidly turning-over organ systems such as the gut and skin epithelium. The second problem is ensuring that every single tumor cell is removed from the marrow before reinfusion. Although it is relatively easy to separate a few tumor cells from 10^7 marrow cells, scaling up to handle 10^{10} cells poses considerable technical difficulties.

SYSTEMIC THERAPY WITH MONOCLONAL ANTIBODIES

T-Cell Leukaemia

The first published clinical study on the therapeutic use of MCAs was in a patient with non-Hodgkin's lymphoma (Nadler et al, 1980). These lymphomas are a heterogenous group of diseases in which neoplastic lymphocytes develop, usually in lymph nodes, but subsequently spread to other organs including the blood and marrow. The patient was a 54 year old man with a one year history of lymph node enlargement with liver and spleen involvement. After extensive chemotherapy with both conventionally used drugs and investigational agents, the patient developed an unresponsive lymphoma-cell leukaemia. A mouse MCA against tumor cells from this patient had been derived a year previously which bound to a limited subset of B cell non-Hodgkin lymphomas. A total of 250 mg purified antibody was given by intravenous infusion to the patient with no untoward side-effects over a 60 hour period. Tumor cells were transiently cleared from the circulation but antibody activity was abrogated by a circulating blocking antigen, presumably shed by tumor cell membranes. Following this study, a total dose of 2 g of the antibody was administered in two courses in the subsequent three weeks, again without significant toxicity. Although there was some transient change in the condition of the patient, the tumor eventually grew out of control and the patient died. Factors contributing to the failure of therapy probably included the large number of tumor cells present, the weak expression on the cell surface of the tumor antigen and the quantity of circulating antigen which blocked tumor cell binding. The mechanism of cell destruction was almost certainly complement cytotoxicity although antibody-coated tumor cells from this patient were approximately three times more adherent to macrophage monolayers than similar cells treated with a non-binding MCA.

Another murine MCA L17F12, which reacts with a normal human T-cell differentiation antigen was used to treat patients with Sezary syndrome (Miller and Levy, 1981a). This rare condition is a lymphoma-cell leukaemia that occurs in about 5% of patients with the T-helper cell lymphoma, mycosis fungoides. Initially, this disease affects the skin, with widespread infiltration in the epidermis. As the disease progresses, it may involve lymph node, viscera and bone marrow. The patient's circulating lymphoma cells were shown to contain large amounts of the antigen recognized by the L17F12 by immunofluorescence. Prior to therapy, circulating antigen was not detectable in the serum using a competition radioimmunoassay. After two courses of intravenous therapy with a total of 7 mg of antibody there was a dramatic fall in the abnormal circulating cell count but this returned to pre-treatment levels within 48 hours. After the first dose of antibody, circulating free antigen and a transient impairment in renal function were noted. The latter suggested immune complex formation. At various times during and after antibody therapy the patient's cells were incubated with the antibody and studied by quantitative immunofluorescence using flow cytometry. Antigenic modulation by leukaemia cells was found transiently, following each course of therapy. A second patient with florid skin manifestations of mycosis fungoides was treated. Transient clinical and histological regressions were noted after antibody therapy (Miller and Levy 1981b).

Acute Lymphoblastic Leukaemia

Four patients with acute lymphoblastic leukaemia, three children and one young adult, were given MCA specific for the common acute lymphoblastic leukaemia antigen. CALLA (Ritz et al, 1981). There was a rapid decrease in circulating blast cells in three patients that began immediately after antibody infusion. Not all the leukaemic cells were cleared and the remaining cells appeared to be resistant to further therapy. Antigenic modulation was again a significant problem and may reflect the nature of the neoplasia treated.

Chronic Lymphoblastic Leukaemia

Two patients with chronic lymphocytic leukaemia (CLL) again extensively treated with chemotherapy, were given IV infusions of a murine MCA to a 65 dalton antigen (T65) on the surface of CLL cells, normal and malignant T cells and thymocytes (Dilman et al, 1982). There was a marked fall in the abnormal circulating cells with a prompt return within several hours to the previously raised level. A serious anaphylactic reaction developed in one patient on the third infusion of 10 mg of MCA with urticaria, diarrhoea, dysphoea and a fall in blood pressure. This responded rapidly to adrenaline, corticosteroids and plasma volume expansion. In vivo binding of antibody was demonstrated with subsequent tumor cell lysis.

Colorectal Carcinoma

There are very few studies of MCA therapy in patients with common solid tumors. A series of MCAs binding to human colorectal carcinoma have been studied (Hurlyn et al, 1979). One of these antibodies (17-1A) has been shown to activate antibody-dependent cell-mediated cytotoxicity by both mouse or human effector cells. The administration of this antibody to mice bearing human colorectal carcinoma xenografts inhibited tumor growth. Furthermore, this antibody perfused through freshly resected human colons containing adenocarcinomas selectively bound to cells of some of these tumors. A phase 1 clinical trial of this antibody in the treatment of gastrointestinal tumors has been performed in four patients who were given 15-200 mg of purified antibody. There was very little objective evidence of tumor regression. In one patient who received autologous MCA into a hepatic artery catheter, the hepatic metastases became smaller and there was a histological change in the infitrating cell type of a resected metastasis. No localization studies with radio-labelled MCA were performed and thus there was no clear evidence of tumor concentration of MCA in vivo.

Malignant Melanoma

A phase 1 evaluation to assess toxicity and kinetics of a MCA specific for a melanoma associated antigen was carried out in two patients with end stage melanoma (Sobol et al, 1982). Both patients had extensive disease and had failed to respond to all conventional chemotherapy. Both patients received 10 mg doses of the antibody intravenously over a two hour period. Subcutaneous tumor biopsies failed to show evidence of in vivo tumor binding. No major toxicity was observed. There was not evidence of tumor cell destruction.

Malignant Glioma

We have studied the use for localization and therapy of a human MCA antibody against glioma (Philips et al, 1983). The antibody was derived from a 21 year old patient presenting with an extensive frontal lobe tumor. At the initial operation a sample of tumor was removed, lymphocytes separated and fused with a human myeloma line, LON-LICR-HMy2. Hybrids were obtained which secreted small amounts of immunoglobulin some of which were found to bind to a variety of glioma cell lines. One such antibody, LGL 1-D6, was prepared in large quantities by growing hybridoma cells in serum-free medium in roller culture. After purification by ammonium sulphate precipitation and ion exchange chromatography, the antibody was labelled with ^{131}I. The patient subsequently relapsed with local recurrence which required the insertion of a catheter to decompress the pressure arising within the cranial cavity. 1 mg purified antibody labelled with 1 mCi of ^{131}I was given intravenously and serial samples collected from the patient's blood, CSF and intra-tumoural fluid. Three weeks later a similar quantity of labelled control human monoclonal IgG was given and similar samples obtained. Whilst the blood and CSF radioactivity were similar, the anti-glioma MCA activity was higher and more prolonged in the tumor fluid, suggesting preferential tumor uptake of MCA. This observation was confirmed by external rectilinear scanning (Fig. 1). A therapeutic dose of 100 uCi of ^{131}I coupled to MCA was subsequently given but no objective response was recorded. The patient had already received whole brain radiation.

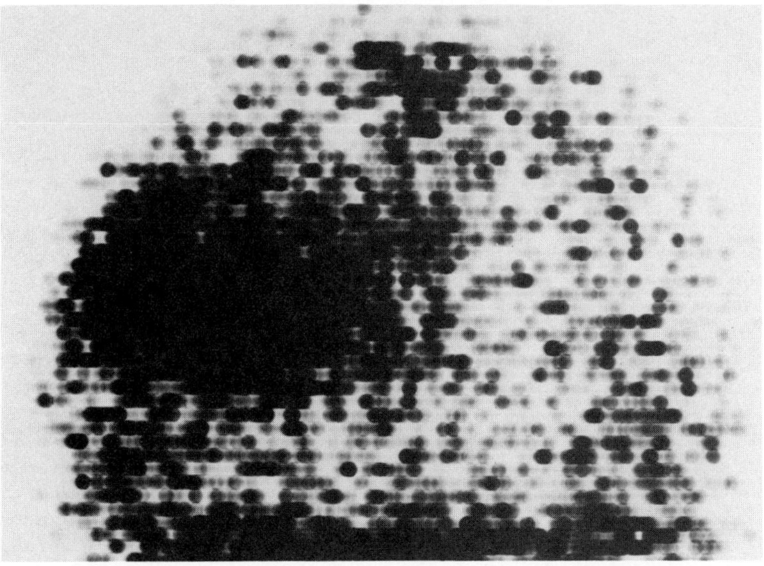

Fig. 1. Radioimmunoscintigraphy using a human MCA of a malignant glioma.

Anti-Idiotype Therapy

The use of a murine MCA to an immunoglobulin idiotype to treat patients with B-cell lymphocytes has recently produced dramatic results (Miller et al, 1982). A normally non-immunoglobulin secreting human malignant B-cell lymphoma was fused to the mouse myeloma cell line NS1. Resulting hybridomas secreted substantial quantities of the human B-lymphocyte surface immunoglobulin. This was then used to raise a monoclonal anti-idiotype antibody. Eight intravenous infusions of this antibody were administered in four weeks with a total dose of 501 mg. Complete tumor regression was observed, which has lasted for over one year. The mechanism for this dramatic response is puzzling. There was no evidence of activation of complement or cellular cytotoxic mechanisms in using the antibody. The authors suggested that the anti-idiotype antibody triggered the observed anti-tumor response by an interaction with the idiotype/anti-idiotype network of the immune system. A disturbance in this network has been implicated in the aetiology of these lymphomas. Several groups are now trying to emulate this success. The surface immunoglobulin idiotype provides a very specific marker for the neoplastic cell clone in surface Ig expressing B-cell lymphoma but cannot be used in the therapy of common epithelial cancers.

ANTIBODIES TO NEW SPECIFICITIES

The fundamental problem of most antibodies tested so far for clinical effectiveness is their lack of specificity and adequate affinity for tumor related proteins. More recently a new approach has been taken using the technique of synthetic peptide immunization. Over the last few years, the genetic basis of malignancy has become clear with the discovery of oncogenes. These are a family of unique sequences of DNA whose abnormal expression is associated with the development of malignant behavior. They were first demonstrated in rapidly transforming RNA viruses. Their significance lies in the discovery that they are not viral in origin, but derive from normal cellular DNA and appear in the retrovirus as a

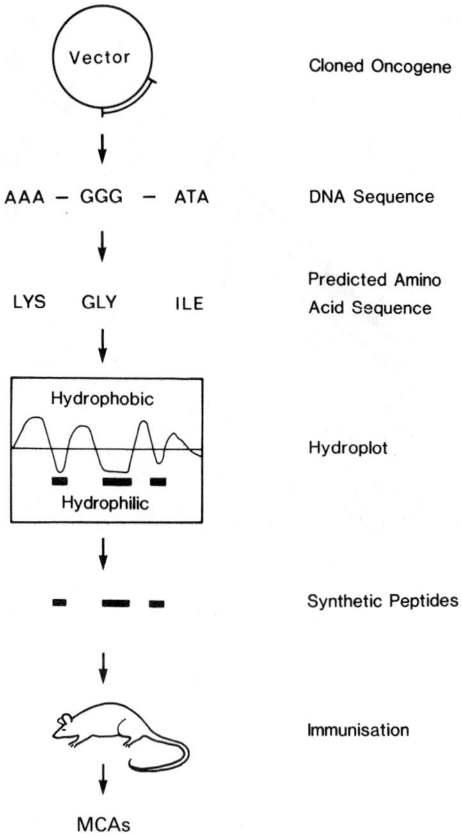

Fig. 2. MCA construction to synthesise peptides.

consequence of recombinant events between the virus and the host's genetic information. The exact mechanisms by which oncogenes cause malignant transformation remain unclear. There is no doubt that they have sequence homology with growth factors and receptors together with functional characteristics pointing to a role in cell cycle control.

Although much is known about the DNA sequences involved in growth control, the isolation and characterization of their encoded proteins is in its infancy. In order to examine the relevance of oncoproteins, MCAs have been constructed to them (Stewart et al, 1986). The DNA sequence of the gene is converted into the putative aminoacid structure and peptides of between 10 and 20 aminoacids long synthesized. The regions chosen for synthesis and immunization are those predicted using a hydropathic plot to be exposed on the intact molecules. These are likely epitopes of the native protein (Fig. 2). One major advantage in utilizing oncoproteins as tumor markers is that unlike the analysis of DNA and RNA which require fresh samples, proteins may be detected by immunohistology, often using formalin fixed wax block material. Such procedures for storing surgical specimens have been routine for many decades and most pathology departments have acquired a huge collection over the years. An example of a tumor stained with a MCA for p62 c-myc is shown in Fig. 3. MCAs against oncogene products are now being intensively investigated for their potential in quantitating oncoprotein on a single cell basis using flow cytometry (Watson et al, 1985); giving information as to tumor prognosis by

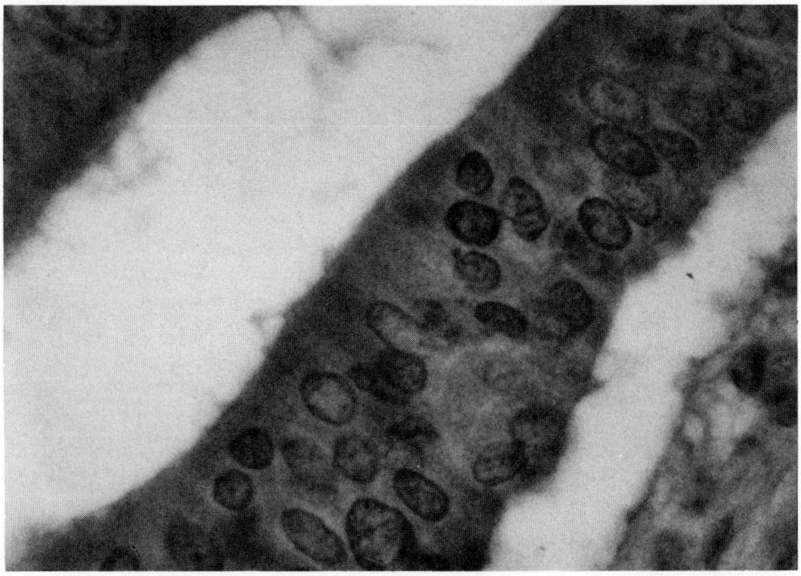

Fig. 3. p62 c-myc detected by immunohistology in a section of colorectal carcinoma using immunoperoxidase and an MCA raised against synthetic peptide.

immunohistology (Sikora et al, 1985); the development of serum assays to determine whether oncoproteins can act as tumor markers and for immunoscintigraphy (Chan et al, 1986). The use of MCA's to study oncogene products will help to unravel the intricate network of molecular control in which cancer and several other diseases begin. Such studies may have profound therapeutic consequences for the future.

REFERENCES

Barratt, G. M., Ryman, B., Begent, R. H., Bagshawe, K. D., 1983, Improved radioimmunodetection of tumors using liposome entrapped antibody, Biochemica et Biophysica acta, 762:154.

Box, P. C., Berenstein, E. H., Siraganian, R. P., 1982, Techniques for enhancing the yield of antigen specific hybridomas in: "Hybridomas and Cancer Diagnosis and Treatment," M.S. Mitchell and H.F. Oettgen, eds., Raven Press, New York.

Bradstock, K. F., Janossy, G., Bollom, F. J., Milstein, C., 1980, Anomolous phenotype in thymic acute lymphoblastic leukaemia, Nature, 284:455.

Brown, J. P., Woodbury, R. G., Hart, C. E., Hellstrom, I., Hellstrom, K. E., 1981, Quantitative analysis of melanoma associated antigen p97 in normal and neoplastic tissues, Proc.Nat.Acad.Sci., 78:539.

Chan, S., Evan, G., Ritson, A., Watson, J., Sikora, K., 1986, Radioimmunolocalization of the c-myc oncogene product in patients with cancer, Brit.J.Cancer., 54:761.

Curry, G. A., 1972, Eighty years of immunotherapy: A review of immunological methods used for the treatment of human cancer, Brit.J.Cancer., 26:141.

Cuttitta, F., Rosen, R., Gazdar, A. F., Minna, J. D., 1981, Monoclonal antibodies that demonstrate specificity for several types of human lung cancer, Proc.Nat.Acad.Sci., 78:4591.

Darras, ed., 1986, "Tumor Markers in Clinical Practice," Blackwell, Oxford.

Dilman, R. O., Collins, H., Royston, I., 1982, T101 monoclonal antibody therapy in chronic lymphocytic leukaemia, in: "Hybridomas in Cancer Diagnosis and Treatment, pp. 151-157, M.S. Mitchell and H.F. Oettgen, eds., Raven Press, New York.

Dippold, W. G., Lloyd, K. O., Lucy, T. C., Old, L. J., 1980, Cell surface antigens of human malignant melanoma, Proc.Nat.Acad.Sci., 77:6114.

Embleton, M. J., Flannery, G. R., Baldwin, R. W., 1982, The effects of monoclonal antibody directed immunotoxins and immunomodulators on cultured osteogenic sarcoma cells, Brit.J.Cancer.

Finan, P. J., Ritson, A., Sikora, K., 1986, Monoclonal antibody immunoscintigraphy for the detection of primary colorectal cancers: a prospective study, Brit.J.of Surgery., 73:177.

Hurlyn, D., Steplewski, Z., Koprowski, H., 1979, Monoclonal antibodies in cell mediated cytotoxicity against human melanoma and colorectal carcinoma, Eur.J.of Immunol., 9:657.

Khaw, B. A., Gold, H. K., Yasuda, T., Haber, E., 1986, Scintigraphic quantification of myocardial necrosis in patients after intravenous injection of myosin-specific antibody, Circulation, 74:501.

Mach, J. P., Buchegger and Karrel, S., 1981, use of radiolabelled monoclonal anti-CEA antibodies for the detection of human carcinomas by external photo-scanning and tomoscinti-scanning, Immunology Today., 239.

Miller, R. A., Levy, R., 1981a, Response of cutaneous T-cell lymphoma to therapy with hybridoma monoclonal antibody, Lancet, 2:226.

Miller, R. A., Levy, R., 1981b, In-vivo effects of murine hybridoma antibody in a patient with T-cell leukaemia, Blood, 58:78.

Miller, R. A., Muloney, D. G., Levy, R., 1982, Treatment of B-cell lymphoma with monoclonal anti-idiotype antibodies, New Engl.J. of Medicine, 306:517.

Nadler, L. M., Ritz, J., Griffin, J. D., Schlossman, S. F., 1983, Diagnosis and treatment of human leukaemias and lymphomas using monoclonal antibodies, Prog.in haematol., 46:187.

Nadler, L. M., Stashenko, P., Hardy, R., Schlossman, S. F., 1980, Serotherapy of a patient with a monoclonal antibody directed against a humanous, lymphoma associated antigen, Cancer Res., 40:3147.

Phillips, J., Alderson, T., Sikora, K., Watson, J. V., 1983, Localization of malignant glioma by radiolabelled monoclonal antibody, J.of Neurol., Neurosurg, Psychiatry, 46:388.

Ritz, J., Clavell, L. A., Schlossman, S. F., 1981, Serotherapy of acute lymphoblastic leukaemia with a monoclonal antibody, Blood 58:361.

Roberts, S., Cheetham, J. C., Reece, A. R., 1987, Generation of an antibody with enhanced affinity and specificity for its antigen by protein engineering, Nature, 328:731.

Ross, B., Langman, R., Trowbridge, I., Becker, R., 1983, Monoclonal antibodies and cancer, Academic Press, New York.

Sikora, K., Lennox, E. S., 1982, Tumor antigens in: "Clinical Aspects of Immunology," P.J. Lachmann and D.K. Peters, eds., Blackwell, Oxford, pp.142-165.

Sikora, K., Alderson, T., Ellis, J., 1982, Human hybridomas from patients with malignant disease, Brit.J.of Cancer, 46:477.

Sikora, K., Wright, R., 1981, Human monoclonal antibodies to lung cancer antigens, Brit.J.Cancer, 43:696.

Sikora, K., Finan, T., Lennox, E., Smedley, H., 1983, The selection of monoclonal antibodies for the localization of colorectal cancer, Brit.J. Cancer, 47:253.

Sikora, K., Evan G., Steward, J., Watson, J., 1985, Detection of the c-myc oncogene product in testicular cancer, Br.J.Cancer, 52:171.

Sobol, R. E., Dillman, R. O., Royston, I., 1982, Phase 1 evaluation of immune monoclonal anti-melanoma antibody in man, in: "Hybridomas in Cancer Diagnosis and Treatment," M.S. Mitchell and H.F. Oettgen, eds., Raven Press, New York.

Smedley, H., Finan, P., Sikora, K., 1983, Localization of metastatic carcinoma by radiolabelled monoclonal antibody, Brit.J.Cancer, 286:262.
Steward, J., Evan G., Watson, J., Sikora, K., 1986, Detection of the c-myc oncogene product in colonic polpys and carcinomas, Br.J.Cancer, 51:1.
Watson, J. V., Evan, G. I., Sikora, K., 1985, Quantitation of oncogene products in archival pathology material, J.Immunol.Methods., 83:170.
Thomas, E. D., Buckner, C. D., Fluornoy, N., 1977, One hundred patients with acute leukaemia treated by chemotherapy, total body irradiation and allogeneic bone marrow transplantation, Blood, 49:511.
Thorpe, P. E., Edwards, D. C., Ross, W. C. J., 1982, Monoclonal antibody toxin conjugates, in: "Monoclonal Antibodies in Clinical Medicine," A.J. McMichael and J.W. Fabre, eds., Academic Press, London.

ANTI-TUMOR EFFECTS OF MONOCLONAL ANTIBODY-RICIN CONJUGATES IN VITRO AND IN VIVO

David C. Blakey and Philip E. Thorpe

Drug Targeting Laboratory
Imperial Cancer Research Fund
Lincoln's Inn Fields, London WC2A 3PX

INTRODUCTION

The idea that antibodies directed against tumor-associated antigens might be used to target a toxic agent specifically to tumor cells and so spare normal tissue from harm has attracted much attention recently. This is mainly because the advent of hybridoma technology for making monoclonal antibodies overcame the problems inherent in using conventionally raised polyclonal antibodies and made this approach much more feasible. Toxins of plant (e.g. ricin, abrin and gelonin) or bacterial (e.g. diphtheria toxin, *pseudomonas aeruginosa* exotoxin A) origin have proved particularly promising for targeting purposes, probably because their extreme potency minimises the number of toxin molecules that have to be delivered to a target cell to kill it. Antibody-toxin conjugates ('immunotoxins') have in many in vitro and in vivo experimental systems been found to kill cancer cells with great potency and specificity.

Here we describe studies carried out with immunotoxins prepared only with the most widely used toxin, ricin. The results are generally applicable, however, to immunotoxins prepared with other toxins. Immunotoxins containing ricin are currently being evaluated clinically both for the removal of malignant cells or T-lymphocytes from bone marrow grafts in vitro and for the treatment of malignant disease in vivo.

The reader is referred to a number of extensive reviews for further information (Thorpe, 1985; Vitetta and Uhr, 1985; Blakey et al, 1987a).

STRUCTURE AND MODE OF ACTION OF RICIN

The plant toxin, ricin, from the beans of *Ricinus communis* is composed of two polypeptide chains, A and B, linked by a single disulphide bond. The A-chain contains the protein synthesis inhibitory activity whereas the B-chain contains binding sites which enable the toxin to bind to cells. Individually the two chains are relatively non-toxic since both are required for the potent cytotoxic action of the toxin.

The first step in the cytotoxic process is that the toxin binds via binding sites in the B-chain, to galactose-terminating glycoproteins and glycolipids which are present on the surface of most cell types. In the second step, the A-chain crosses a cell membrane to reach the cytosol. The mechanism involved in this translocation and the site at which it takes place have not yet been elucidated. It appears, however, that the B-chain portion of the toxin assists the A-chain in its passage across the membrane. Having reached the cytosol of the cell the A-chain then catalytically inactivates the 60S ribosomal subunit in a way that decreases its capacity to bind elongation factor, EF-2 (Nolan et al, 1976a). Protein synthesis is thus terminated resulting in the death of the cell. Studies by Eiklid and his colleagues (1980) demonstrated that possibly as few as one A-chain molecule entering the cytosol is sufficient to kill the cell. It is this extreme potency of the toxin molecule which makes it an attractive candidate for targeting to tumor cells.

CYTOTOXIC PROPERTIES OF IMMUNOTOXINS

Ricin Immunotoxins

Immunotoxins have been prepared by linking intact ricin to antibodies. They are almost always extremely potent cytotoxic agents in vitro for cells with the appropriate target antigen, often matching or even surpassing the native toxin in potency (Thorpe, 1985). Intact ricin immunotoxins, however, suffer from a lack of complete specificity since they can also bind to galactose residues on non-target cells by means of the binding sites present on the B-chain. In vitro, this non-specific binding can be blocked with high concentrations of free galactose or lactose (Youle and Neville, 1980; Thorpe et al, 1982) but such a blockade in vivo is of little value since these sugars are rapidly excreted. As a result, intact ricin immunotoxins have thus far only been used clinically for purging bone marrow in vitro (Filipovich et al, 1985). A method for attenuating the non-specific binding properties of intact ricin immunotoxins has been devised by Thorpe and colleagues (1984). Immunotoxins were prepared by attaching ricin to antibody via a short linkage. A significant proportion of the immunotoxin preparation was unable to bind to galactose residues probably because of steric hindrance of the binding sites on the B-chain by the antibody component itself. The 'Blocked' ricin immunotoxins retained their potent cytotoxic effects but unlike other ricin-antibody conjugates, showed excellent specificity. In the future it may be possible to eliminate non-specific toxicity of ricin immunotoxins by chemically blocking the B-chain binding sites or deleting them from the ricin gene using recombinant DNA technology.

Ricin A-Chain Immunotoxins

The most widely used approach to prepare immunotoxins is to attach only the inhibitory A-chain subunit of ricin to the antibody. A-chain immunotoxins show virtually complete specificity in their cytotoxic effects upon target cells in vitro because they do not have a B-chain by which to bind to cells nonspecifically. Unfortunately, however, they have variable cytotoxic potency. Some ricin A-chain immunotoxins are as toxic to their target cells as their intact toxin counterparts whereas others are only weakly cytotoxic. This variability appears to depend in part on the molecule recognized by the antibody. Some antigens appear to deliver the immunotoxin to sites in the cell from which the A-chain can escape to the cytosol whereas other antigens do not. A second factor is that cells with high density of the target antigen are more sensitive to immunotoxins than cells with low densities of the same antigen (Laurent et al, 1986a). Other factors which affect potency include the affinity of the antibody, its valency, and the cell type on which the target antigen is present (reviewed by Blakey et al, 1987a). The cytotoxic potency of weakly effective ricin A-chain immunotoxins can often be potentiated in vitro using lysosomotrophic amines (Casellas et al, 1984), carboxylic ionophores (Casellas et al, 1984) or the B-chain of ricin (McIntosh et al, 1983; Vitetta et al, 1984).

Ricin A-chain immunotoxins are generally prepared by modifying the antibody with a heterobifunctional coupling reagent (e.g. N-succinimidyl 3-(2-pyridyldithio) proprionate (SPDP)) which introduces an activated disulfide group and then mixing the antibody derivative with freshly reduced ricin A-chain. A thiol-disulfide exchange reaction then takes place which leaves the A-chain attached to the antibody by a reducible disulphide bond (Fig. 1) (Wawrzynczak and Thorpe, 1987). The introduction of this disulphide bond is important, since its reduction to release free A-chain appears to be an important step in the cytotoxic process of the immunotoxin. This conclusion comes from the fact that ricin A-chain immunotoxins prepared with irreducible thioether linkages between the A-chain and antibody are consistently less potent than their disulphide bonded counterparts at inhibiting protein synthesis in target cells (reviewed by Wawrzynczak and Thorpe, 1987).

An example of a highly effective ricin A-chain immunotoxin prepared in this way which we have extensively studied in our laboratory is that made from the mouse monoclonal antibody (OX7) against the Thyl.1 antigen. It binds to the Thyl.1 antigen present on T-lymphocytes and most T-cell lymphomas originating in the AKR strain of mice. The immunotoxin inhibits 50% of protein synthesis in AKR-A lymphoma cells in tissue culture at a concentration (the IC50) of 10^{-11}M (Fig. 2). Its effect was specific since antibody alone did not inhibit protein synthesis and free A-chain only inhibited protein synthesis at a concentration 100,000 fold higher than it did as an immunotoxin. Furthermore, a ricin A-chain immunotoxin prepared with a mouse monoclonal antibody that did not bind to the AKR-A cells had no effect on protein synthesis at concentrations as high as 10^{-8}M. Disappointingly, however, the same immunotoxin elicited only a weak antitumor effect when administered to mice bearing AKR-A tumor cells in vivo (Blakey et al, 1987b). The tumor test system we used was to inject one million AKR-A lymphoma cells into the peritoneal cavity of nude mice and one or three days later give them a single intravenous injection of the immunotoxin containing 10μg of ricin A-chain (equivalent to 1/10 of the LD50 dose). The anti-Thyl.1-ricin A-chain immunotoxin extended the median survival time of the tumor bearing mice by 6 days compared to control groups and from calibration experiments in which graded numbers of tumor cells were injected into mice, this was calculated to correspond to the killing of about 99% of the tumor cells in the peritoneal cavity (Fig. 3). Again the effect was specific. Antibody alone, A-chain alone, mixtures of the two and control immunotoxins of irrelevant specificity were without protective effect. In a number of other experimental systems, ricin A-chain immunotoxins which are potent and specific in vitro also only produced rather small antitumor effects in vivo (Jansen et al, 1984; Seto et al, 1982; Kishida et al, 1983) suggesting that factors operate in vivo which reduce the potency of ricin A-chain immunotoxins.

FACTORS THAT EFFECT THE POTENCY OF RICIN A IMMUNOTOXINS IN VIVO

Contaminating Free Antibody

Ricin A-chain immunotoxins used in most experimental studies and in the clinical trials to date have been contaminated with unreacted free antibody. This is because the antibody (150 kDa) and the immunotoxin containing one molecule of A-chain and one of antibody (180 kDa) are similar in molecular weight and are not effectively separated during the gel filtration step used to remove free A-chain from the immunotoxin

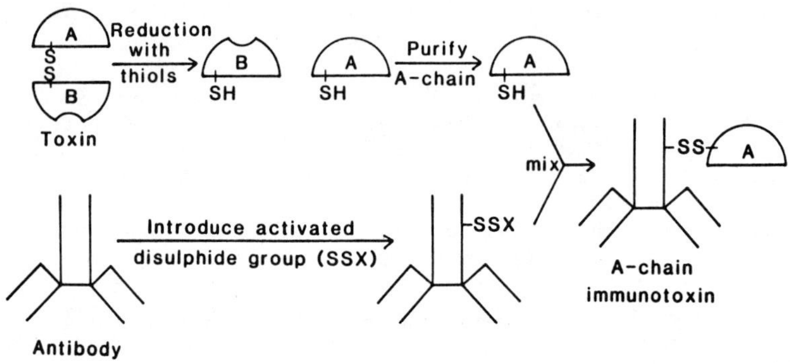

Fig. 1. Preparation of A-chain immunotoxin.

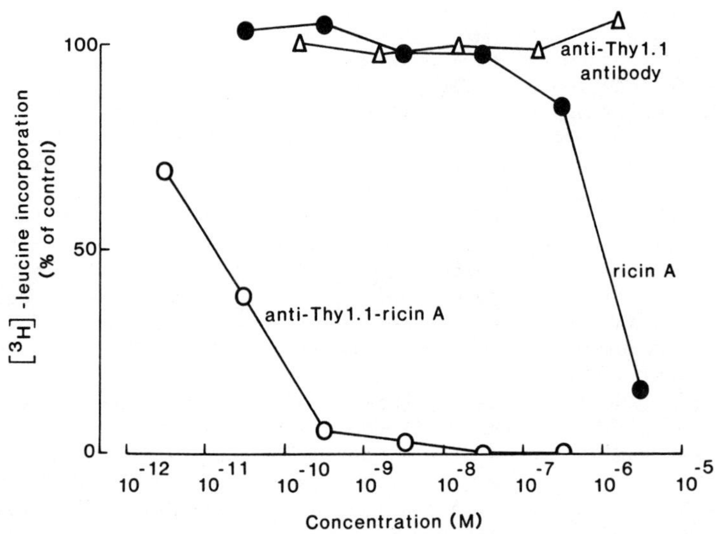

Fig. 2. Cytotoxicity of an anti-Thyl.l-ricin A-chain immunotoxin to AKR-A lymphoma cells in vitro.
AKR-A lymphoma cells were cultured for 48h at 37°C in the presence of anti-Thyl.l-ricin A-chain immunotoxin (o), anti-Thyl.l antibody (Δ) or ricin A (●). The [^{3}H] leucine incorporation by the cells during the final 24 h period of culture was measured (Thorpe et al, 1984).

preparation. This contaminating free antibody in the immunotoxin preparation can compete for binding to the target antigen and reduce the antitumor effectiveness of the immunotoxin. For example, in our tumor model system an anti-Thyl.l-ricin A-chain immunotoxin preparation containing 20% free antibody only extended the median survival time of the tumor bearing mice by 6 days and all the animals developed tumor in the peritoneal cavity. In contrast only 50% of tumor bearing animals developed tumor in the peritoneal cavity after treatment with an identical immunotoxin from which the free antibody had been removed (Blakey and Thorpe, unpublished results). This effect is probably due to the fact that the ricin A-chain immunotoxin is rapidly cleared from the blood in vivo (see below) whereas the antibody persists for prolonged periods. As a result, the concentration of the contaminating antibody soon exceeds that of the

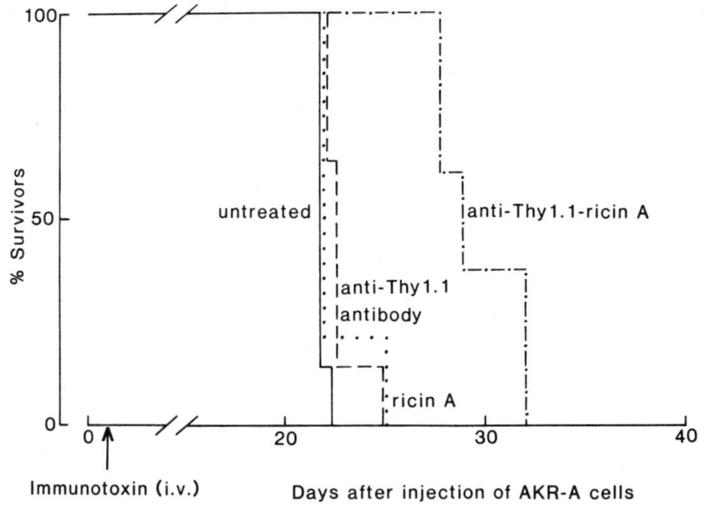

Fig. 3. Antitumor effect of an anti-Thyl.l-ricin A-chain immunotoxin. Groups of 10 nu/nu mice received intraperitoneal injection of 10^6 AKR-A lymphoma cells (approximately 10^5 lethal doses) and 1 day later were given a single intravenous injection of the following: 50μg anti-Thyl.l-ricin A-chain immunotoxin (—·—·), 50 μg anti-Thyl.l antibody (---), 50μg A-chain (....) or diluent (——). Antitumor effects were judged from the survival time of the mice.

immunotoxin after injection. It therefore diffuses in greater amount into the peritoneal cavity and masks the antigens on the tumor cells from the immunotoxin.

A simple procedure was recently developed in our laboratory (Knowles and Thorpe, 1987) for removing free antibody from ricin A-chain immunotoxin preparations. The procedure involves affinity chromatography on Cibacron blue F3GA coupled to Sepharose (blue Sepharose). The antibody does not bind to the gel whereas the immunotoxin is bound by its A-chain moiety and can be recovered in high yield and purity using mild elution conditions.

Liver Entrapment

Ricin A-chain contains mannose and fucose residues which are recognized by receptors present on the non-parenchymal (Kupffer and sinusoidal endothelial cells) and parenchymal cells of the liver (Skilleter and Foxwell, 1986; Blakey and Thorpe, 1986). This recognition results in the rapid clearance of ricin A-chain immunotoxins from the bloodstream by the liver (Blakey et al, 1987c) and reduces the amount of immunotoxin available to reach the tumor site. This problem can be overcome by chemically modifying ('deglycosylating') the mannose and fucose residues on the A-chain with a mixture of sodium metaperiodate and sodium cyanoborohydride (Thorpe et al, 1985). Deglycosylated ricin A-chain, unlike its native counterpart, is not taken up by liver cells in vitro or in vivo (Blakey and Thorpe, 1986) and, consequently, immunotoxins prepared with the deglycosylated ricin A-chain are more slowly cleared from the blood stream than native ricin A-chain immunotoxins (Fig. 4) (Blakey et al, 1987c). Ten minutes after injection, only 12% of the deglycosylated ricin A-chain immunotoxin was taken up by the liver as compared with 30% of the native ricin A-chain immunotoxin. Nevertheless, the amount of the deglycosylated ricin A-chain immunotoxin taken up by the liver significantly exceeded that of the native antibody (5% of injected dose). Studies in vitro indicate that the uptake of the deglycosylated ricin A-chain immunotoxin is neither due to recognition of carbohydrate on the A-chain or antibody nor to recognition of the Fc portion of the antibody (Blakey et al, unpublished results). Possibly attachment of the A-chain to the antibody exposes hidden sites in the A-chain which are recognized by receptors on liver cells.

Deglycosylation does not effect the cytotoxic potency of ricin A-chain immunotoxins in vitro (Blakey et al, 1987c). In the AKR-A tumor model system the deglycosylated ricin A-chain immunotoxin produced a better antitumor effect than the native ricin A-chain immunotoxin (Fig. 5) suggesting that more deglycosylated ricin A-chain immunotoxin reaches the tumor site as a result of its longer blood half life.

Instability

There have been conflicting reports concerning the stability of the disulphide bond between the antibody and ricin A-chain (Blakey et al, 1987c; Worrell et al, 1986; Jansen et al, 1984; Bourrie et al, 1986) which might possibly be explained by differences in the ability of a particular antibody to shield the disulfide bond from attack by reducing agents. In our laboratory, ricin A-chain immunotoxins prepared with the most widely used heterobifunctional coupling reagents, SPDP and 2-iminothiolane have consistently broken down (t1/2 = 6 to 8h) in the bloodstream of mice to release free antibody (Fig. 6a) (Blakey et al, 1987c). Instability of immunotoxins is a problem for two reasons. Firstly, it accelerates the rate of disappearance of the immunotoxin from the bloodstream and , secondly, it results in the release of free antibody which, as described above, can compete for the target antigen and reduce the antitumor effect of the immunotoxin. We have prepared a new coupling reagent (SMPT) which introduces a hindered disulphide bond between the antibody and A-chain which is more resistant to reduction. A deglycosylated ricin A-chain

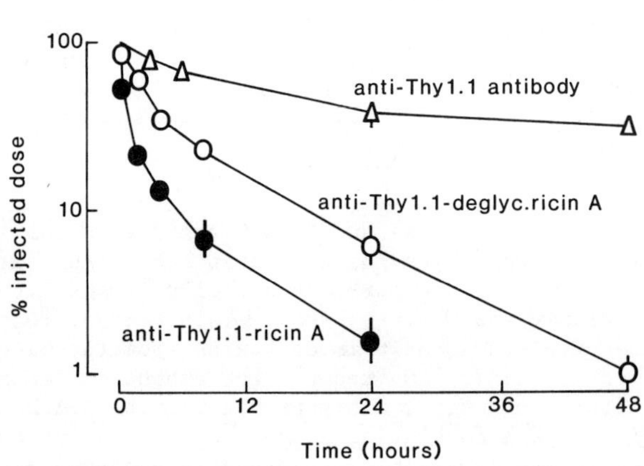

Fig. 4. Blood clearance rates of a ricin A-chain immunotoxin and a deglycosylated ricin A-chain immunotoxin in mice.
Mice were given injections (i.v.) of radioiodinated anti-Thyl.1 immunotoxins and blood samples were taken at various periods of time later. The percentage of the injected dose at the various time points that corresponded to intact anti-Thyl.1 antibody (Δ), deglycosylated ricin A-chain immunotoxin (O) or ricin A-chain immunotoxin (●) was determined by scanning autoradiographs of SDS gels on which the plasma samples taken from the mice had been separated (Blakey et al, 1987).

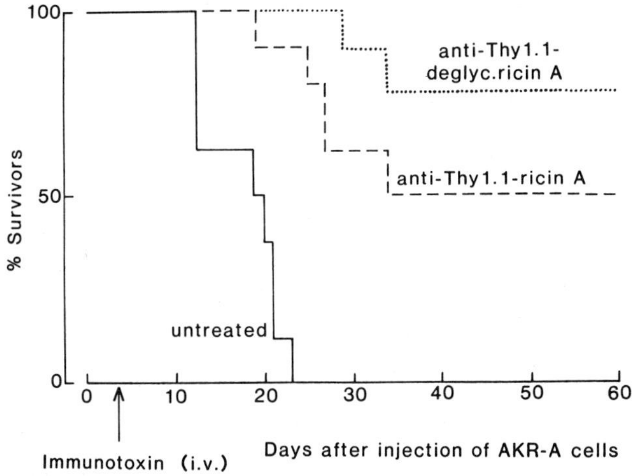

Fig. 5. Improved antitumor effect an anti-Thyl.1-deglycosylated ricin A immunotoxin.
Groups of 8 or 10 nu/nu mice received intraperitoneal injections of 10^6 AKR-A cells and 3 days later were given a single injection (i.v.) of the following: 50μg anti-Thyl.1-ricin A-chain immunotoxin (---), 50μg anti-Thyl.1-deglycosylated ricin A-chain immunotoxin (...) or diluent (——).

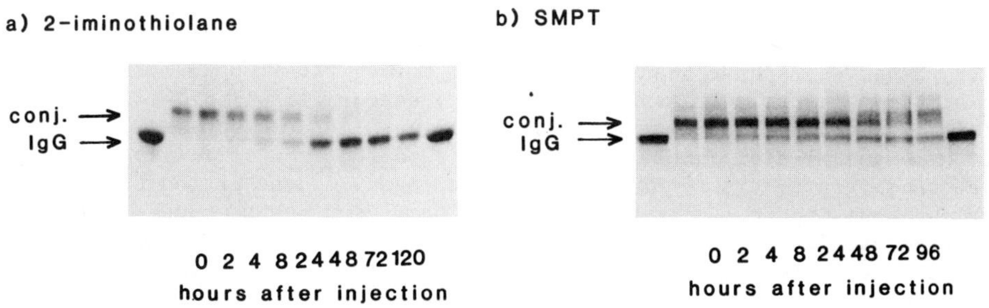

Fig. 6. Stability of 2-iminothiolane and SMPT linked immunotoxins in the bloodstream of mice.
Autoradiographs of SDS gels of plasma samples taken from mice at various times after injection of an unstable 2-iminothiolane linked deglycosylated ricin A-chain immunotoxin (a) or a more stable SMPT linked deglycosylated ricin A-chain immunotoxin (b). In both instances immunotoxins were radiolabeled with ^{125}I.

immunotoxin prepared with the SMPT coupling reagent was 6-9 times more stable in the bloodstream of mice than its 2-iminothiolane linked counterpart as judged by the release of free antibody (Fig. 6b) (Thorpe et al, in press). Consequently, the SMPT linked immunotoxin had a longer blood half-life than the 2-iminothiolane linked immunotoxin. Forty-eight hours after injection 10% of the SMPT-linked immunotoxin remained in the bloodstream of mice as compared with 1% of a 2IT-linked immunotoxin. Since both immunotoxins were equally effective at inhibiting protein synthesis in vitro, the SMPT linked immunotoxin would thus be expected to have improved antitumor activity.

Resistant Tumor Cells

Resistant tumor cells may emerge which lack the target antigen and so cannot be killed by the immunotoxin. This is routinely the cause of death in AKR-A tumor-bearing mice treated with SMPT-linked deglycosylated ricin A-chain immunotoxins. Heterogeneity of antigen expression has been observed with many types of tumor cells (Woodruff, 1983) and so is almost certain to be a common problem with immunotoxin therapy. By using a cocktail of immunotoxins directed against a variety of tumor-associated antigens it may be possible to overcome this problem. Alternatively, immunotoxin therapy could be combined with other forms of therapy which do not rely on antigen expression of a target antigen for their effectiveness. Tumor cells may also emerge which are resistant to the action of the toxin molecule. If so, it will be necessary to use a cocktail of immunotoxins made from toxic components whose mode of action does not overlap.

Access In Solid Tumors

Experiments done in our laboratory have shown that AKR-A lymphoma cells growing as a solid tumor subcutaneously in mice are less responsive to intravenously administered immunotoxin than the same cells growing free in the peritoneal cavity (Thorpe, unpublished results). Improved access of immunotoxins to tumor cells growing as a solid mass may be achieved either by the use of smaller antibody fragments or by co-administrating a vasodilating agent to increase vasculature permeability. Alternatively, it may be possible to target immunotoxins to the vasculature of a solid tumor and so kill the tumor cells indirectly by starving them of nutrients.

CLINICAL USE OF IMMUNOTOXINS IN VIVO

Immunotoxins have been used clinically in vitro in bone marrow transplantation and the results have been quite encouraging (reviewed in Blakey et al, 1987a). Filipovitch and her colleagues (1985) have used a cocktail of three anti-T-cell immunotoxins to purge bone marrow of T-lymphocytes that cause graft-versus-host disease in allogeneic bone marrow transplantation. Only two out of eleven patients receiving purged bone marrow from histocompatability matched donors developed stage II graft-versus-host disease. Unfortunately, the elimination of T-cells using this approach has resulted in an increased incidence of graft failure possibly because a few remaining T-cells in the graft are needed to suppress residual host immunity. In autologous bone marrow transplantation a portion of the patient's own bone marrow is removed before giving the patient intensive radiochemotherapy. Gorin and his colleagues (1985) have used an anti-T-cell immunotoxin to remove leukaemia cells from the bone marrow of four patients with T-ALL or T-cell lymphoma prior to its reengraftment. All four patients showed prompt engraftment and hematopoietic recovery was normal. One patient is in complete remission eleven months after transplantation.

However, by far the greatest use of immunotoxins is likely to be for in vivo therapy in which the immunotoxin is given directly to the patient. There have been three reports so far describing the in vivo use of ricin A-chain immunotoxins in man. Firstly, twenty-two patients with metastatic malignant melanoma have been treated with a murine monoclonal anti-melanoma antibody-ricin A-chain immunotoxin in a phase 1 trial in the USA. Half of those treated showed some response and one patient had a complete remission (Spitler et al, 1987). Side effects of the treatment included hypoalbuminaemia, oedema and allergic reactions but they were generally transient and reversible. Secondly, two patients with leukemia have been treated with an anti-CD5 antibody-ricin A-chain immunotoxin in a phase 1 trial in France. Both patients showed a transient decrease in the number of tumor cells in the bloodstream following infusion of the immunotoxin and there were no adverse side effects (Laurent et al, 1986b). Finally, 15 patients who had steroid-resistant graft versus host disease after allogeneic bone marrow transplantation were treated with an anti-CD5 antibody-ricin A-chain immunotoxin (Byers et al, 1987). Twelve of the patients showed significant reduction in their disease in at least one organ at 7 days and at 28-40 days, 3 of 10 patients had complete remission of their disease.

Encouraging though these results are, it must be stressed that the immunotoxins in these clinical trials contained native ricin A-chain and contaminating free antibody and were prepared with the unstable SPDP reagent. As described above, this would have lead to rapid blood clearance of the immunotoxins, instability and competition of the free antibody for the target antigen. It seems likely that in the future, purified immunotoxins prepared with deglycosylated ricin A-chain and new, more stable, coupling reagents, will have improved therapeutic activity in the clinic.

REFERENCES

Blakey, D. C., and Thorpe, P. E., 1986, Effect of chemical deglycosylation on the in vivo fate of ricin A-chain, Cancer Drug Delivery, 3:189.

Blakey, D. C., Wawrzynczak, E. J., Wallace, P. M., and Thorpe, P. E., 1987a, Antibody toxin conjugates: a perspective in: "Progress in Allergy," H. Waldmann, ed., S. Karger AG., Basel, in press.

Blakey, D. C., Wawrzynczak, E. J., Stirpe, F., and Thorpe, P. E., 1987b, Anti-tumor activity of a panel of anti-Thy 1.1 immunotoxins made with different ribosome inactivating proteins in: "UCLA Proceedings Volume: Membrane Mediated Cytotoxicity," B. Benjamin and R. J. Collier eds., Alan R. Liss, Inc., New York.

Blakey, D. C., Watson, G. J., Knowles, P. P., and Thorpe, P. E., 1987c, Effect of chemical deglycosylation of ricin A-chain on the in vivo fate and cytotoxic activity of an immunotoxin composed of ricin A-chain and anti-Thy 1.1. antibody, Cancer Res., 47:947.

Bourrie, B. J. P., Casellas, P., Blythman, H. E., and Jansen, F. K., 1986, Study of the plasma clearance of antibody-ricin A-chain immunotoxins. Evidence for specific recognition sites on the A-chain that mediate rapid clearance of the immunotoxins, Eur.J.Biochem., 155:1.

Byers, V., Kernan, N., Henslee, J., Gingrich, R., Frame, J. Waid, J., Messino, M., Blazer, B., Mischak, R., O'Reilly, R., Thompson, J., Dupont, B., and Scannon, P., 1987, A phase 1 study using pan T lymphocyte-ricin A chain immunotoxin to treat graft versus host disease in: "Second International Conference on Monoclonal Antibody Immunoconjugates for Cancer," I. Royston, R.O. Dillman eds., USCD Cancer Center, San Diego, in press.

Casellas, P., Bourrie, B. J. P., Gros, P., and Jansen, F. K., 1984, Kinetics of cytotoxicity induced by immunotoxins: enhancement by lysosomotrophic amines and carboxylic ionophores, J.Biol.Chem., 259:9359.

Eiklid, K., Olsnes, S., and Pihl, A., 1980, Entry of lethal doses of abrin, ricin and modeccin into the cytosol of HeLa cells, Exp.Cell Res., 126:321.

Filipovich, A. H., Vallera, D. A., Youle, R. J., Neville, D. M., and Kersey, J. H., 1985, Ex vivo T-cell depletion with immunotoxins in allogeneic bone marrow transplantation: the pilot clinical study for prevention of graft-versus-host disease, Transpl.Proc., 17:442.

Gorin, N. C., Douay, L., Laporte, J. P., Lopez, M., Zittoun, R., Rio, B., David, R., Stachwiak, J., Jansen, J., Cazellas, P., Poncelet, P., Liance, M. C., Voisin, G. A., Salmon, C., Le Blanc, G., Deloux, J.,

Najman, A., and Duhamel, G., 1985, Autologous bone marrow transplantation with marrow decontaminated by immunotoxin T101 in the treatment of leukemia and lymphoma: first clinical observations, Cancer Treat. Rep., 69:953.

Jansen, F. K., Blythman, H. E., Bourrie, B., Carriere, D., Casellas, P., Dussossoy, D., Gros, O., Laurent, J. C., Liance, M. C., Poncelet, P., Richer, G., and Vidal, H., 1984, Significance of the kinetics of immunotoxin cytotoxicity in: "Receptor Mediated Targeting of Drugs," G. Gregoriadis, G. Poste, J. Senior and A. Trouet, eds., Plenum Press, New York.

Kishida, K., Masuho, Y., Saito, M., Hara, T., and Fuji, H., 1983, Ricin A-chain conjugated with monoclonal anti-L1210 antibody. In vitro and in vivo antitumor activity, Cancer Immunol.Immunother., 16:93.

Knowles, P. P., and Thorpe, P. E., 1987, Purification of immunotoxins containing ricin A-chain and abrin A-chain using Blue Sepharose Cl 6B, Anal.Biochem., 160:440.

Laurent, G., Kuhlein, E., Casellas, P., Canat, X., Carayon, P., Poncelet, P., Correll, S., Rigal, F., and Jansen, F. K., 1986a, Determination of sensitivity of fresh leukemia cells to immunotoxins, Cancer Res., 46:2289.

Laurent, G., Pris, J., Farcet, J. P., Crayon, P., Blythman, H., Casellas, P., Poncelet, P., and Jansen, F. K., 1986b, Effect of therapy with T101 ricin A-chain immunotoxin in two leukemia patients, Blood, 67:1680.

McIntosh, D. P., Edwards, D. C., Cumber, A. J., Parnell, G. D., Dean, C. J., Ross, W. C. J., and Forrester, J. A., 1983, ricin B-chain converts a non-cytotoxic antibody-ricin A chain conjugate into a potent and specific cytotoxic agent. FEBS Lett., 164:17.

Nolan, R. D., Grasmuk, H., and Drews, J., 1976, The binding of tritiated elongation factor 1 and 2 to the ribosomes from Krebs II mouse ascites cells. The influence of various antibiotics and toxins, Eur.J. Biochem., 64:69.

Seto, M., Umemoto, N., Saito, M., Masuho, Y., Hara, T., and Takahashi, T., 1982, Monoclonal anti-MM46 antibody:ricin A chain conjugate: In vitro and in vivo antitumor activity, Cancer Res., 42:5209.

Skilleter, D. N., and Foxwell, B. M. J., 1986, Selective uptake of ricin A-chain by hepatic non-parenchymal cells in vitro. Importance of mannose oligosaccharides in the toxin, FEBS Lett., 196:344.

Spitler, L. E., del Rio, M., Khentigan, A., Wedel, N. I., Brophy, N. A., Miller, L. L., Harkonen, W. S., Rosendorf, L. L., Lee, H. M., Mischak, R. P., Kawahata, R. T., Stoudemire, J. B., Fradkin, L. B., Bautista, E. E., and Scannon, P. J., 1987, Therapy of patients with malignant melanoma using a monoclonal anti-melanoma antibody-ricin A-chain immunotoxin, Cancer Res., 47:1717.

Thorpe, P. E., 1985, Antibody carriers of cytotoxic agents in cancer therapy in: "Monoclonal Antibodies '84, Biological and Clinical Applications," A. Pinchera, G. Doria, F. Dammacco and A. Bargellesi, eds., Edritice Kurtis s.r. l., Milan.

Thorpe, P. E., Mason, D. W., Brown, A. N. F., Simmonds, S. J., Ross, W. C. J., Cumber, A. J., and Forrester, J. A., 1982, Selective killing of malignant cells in a leukaemic rat bone marrow using an antibody-ricin conjugate, Nature, 297:594.

Thorpe, P. E., Ross, W. C. J., Brown, A. N. F., Myers, C. D., Cumber, A. J., Foxwell, B. M. J., and Forrester, J. A., 1984, Blockade of the galactose-binding sites of ricin by its linkage to antibody: specific cytotoxic effects of the conjugate, Eur.J.Biochem., 140:63.

Thorpe, P. E., Detre, S. I., Foxwell, B. M. J., Brown, A. N. F., Skilleter, D. N., Wilson, G., Forrester, J. A., and Stirpe, F., 1985, Modification of the carbohydrate in ricin with metaperiodate-cyanoborohydride mixtures. Effects on toxicity and in vivo distribution, Eur.J.Biochem., 147:197.

Vitetta, E. S., and Uhr, J. W., 1985, Immunotoxins, Ann.Rev.Immunol., 3:197.
Vitetta, E. S., Fulton, R. J., and Uhr, J. W., 1984, Cytotoxicity of cell-reactive immunotoxin containing ricin A chain is potentiated by an anti-immunotoxin containing ricin B chain, J.Exp.Med., 160:341.
Wawrzynczak, E. J., and Thorpe, P. E., 1987, Methods for preparing immunotoxins: effects of the linkage on activity and stability in: "Immunoconjugates: Antibody Conjugates in Radioimaging and Therapy of Cancer," C.W. Vogel, ed, Oxford University Press, New York, in press.
Woodruff, M. F. A., 1983, Cellular heterogeneity in tumors, Br.J.Cancer., 47:589.
Worrell, N. R., Cumber, A. J., Parnell, G. D. Ross, W. C. J., and Forrester, J. A., 1986, Fate of an antibody-ricin A-chain conjugate administered to normal rates, Biochem.Pharmacol., 35:417.
Youle, R. J., and Neville, D. M., 1980, Anti-Thyl.2 monoclonal antibody linked to ricin is a potent cell-type-specific toxin, Proc.Natl. Acad.Sci.USA., 77:5483.

LIMITATIONS IN LOCALIZING AND KILLING TUMORS USING RADIOLABELLED ANTIBODIES

A.R. Bradwell, P.W. Dykes and A.T.M. Vaughan

Immunodiagnostic Research Laboratory,
Department of Immunology, The Medical School
University of Birmingham, B15 2TJ, UK

SUMMARY

Radiolabelled antibodies against tumors are being studied intensely for their use in tumor detection and tumor destruction. Although a host of different tumor types can be localized, results at present are no better than other scanning techniques. In addition no tumor has been eradicated using labelled antibodies. Dramatic improvements will depend upon a fuller understanding of tumor cell biology and optimizing all the parameters involved.

INTRODUCTION

Localizing tumors with radiolabelled antibodies (radioimmuno-localization - RAIL) has not been as useful as was originally hoped (Goldenberg et al, 1987). In order to understand the problems and limitations of the technique and possibilities of improvements we developed a mathematical model. This was subsequently modified to ascertain the limitations in using radiolabelled antibodies for tumor killing. These models are described together with the basis for their construction.

MODEL FOR TUMOR LOCALIZATION

Eleven parameters have been used which comprise the major factors involved (Fig. 1). It can be seen that the parameters that cannot be altered are, (1) the tumor area, (2) the tumor depth, (3) the patient's body thickness at the site of the tumor and (4) the signal to noise ratio seen in the final picture (normally about three). The uptake ratio (5) of the antibody onto the tumor compared with the surrounding normal tissue is the most important single determinant of scan sensitivity. Reported values in patients average less than five and are usually nearer two (Table). The influence of the uptake ratio in detecting different sized lesions at various depths as determined by the model is shown in Fig. 2.

The model indicates that, next to uptake ratios, the most important factor that can be improved is the absolute count rate of the tumor locating isotope - (6). When using iodine-131 (I-131), count rates are

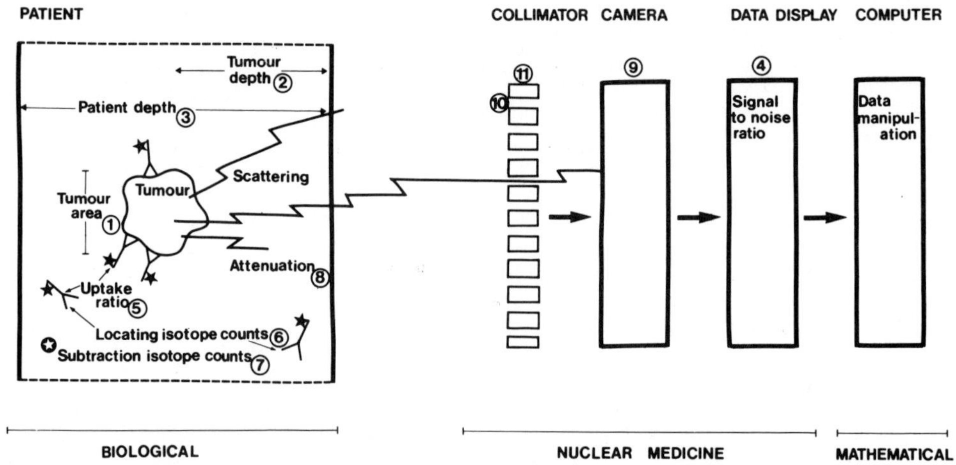

Fig. 1. Parameters used in the model for the radioimmunodetection of tumors - also see text. (Units 1, 2 and 3 in cm^2. 4, signal to noise ratio. 5, Tumor to normal tissue uptake ratio. 6 and 7, counts/cm^2. 8, per cm.9, Camera resolution - FWHM in mm. 10, Hole diameter in mm. 11, Hole depth in mm).

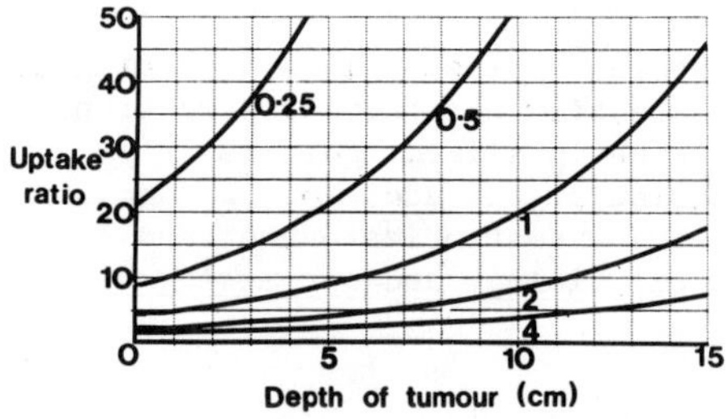

Fig. 2. Relationship between uptake ratio (5) and tumor depth (2) with contours on tumor area (1). Other parameters from fig. 1: 3 = 20; 4 = 3; 6 = 300; 8 = 0.127 (indium-111); 9 = 4.8; 10 = 4.3; 11 = 50.

usually low compared with other isotopic scanning methods and in our hands often less than 100 per cm^2 in the image. Trebling of this rate would improve the clarity of the image and be equivalent to doubling the uptake ratio. A tenfold increase would be equivalent to a threefold rise in uptake ratio but clearly there is a dose limitation and each isotope has its own problems.

Technetium-99m (99m-Tc) has frequently been used as the subtracting isotope (Fig. 1; factor (7) to mask the variation in vascular supply to different tissues. This assists the detection of small tumors with low uptake ratios. On theoretical grounds such a process must add noise (Bradwell et al, 1985) and hence reduce the resolution of the scans. It is important to keep the count rates from the subtracting isotope as high as possible to minimize noise.

As regards the nuclear medicine aspects (Fig. 1) factors such as gamma ray scattering in the body, camera blurring and sensitivity are physical limitations inherent in the present systems. Computerised tomography is potentially advantageous (Berche et al, 1982) providing count rates are high but this tends not to be the case in RAIL because of the high dosimetry of the isotopes.

Table 1. Clinical Studies with Radiolabelled Antibodies

antibody type	tumour type	uptake ratio (U)	% injected activity per gm tumour	reference
AP	mixed	2.5	-	Goldenberg 1978
Pab	germ cell	40*	-	Goldenberg 1980
Mab	kidney	2.3	-	Ghose 1980
Pab	colon	2.5	-	Dykes 1980
AP	colon	3.6	0.0033	Mach 1980
Mab	mixed	4	0.0026	Mach 1981
Mab	colon	2.3	0.005	Farrands 1982
AP	hepatoma	5	-	Ishii 1983
Mab	colon	2-5	-	Mach 1983
Mab	melanoma	4.3	0.0057-0.01	Larson 1983
Mab	breast	4	-	Rainsbury 1983
Mab	colon	2.5	-	Armitage 1984
Mab	colon	3	-	Chatal 1984
Mab	breast	>2	-	Thompson 1984
Mab	melanoma	2.3	0.0036	Buraggi 1985
Fab_2	melanoma	2.4	0.0017	Buraggi 1985
Mab	ovary	2.2	-	Epenetos 1986
Mab	various	1.22-35.8$	0.015	Epenetos 1986
Mab	bone	2.8*	-	Armitage 1986

Pab = Polyclonal antibody. AP = Affinity purified polyclonal. Mab = Monoclonal. U = Tumour to normal tissue ratio in surgically resected samples. * = One sample only. $ = Sample taken 12 days post-injection

Potential Areas for Improvement

Most work has centered on attempts to improve antibody uptake ratios with, as yet, relatively little success. In this respect the use of monoclonal antibodies has shown little advantage over polyclonal antisera (Table). Monoclonal antibodies have several problems. They may be too specific; bind unexpectedly to similar determinants on other antigens; have low affinity and cross react with human cellular Fc receptors so producing false positive localization. These factors may explain why it has proved difficult to improve on affinity purified polyclonal antisera. The use of antibody fragments, second antibodies, alternative isotopes (such as indium-111 and iodine-123) and different cameras has had little overall effect on results (Bradwell et al, 1985).

In spite of this pessimism most reported series suggest that a few patients benefit from the scans. Such patients are serum marker positive but clinically in remission by all criteria except the antibody scan. Provided there is useful treatment for such patients there is a small role for RAIL. However, for there to be dramatic improvements in their value, tumor uptake ratios must be improved. This will depend on greater understanding of cell biology, particularly in relation to cell surface antigens. Only then will the exquisite biochemical specificity of antibodies find use in tumor localization.

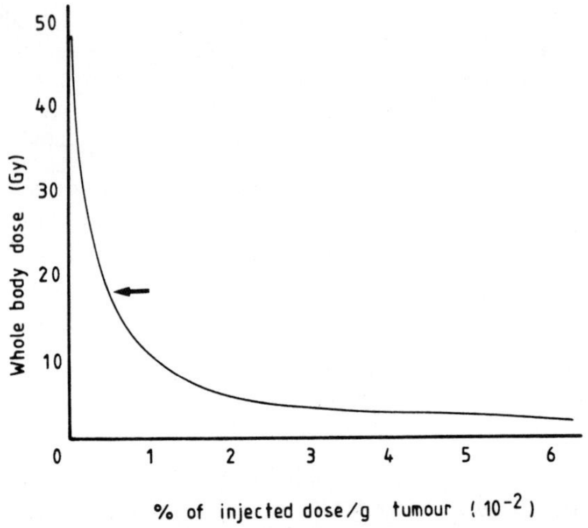

Fig. 3. The whole body dose resulting from the delivery of a 60 Gy tumour dose in one week by I-131 labelled antibody in relation to the percentage of injected dose in the tumour. A half-time for tumour uptake of 8 hours is taken and activity is eliminated from the body and tumour with a half-time of 3 days. The arrow marks the currently attainable tumour uptake. A similar curve is obtained for yttrium-90 labelled antibody for which there is no excretion of activity from the body or tumour.

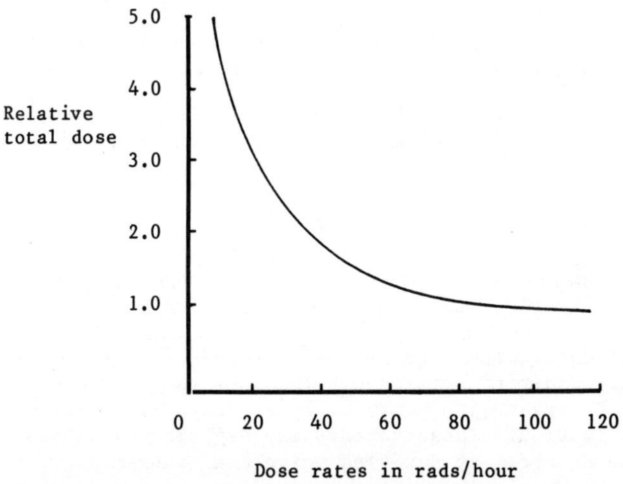

Fig. 4. Relative increase in total dose required for continuous irradiation at various dose rates to give the same effect as 60 Gy of a single exposure of 1.5 hours.

MODEL FOR TUMOR KILLING

There have been many articles proposing the destruction of tumors using radiolabelled antibodies. Most suggest that antibody therapy may follow directly from the success of antibody scanning programmes but this extrapolation depends upon considerable improvements in tumor uptake ratios. We have therefore constructed a model to assess these theoretical and practical limitations.

This model is based upon antibody uptake figures reported in the literature (Table) and indicates that the amount of labelled antibodies located in human tumors is approximately 0.005% of the i/v injected dose per gm of tumor. We have used this to assess the tumor dose. The calculation was based on a compartmental model evaluating antibodies labelled with either iodine-131 or yttrium-90 (Vaughan et al, 1987). We assumed an initial uniform distribution of activity throughout the body. Accumulation and loss of activity by the tumor and loss from the whole body, was calculated using rate constants obtained from published work (Table). If we consider a 60 Gy dose in one week is sufficient for tumor sterilisation (Ellis, 1968) the model can be used to calculate associated whole body doses.

The calculations show that for an I-131 labelled antibody to deliver the required dose the model indicates that 17 Gy would be absorbed by the whole body. Since the maximum reasonable whole body dose is 2 Gy, successful tumor therapy with the available distribution of antibody is impossible. Increased tumor retention of iodine activity by a factor of two, as reported by Ettinger (1982) only allowed a reduction in the total body dose of 2 Gy. The figure indicates that the percentage of injected dose within the tumor must be increased at least ten fold if lethal whole body irradiation is to be avoided. Even using a radiolabel with one of the best set of physical characteristics (yttrium-90) the whole body dose would still be 14.3 Gy for a 60 Gy tumor dose. There are no isotopes that are obviously physically superior.

Potential Areas for Improvement

The model also permits assessment of manipulations that might improve the situation. For instance, our model indicates it is possible to halve the whole body dose by trebling the excretion rate. This might favor the use of antibody fragments such as $F[ab]_2$ (Burraggi et al, 1985) or a clearing second antibody (Goodwin et al, 1984) but in practice there is a reduction of labelled antibody in the tumor so the reduction in whole body dose tends to be small. Our calculations (Fig. 3) show that there is a non linear relationship between tumor uptake and whole body dose. Considerable increases in specific tumor uptake are essential before the whole body dose is reduced to a tolerable level. This may not be generally appreciated. It should also be added that this is the most optimistic situation. First, if there is specific organ accumulation of activity, such as in the bladder, local toxicity will further limit the injected dose. Second, the dose rate needs to be several times that presently achieved, ideally greater than 30 centi-Greys per hour (Fig. 4). In practice dose rates are much less even shortly after isotope injection (Ettinger et al, 1982).

Specific situations, such as regional therapy, have not been considered in this model. It is unlikely that such an approach will prove useful since the main purpose of systemic cancer therapy is to remove occult systemic disease. The question of drug antibody conjugates is another matter. In our view this has less immediate appeal since success depends upon an adequate blood supply to all tumor cells and each must express the target antigen. This is most unlikely.

For these reasons it seems unlikely that in-vivo radiolabelled antibodies will have any significant impact on in vivo cancer localization and therapy in the immediate future.

Acknowledgements

We would like to thank the Cancer Research Campaign for their support of this work. The figures have been reproduced with the kind permission of

Pitman Publishing Lit. (Figs. 1-2) and the Lancet (Fig. 3). Figure 4 has been adapted from the British Journal of Radiology, 1963, 34:123-129.

REFERENCES

Armitage, N. C., Perkins, A. C., Pimm, M. V., Farrands, P.A., Baldwin, R. W., and Hardcastle, J. D., 1984, The localization of an anti-tumor monoclonal antibody (791T/36) in gastrointestinal tumors, Br.J.Surg., 71:407.

Armitage, N. C., Perkins, A. C., Pimm, M. V., Wastie, M., Hopkins, J. S., Dowling, F., Baldwin, R. W., and Hardcastle, J. D., 1986, Imaging of bone tumors using a monoclonal antibody raised against human osteosarcoma, Cancer, 58:37.

Berche, C., Mach, J-P., Lumbroso, J.-D., Langlais, C., Aubry, F., Buchegger, F., Carrel, S., Rougier, P., Parmentier, C., and Tubiana, M., 1982, Tomoscintigraphy for detecting gastrointestinal and medullary thyroid cancers: first clinical results using radiolabelled monoclonal antibodies against carcinoembryonic antigen, Brit.Med.J., 285:1447.

Bradwell, A. R., Fairweather, D. S., Dykes, P. W., Keeling, A., Vaughan, A., and Taylor, J., 1985, Limiting factors in the localization of tumors with radiolabelled antibodies, Imm.Today, 6:163.

Buraggi, G. L., Callegaro, L., Marianai, G., Turrin, A., Cascinelli, N., Attili, A., Bombardieri, E., Terno, G., Plassio, G., Dovis, M., Mazzuca, N., Natali, P. G., Scassallati, G. A., Rosa, U., and Ferrone, S., 1985, Imaging with I-131 labelled monoclonal antibodies to a high-molecular-weight melanoma-associated antigen in patients with melanoma: Efficacy of whole immunoglobulin and its F(ab')2 fragments. Cancer Res.,45:3378.

Chatal, J. F., Saccavini, J. C., Fumoleau, P., Douillard, J-Y., Curtet, C., Kremer, M., LeMevel, B. Q., and Koprowski, H., 1984, Immunoscintigraphy of colon carcinoma, J.Nucl.Med., 25:307.

Ellis, F., 1968, The relationship of biological effect to dose-time fractionation factors in radiotherapy, Current Topics in Radiation Research, 4:359.

Epenetos, A. A., Carr, D., Johnson, P. M., Bodmer, W. F., and Lavender, J. P., 1986, Antibody guided radiolocalization of tumors in patients with testicular or ovarian cancer using two radioiodinated monoclonal antibodies to placental alkaline phosphatase, Br.J.Radiol., 59:117.

Epenetos, A. A., Snook, D., Durban, H., Johnson, P. M., and Taylor-Papadimitriou, J., 1986, Limitations of radiolabelled monoclonal antibodies for localization of human neoplasms, Cancer Res., 46:3183.

Ettinger, D. S., Order, S. E., Wharam, M. D., Parker, M. K., Klein, J. L., and Leichner, P. K., 1982, Phase I-II study of isotopic immunoglobulin therapy for primary liver cancer, Cancer Treat Rep., 66:289.

Farrands, P. A., Perkins, A. C., Pimm, M. V., Hardy, J. D., Embleton, M. J., Baldwin, R. W., and Hardcastle, J. D., 1982, Radioimmunodetection of human colorectal cancers by an anti-tumor monoclonal antibody, Lancet II: 398-400.

Ghose, T., Norvell, S. T., Aquino, J., Belitsky, P., Tai, J., Guclu, A., and Blair, A. H., 1980, Localization of 131-I-labelled antibodies in human renal cell carcinomas and in a mouse hepatoma and correlation with tumor detection by photoscanning, Cancer Res., 40:3018.

Goldenberg, D. M., DeLand, F. H., Kim, E., Bennett, S., Primus, F. J., van Nagell, J. R., Estes, N., DeSimone, P., and Rayburn, P., 1978, Use of radiolabelled antibodies to carcinoembryonic antigen for the detection and localization of diverse cancers by external photoscanning, N.Engl.J.Med., 298:1384.

Goldenberg, D. M., Kim, E. E., Deland, F. H., van Nagell, J. R., and Javadpour, N., 1980, Clinical radioimmunodetection of cancer with radioactive antibodies to human chorionic gonadotrophin, <u>Science</u> 208:1284.

Goodwin, D., Meares, C., Diamanti, C., McCall, M., Lai, X., Torti, F., McTigue, M., and Martin, B., 1984, Use of specific antibody for rapid clearance of ciruclating blood background from radiolabelled tumor imaging proteins, <u>Eur.J.Nucl.Med.</u>, 9:209.

Ishii, N., Nakata, K., Munehisa, T., Koji, T., Nishi, S., and Hirai, H., 1983, Radioimmunodetection of cancer using radiolabelled antibodies to alphafetoprotein, <u>Prot.Biol.Fluids.</u>, 31:305.

Larson, S. M., Brown, J. P., Wright, P. W., Carrasquillo, J. A., Hellstrom, I., and Hellstrom, K. E., 1983, Imaging of melanoma with I-131-labelled monoclonal antibodies, <u>J.Nucl.Med.</u>, 24:123.

Leichner, P. K., Klein, J. L., Siegelman, S. S., Ettinger, D. S., & Order, S. E., 1983, Dosimetry of I-131-labelled antiferritin in hepatoma: Specific activities in the tumor and liver, <u>Cancer Treat.Rep.</u>, 67:647.

Mach, J-P., Buchegger, F., Forni, M., Ritschard, J., Berche, C., Lumbroso, J. D., Schreyer, M., Giardet, C., Accolla, R. S., and Carrel, S., 1981, Use of radiolabelled monoclonal anti-CEA antibodies for the detection of human carcinomas by external photoscanning and tomoscintigraphy, <u>Immunol.Today.</u>, 2:239.

Mach, J-P., Chatal, J. F., Lumbrosco, J. D., Buchegger, F., Forni, M., Ritschard, J., Berche, C., Douillard, J-Y., Carrel, S., Herlyn, M., Steplewski, Z. and Koprowski, H., 1983, Tumour localisation in patients by radiolabelled monoclonal antibodies against colon carcinoma. <u>Cancer Res.</u>, 43:5593.

Mach, J-P., Carrel, S., Forni, M., Ritschard, J., Donath, A., and Alberto, P., 1980, Tumor localization of radiolabelled antibodies against carcinoembryonic antigen in patients with carcinoma, <u>N.Engl.J.Med.</u>, 303:5.

Rainsbury, R. M., Ott, R. J., Westwood, J. H., Kalirai, T. S., Coombes, R. C., McCready, V. R., Neville, A. M., and Gazet, J-C., 1983, Location of metastatic breast carcinoma by a monoclonal antibody chelate labelled with indium-III, <u>Lancet</u> II:934.

Thompson, C. H., Lichtenstein, M., Stacker, S. A., Leyden, M. J., Salehi, N, Andrews, J. T., and McKenzie, I. F. C., 1984, Immunoscintigraphy for detection of lymph node metastases from breast cancer, <u>Lancet</u> II: 1245.

Vaughan, A. T. M., Anderson, P., Bradwell, A. R., Dykes, P. W., and Anderson, P., 1987, Limitations to the killing of tumors using radiolabelled antibodies, <u>Br.J.of Radiology</u>, 60:567.

INDEX